CUTTING TOOL MATERIALS

Proceedings of an International Conference
15-17 September 1980
Ft. Mitchell, Kentucky

Sponsored by the
Machinability Activity
of the
Mechanical Working and Forming Division
American Society for Metals
and the
Society of Manufacturing Engineers,
Iron and Steel Institute of Japan,
and
Collège Internationale pour l'Etude Scientifique
des Techniques de Production Mécanique

Organized and compiled under the direction of
FRANK W. GORSLER
General Electric Company

MATERIALS/METALWORKING TECHNOLOGY SERIES

American Society for Metals
Metals Park, Ohio 44073

ISBN: 0-87170-111-1

PRINTED IN THE UNITED STATES OF AMERICA

CONFERENCE ORGANIZING COMMITTEE

FRANK W. GORSLER
General Electric Company
Conference General Chairman

GREGG B. BALKO
Society of Manufacturing Engineers

MOHAN CHAUDHARI
Grey Iron Research Institute

JAY HAZRA
General Electric Company

ROBERT JAGERS
Valeron Corporation

ROGER A. JOSEPH
Republic Steel Corporation

RANGA KOMANDURI
General Electric Company

JOHN C. MILLER
Reynolds Metals Company

EDWARD J. MYERS
American Society for Metals

S. RAMALINGAM
Georgia Institute of Technology

VINOD K. SARIN
GTE Laboratories, Inc.

MYRON J. SCHMENK
Cincinnati Milacron

THOMAS J. SKINGLE
Acme-Cleveland Development Company

ROBERT W. THOMPSON
Inland Steel Company

VIJAY A. TIPNIS
Tipnis Associates, Inc.

CHARLES C. TOBIN
Caterpillar Tractor Company

FOREWORD

The papers in these proceedings were presented at the International Conference on Cutting Tool Materials, held September 15-17, 1980, in Ft. Mitchell, Kentucky. This conference was made possible through the effort and cooperation of the American Society for Metals, the Society of Manufacturing Engineers, CIRP, and the Iron and Steel Institute of Japan.

This is a continuation of a series of conferences devoted to the general theme of machinability. To date this series consists of:

The Influence of Metallurgy on Machinability, held October 14-16, 1975, in Chicago, Illinois; proceedings published by ASM

The Influence of Metallurgy on Hole Making Operations, held May 19-20, 1977, in Boston, Massachusetts; proceedings published by ASM

The International Symposium on Influence of Metallurgy on Machinability of Steel, held September 26-28, 1977, in Tokyo, Japan; proceedings published by ISIJ

Machinability Testing and Utilization of Machining Data, held September 12-13, 1978, in Oak Brook, Illinois; proceedings published by ASM.

The objective of this particular conference was to bring together researchers, manufacturers, and users of cutting tools to focus on common problems relating to their design, production, quality control, and application. The format consisted of technical presentations grouped by cutting tool material type and an international panel of users and producers. Panel discussions as well as questions and answers from the speakers were recorded for publication herein.

Frank W. Gorsler
Conference General Chairman

Robert W. Thompson
Chairman, Machinability Activity

CONTENTS

SESSION I

AVAILABILITY, PROPERTIES, WEAR OF CUTTING TOOL MATERIALS

FUTURE MATERIALS AVAILABILITY AND PROBABLE ALTERNATIVES

F. X. Cook
Emory Ayers Associates Incorporated

Cemented carbide producers and users are perhaps more affected by supply concerns since every one of their principal materials is plagued by near-term, if not long-term questions. In addition, the availability situation has been exacerbated by dramatic price increases. For example, tantalum and columbium prices have risen 400% in the past 18 months.

OUTLINE OF PRESENTATION

The paper would cover three areas, with most of the time and emphasis being placed on the first two:

1. An assessment of the availability of the five key materials: tungsten, cobalt, tantalum, columbium and molybdenum, including:
   - Current supply and demand balance assessment
     - Background
     - Price Effects
     - Explanation of the major causes
   - Supply forecast
     - Key areas of the world
     - Trouble spots
     - Possibilities and probability of shortages
     - Scrap availability
2. Probable future trends in materials usage by cemented carbide producers
   - Cemented carbide and user market situation today
     - End product trends
     - Effect of CC price increases
   - The influence and motivation of cemented carbide producers
   - Future trends, e.g.,
     - Substitution in design stages for carbide mixes
     - Product/market share trends, i.e., product-for-product substitution
3. Possibilities for alternative materials

RESEARCH PROGRAM

During the next eight months, we will be conducting on-going research in materials availability. In addition, it is very likely that we will be conducting two specific projects on tantalum and columbium which should prove detailed and very up-to-date information in time for the presentation:

1. Supply study, world wide, of tantalum and columbium, e.g.,
   - Miners' exploration and development plans
   - Political problems/economic limitations
2. Survey of leading cemented carbide producers and end users
   - Price effects
   - Materials usage
   - Substitution trends
3. Interviews with other experts
   - R & D
   - Universities
   - Government

DISCUSSION

Howard Kline (Caterpillar Tractor Co.)

Are the private (cartel) investors controlling the prices of strategic Material?

Frank Cook

If we define a cartel as a group of producers that collaborate to conciously fix prices then we have not found that, NO. As an offhand remark I would say that there was a lot of pent up hedging, for example, for cobalt which was galvanized by the political instability in Zaire, so that the price of $25 per pound is not in my opinion all due to increased mining costs.

John Knox (Vermont American Corp.)

If I may, I would like to add some specifics on the cobalt supply problem for the carbide producers. The Cemented Carbide Producers Association has a committee keeping track of the cobalt problem. Our problem is not just the supply of cobalt but the supply of fine cobalt powder. There is no powder in the Federal Stockpile and until June of this year no production facility for making it in the U.S. Credit is due to Washington bureaucrats, or as they prefer - senior civil servants, for having pursuaded The Belgian Producer to put cobalt power plant in North Carolina. However, this plant is still dependent on Zaire for feedstock.

There is enough cobalt ore in U.S. deposits recoverable at present prices to supply 25% of total requirements. The senior civil servants are fully aware of the criticality of cobalt and other materials and have programs worked out to handle the cobalt situation. However, they require the approval of the higher level political appointees and the Congress. Capturing and holding the attention of these people long enough to get concrete action is extremely difficult. Anything we can do as users of critical materials to educate the the Congress would be usefull.

IDENTIFICATION OF THE DOMINANT WEAR MECHANISM IN SPECIFIC TOOL-WORK SYSTEMS

P.K. Wright*, A. Bagchi* and J.G. Horne†

*Carnegie-Mellon Univ., Dept. of Mech. Engr., Pittsburgh, PA
†International Harvester Comp., Sci. & Tech., Hinsdale, IL

ABSTRACT

Micrographs are presented which show the main forms of cutting tool wear in machining. The broad concern of the research is to identify which of these possible mechanisms have the dominant influence on tool performance, for a given tool-work combination. It is proposed that the commercially relevant measure of tool performance is the volume of material worn away from the nose-radius region of the tool. A general equation is presented stating that this is affected by adhesion, abrasion, oxidation/adhesion wear, diffusion and plastic deformation of the edge. Examples are presented to show cutting conditions when one of these dominates. The elevated-temperature, high-strain-rate shear flow strength of the workmaterial has been found to be important. This controls the localized stress and temperature distribution at the edge.

1. INTRODUCTION

A great deal of metallurgical research is now being carried out to identify specific tool wear mechanisms. Such a mechanistic approach, often employing detailed metallography, provides valuable information and improves upon the past descriptions of tool performance solely in terms of flank and crater wear. At the same time, however, it is leading to a proliferation of proposed wear mechanisms and classifications.

The broad propositions of this paper are that: i) most of these proposed mechanisms can be found to play some role during a particular cutting operation but that ii) the key question to address is: for a particular tool-work combination and cutting speed, which are the wear mechanisms which have the *dominant* influence on tool life? Following a brief description of the major wear processes, two examples are discussed that illustrate how the dominating mechanism varies according to chip-flow pattern over the tool and the temperature distribution around the cutting edge region.

2. WEAR MECHANISMS

In everyday tribological studies, adhesive and abrasive wear are the most commonly encountered mechanisms. However, when dealing with metal cutting, the elevated temperatures and high stresses imposed on the tool influence the nature of these two basic mechanisms and also bring other processes into play. The review below outlines *seven* mechanisms that, depending on circumstances, can individually dominate cutting tool performance. The micrographs come from recent semi-orthogonal turning work on a wide variety of workmaterials ranging from low carbon to alloy steels and aerospace alloys. The tool materials have been M34 high-speed steel, a variety of carbide inserts and ceramic tools. After each test the tools have been examined metallographically and the wear mechanisms identified by optical and scanning electron microscopy.

2.1. Adhesive Wear (Attrition)

This general description of wear mechanisms concentrates on events that occur at commercially relevant speeds. Hence, while adhesive wear is strongly associated with built-up-edge formation (as shown in Figures 18 and 19 of reference [1]), the latter rarely occurs at the speeds attainable with cemented carbide turning tools. This is not to say that adhesive wear does not occur at higher rates of metal removal. Recent work [2] has re-emphasized that the chip cleans the tool rake face and cutting edge regions so that 'naked' surfaces free from organic contaminants and oxides are created. As the cleaning action proceeds the opportunity for adhesive wear increases, particularly in view of the intimate contact between chip and tool. Archard's general work on adhesive wear gives a basis for discussing the basic parameters: for like materials the rate at which a volume of material (dv) is removed with sliding distance (dL) is given by

$$dv/dL = K P_n / 3H \quad (1)$$

where H is the hardness of the materials, P_n the normal load and K the probability that a wear junction will arise.

In cutting there are special circumstances that affect this equation. In particular the normal load, P_n, exceeds the yield strength of the chip hence leading to an interface where the real area of contact is equal to the apparent area. Once the chip cleans the oxide layers from the tool there is potential for the chip-tool interface to become one continuous adhesive junction. At this stage the value of K then depends solely on the cohesive strength of the tool material. The above equation relates to like materials and indicates the expected trend of wear rate, decreasing with increased hardness. It has also been found that for unlike wear pairs the ratio of wear of the harder to that of the softer material is $(H_S/H_H)^2$. During high speed cutting, especially with aerospace materials that have high elevated-temperature yield strength, the ratio H_S/H_H is reduced because the tool material becomes thermally weakened by the heat generated during cutting. This increases the likelihood of adhesive wear. Adhesive junction formation is evident in Figure 1 which shows a section through the flank surface of a high-speed steel tool that has been used to machine A.I.S.I. type 304 stainless steel.

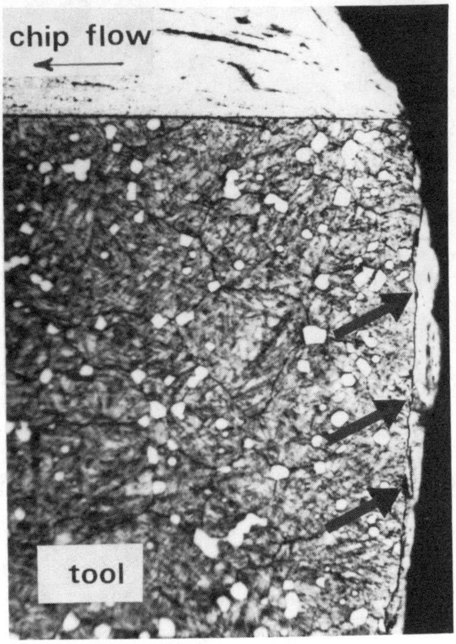

Fig. 1. Adhesive Wear Junctions (Arrowed) on Flank Face of M34 High-Speed Steel Tool used to Machine Stainless Steel (x1000).

With the more brittle cemented carbide tools, the adhesive junction can result in a whole carbide particle being 'plucked' from the tool surface to be carried off in the chip. Trent has termed this process attrition wear [3].

2.2. Abrasive Wear

Abrasive wear involves the removal of tool material by a scoring action of hard phase inclusions in the chip. Recent work [4] has described in detail the mechanisms involved in such a process and a typical micrograph is presented in Figure 2. In this case machining has been carried out on a titanium

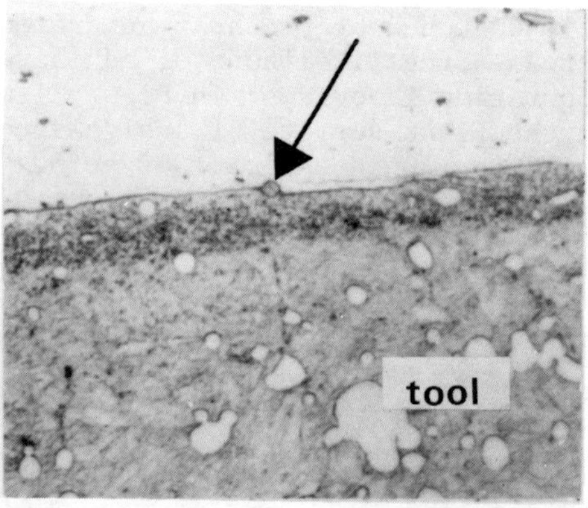

Fig. 2. Gray TiC Particle (Arrowed) in Titanium Stabilized Stainless Steel (AISI grade 316) Scoring a Groove in the Tool Rake Face (x1500).

stabilized stainless steel and thus the chip material contains TiC particles having a hardness of ~2400 VPN. It has been shown that these score grooves into the softer martensitic matrix of the tool, removing thin layers of the latter as a 'micro-chip' or 'prow'. Figure 2 shows a groove of lost tool material in the wake of the particle. With high-speed steel tools such a process has been found on both the flank face and rake face. In the thermally weakened crater area of the latter the abrasive wear process appears to have greater effects. Again this is to be expected since, using the same terminology as in equation (1), the abrasive wear rate is

$$dv/dL = G P_n/H \qquad (2)$$

where G is a geometric factor depending on the shape of the particle. Thus as the tool hardness decreases, due to temperature, the likelihood of an abrasive interaction increases. The same form of abrasive wear has also been proposed for cemented carbide tools where in the crater regions temperatures exceed the ductile-brittle transition temperature. Machining experiments with materials of controlled microstructure have shown a clear correlation between increased tool wear and increased amounts of abrasive particles in the workmaterial [5].

2.3. Fracture and Chipping

The inherent brittleness of cemented carbide tools renders them susceptible to severe damage by cracking if sudden loads or thermal gradients are applied to their edge. While such difficulties are generally associated with milling operations and other interrupted cuts, recent interest in very high speed machining has focussed attention on the starting conditions as the tool encounters the bar shoulder. In fact, in any commercial situation, unpredictable tool breakage cannot be tolerated and it is assumed here that steps will always be taken to avoid this situation. Catastrophic fracture is not therefore considered as a major influence in the steady-state wear problems below.

2.4. Wear by Oxidation and Subsequent Adhesion

One aspect of the present research work is direct observations of the chip-tool interface using transparent sapphire cutting tools. Until recently these have been used to machine only soft materials such as aluminum but, using a modified design, it is anticipated that these may be used to cut low carbon steel. The ability to observe the interface has enabled the features at the perimeter of the chip-tool contact area to be more closely studied [5]. Figure 3a, from machining aluminum, shows that the contact zone may be divided into a clear area close to the cutting edge (zone 1) and a region of substantial transfer around the perimeter of the chip-tool contact region (zone 2). It has been proposed that the perimeter transfer arises from the formation of a strong bond between the aluminum oxide (sapphire) tool and the oxide that forms on the underside of the chip. The transfer is a result of this adhesive process.

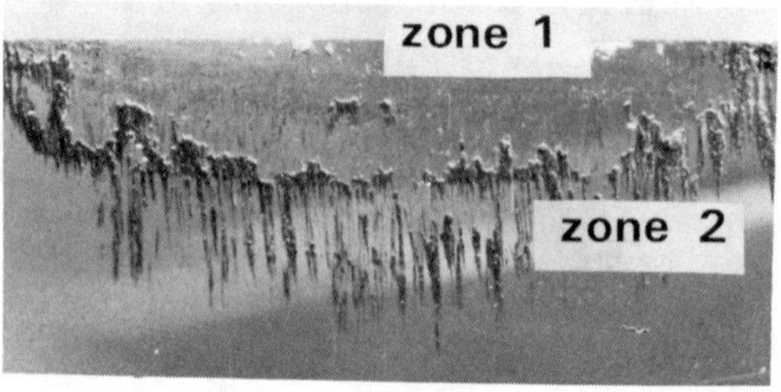

3(a)

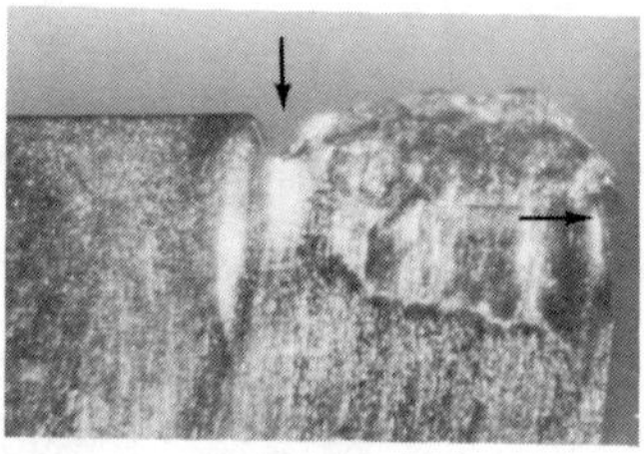

3(b)

Fig. 3. (a) Contact Area After Planing Aluminum with Sapphire Tool. (b) Grooves (Arrowed) at Extremity of Depth-of-Cut and Nose Radius After Machining Steel. Both the Zone 2 Transfer and These Wear Grooves Arise by Oxidation and Subsequent Adhesion.

At higher cutting speeds it has been found that oxygen cannot penetrate to the rear of the chip-tool contact area and transfer continues to form only at the sides. In semi-orthogonal turning tools, such transfer has been observed to form at the extremity of the depth of cut and immediately behind the nose radius. After long cutting times, deep notches have been observed in both these regions and an example is shown in Figure 3b after machining a medium carbon steel. The evidence shows that the continuous formation and breaking down of adhesive junctions between oxides also causes the underlying tool material to be occasionally removed.

2.5. Diffusion Wear

Mechanisms such as adhesion and abrasion are perceived as being enhanced by elevated temperature, but not as critically dependent on a temperature rise as diffusion or plastic deformation. Diffusion involves transport of tool material atoms into the moving chip and, as shown by Trent [3], is predominantly controlled by the solubility of the carbide phase in the chip material. It has been proposed that this is the major cause of crater wear at high speeds. Figure 4 shows how the diffusion wear process can lead to the degeneration

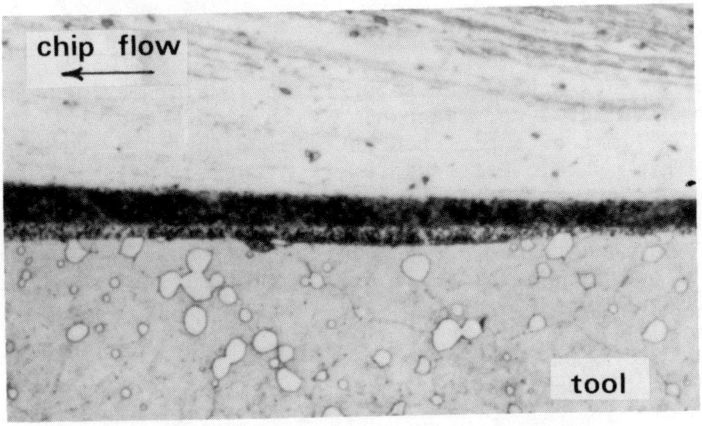

Fig. 4. Degeneration of Tool Material to Form Darker Etching Layer is Evidence of Diffusion and Solution of Tool Material (x1500).

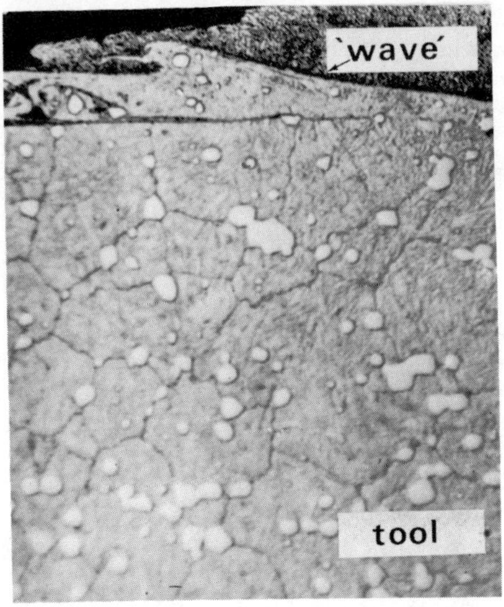

Fig. 5. Superficial Plastic Deformation of Tool Rake Face Surface at Rear of Crater. The Chip Carries Tool Material Away in a Characteristic 'Wave' Formation. Machining Medium Carbon Steel (x1200).

of the tool material. In this case a dark etching layer has been formed by the diffusion of tool material elements into the lighter etching chip material. In this work such areas were also observed on the flank face of the tool suggesting that wear in this area can be partially controlled by diffusion.

2.6. Superficial Plastic Deformation

This process has been shown to be a major influence in crater wear rate, when machining with high-speed steel tools near the limit of their endurance [1]. The chip, deformed at a very high strain rate, can exert a sufficient shear stress onto the surface layers of the tool to deform the latter at a low strain rate and 'drag' tool material from the crater region. In many respects such a mechanism, relying on intimate contact and tangential shear of the tool material, can be regarded as a gross form of adhesive wear (Figure 5).

2.7. Plastic Deformation of the Cutting Edge

The most serious form of wear involves thermal weakening of the cutting edge region and deformation under the applied normal load. Once this deformation occurs the blunt edge causes an additional heat source as it rubs on the machined bar, further weakening the material and leading to plastic collapse. The deformation of a high-speed steel tool is shown in Figure 6 and similar deformation has been found in carbides [2]. At high speeds this is one of the most dominant wear mechanisms as described below.

3. TOOL PERFORMANCE

Tool performance is usually judged in terms of crater wear or flank wear. In fact there is now growing evidence to suggest that a large crater in the rake face of the tool may not necessarily impair tool performance. Indeed in some experiments it has been shown that a deep crater in the tool increases the effective rake angle in much the same way as a controlled-contact design or an insert with a sintered in chip-breaker.

A point of emphasis in this paper is that the wear around the nose radius is of most significance commercially. Wear in this region reduces the pre-set depth-of-cut, thus making

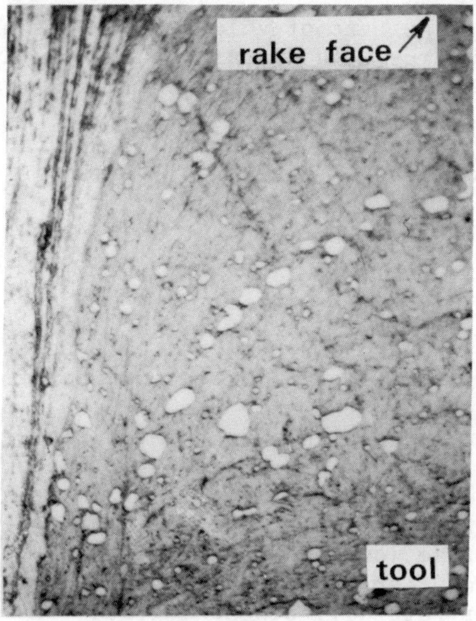

Fig. 6. Section Directly Through Nose Radius of Tool used to Machine Medium Carbon Steel at Limit of Performance Shows Plastic Deformation and Edge Rounding.

the turned component oversize. Despite this obvious fact research workers concern themselves predominantly with crater wear and flank wear mechanisms. While the features on the flank generally resemble the wear land on the nose radius, the latter is frequently deeper and may also be affected by the groove due to the oxidation-adhesion wear process (2.4). An excellent example of this is shown in Figure 7.

Fig. 7. Nose Radius Wear on ISO Grade P20 Insert Used to Cut 0.35% C Steel. Both a Wear Land and a Groove (Arrowed) Due to the Oxidation and Subsequent Adhesion Mechanism are Evident.

Considering the rate of volumetric loss of tool material from the nose radius with distance machined $(dv/dL)_{nr}$ the wear rate will be written in the schematic form,

$$(dv/dL)_{nr} = \text{function}\left[\frac{KP_n}{H^*}\ ;\ \frac{GP_nV'}{H}\ ;\ \frac{\xi_1,\xi_2}{\sigma_c}\ ;\ A_o\exp\left(\frac{-E}{RT}\right)\ ;\ Y^*\right] \quad (3)$$

The five functional terms state that nose radius wear depends on the mechanisms of adhesion, abrasion, oxidation, diffusion and gross plastic deformation. In the first term H^* denotes the difference in hardness between tool and chip (the adhesion couple) at the prevailing flank face temperature. This term has been presented in the same form as equation (1) and clearly if H^* remains high, and the tool is kept as sharp as possible so as to keep P_n low, then the adhesive wear contribution to the wear rate will be low. In the second term H (as in equation 2) is the hardness of the tool and ideally this should not be lowered by thermal effects. V' is the volume fraction of abrasive particles. In the oxidation-adhesion term, wear rate will depend initially upon how readily oxides form on both chip and tool ξ_1 and ξ_2. Once a bond is formed between such oxides, tool loss by adhesion is then primarily controlled by the cohesive strength of the tool material, σ_c, which ideally should be high. The fourth term is concerned with the diffusion from the nose radius region into the adjacent workmaterial and thus appears in Arrhenius form with A_o a constant, E the activation energy and T temperature. Finally Y^* conveys the difference between the elevated-temperature, compressive yield strength of the tool edge and normal stress acting on the rake face. The latter is related to the shear strength of material in the primary deformation zone of chip formation and also the material flow patterns around the edge.

4. VARIATION OF THE DOMINANT WEAR MECHANISM WITH MATERIAL PROPERTIES

The parameters in equation (3) relate to the machinability of a particular workmaterial and tool performance. Since the wear rate, $(dv/dL)_{nr}$, is a measure of tool life, T, the ultimate aim of the present research is the ability to rewrite equation (3) in Taylor form ($T = AV^{-B}$, where V is velocity and A,B are constants) so that each of the five main wear mechanism parameters are expressed in terms of a quantitative exponent. The ranking of these exponents would then be related to the dominant wear mechanism for a specific tool-work system.

4.1. Order of Influence

In order to achieve the above, the five wear rates must be correlated with the physical and mechanical properties of both tool and workmaterial. The 'first-order' variables which control the nature of these materials are chemical composition and microstructure. While it would be of major benefit to be able to express tool life solely in terms of composition, the evidence suggests that many factors control wear. Thus current experiments are being carried out with reference to 'second-order' parameters such as cutting temperatures and the stresses acting on the tool. For the tool material, hot-hardness has been identified as one of the most important variables, influencing particularly, the mechanisms of adhesion, abrasion and plastic deformation of the edge. For the workmaterials, the elevated-temperature, high-strain-rate shear flow strength of the material is considered to be one of the most important parameters. The proportion of abrasive particles in the workpiece, and the solid solubility of tool material elements in the workmaterial, are also dominant influences in certain cutting conditions. Some other variables, of higher order, which also affect machinability are tool geometry, vibration and lubrication. These orders of influence are summarized in Table I and further discussed in relation to the two examples of the dominant wear mechanism approach given below. Again, these examples are concerned with wear of the nose radius region.

4.2. Workmaterials That Promote Abrasion

The detailed mechanism of abrasion has been discussed in previous work [1,4], comparing two stainless steels of similar bulk properties but different inclusion content. One grade, A.I.S.I. type 316, was titanium stabilized and contained hard abrasive particles whereas the other grade did not contain such particles. Figure 8 shows how the rates of flank wear, similar to nose radius wear for this case, vary with time. The abrasive particles clearly promote a higher wear rate for all the cutting time range shown, but have more influence in the first three minutes. The evidence indicates that the abrasive particles are particularly active while the cutting edge is sharp, but that as the land increases they are more likely to be embedded in the tool and remove less material. Despite the fact that abrasion plays a dominant role in this example, and is prominent in terms of equation (3), it should be noted that there is still *some* wear of the tools used to machine the non-abrasive-carrying steel. The micrograph in Figure 1 is a section through a tool used to machine the A.I.S.I. grade 304 steel and indicates that adhesive wear is important in this case.

Table I. Significance of parameters in assessing machinability. The second order effects (II) are indirect measures of chemical composition and microstructure but they can be investigated in machining experiments.

I	Chemical composition and microstructure of tool and workmaterial.
II	a) Cutting temperatures in relation to tool hot-hardness
	b) Elevated temperature, high-strain-rate, flow strength of work material
	c) Normal stress acting on region behind tool edge
	d) Adhesive strength of chip and tool around cutting edge
	e) Mutual solubility between chip and tool
	f) Volume fraction of hard abrasives in workmaterial
III	a) Tool geometry and nose radius
	b) Machine tool vibration
	c) Cutting fluids as lubricants and coolants

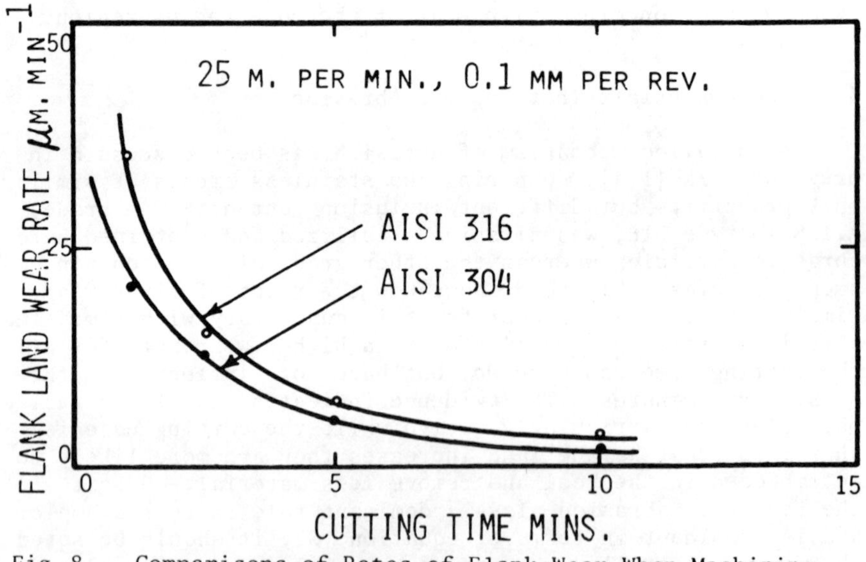

Fig. 8. Comparisons of Rates of Flank Wear When Machining Stainless Steel that Contains TiC Particles (316) and a Stainless Steel Without Such particles (304).

4.3. Workmaterials That Promote Plastic Deformation of the Tool Edge

Comparisons have been made between a medium carbon steel (0.45% C), an austenitic manganese steel (1.46% C, 13.6% Mn) and commercially pure nickel. These caused plastic collapse of cemented carbide tools at very different speeds, despite the fact that the room-temperature hardnesses were fairly similar. Such observations focus attention on the elevated-temperature, high-strain-rate, shear flow strength of the workmaterial. The flow strength in the primary deformation zone, τ_p, has been calculated by resolving forces onto the shear plane, and the mean primary zone temperature, θ_p, has been found from the rate of work done [6]. In addition, the forces have been resolved onto the contact area in order to calculate the mean normal stress σ_N. Table II summarizes this data for a number of materials used in the present work.

Table II. Comparison of machining data at a speed of 60 m/min showing primary shear zone stress (τ_p) and temperature ($\theta_p°$) and normal stress on rake face σ_N.

Material	τ_p,MPa	θ_p°C	σ_N,MPa
Austenitic Mn Steel	1100	470	1005
Comm. Pure Nickel	900	627	371
Stainless Steel (316)	865	378	348
Medium Carbon Steel	725	290	278
Comm. Pure Copper	470	180	192

It is of particular interest that the ranking of the normal stress acting on the rake face also ranks these materials in terms of machinability. The high elevated temperature strength of the austenitic manganese steel also influences the features in the secondary shear zone. In Figure 9, a section through a 'quick-stop' sample, intense shear strains are evident around the edge of the carbide tool. Previous 'quick-stops' on copper [2] and work with the transparent tool [5] indicate that the flow patterns with softer materials are markedly different. In these cases the adhesion between the chip and the part of the rake face immediately behind the edge is much less and sliding rather than seizure occurs. As a result the tool temperature profiles that arise with the materials listed in Table II differ in form. With the difficult-to-machine alloys, edge temperatures are not a great deal less

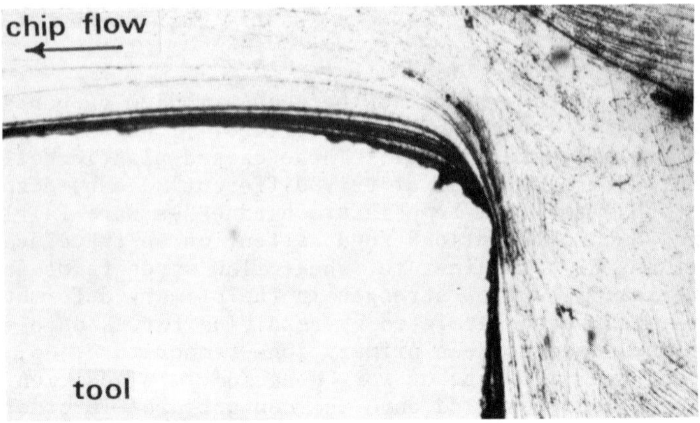

Fig. 9. Intense Seizure Flow Pattern Around Edge of ISO grade P20 Tool Machining Austenitic Manganese Steel. 'Quick-Stop' Section (x400).

than the maximum temperature further along the rake face. Such temperature profiles can be obtained using a recently developed metallographic method [1,3,6] and the results for nickel show that thermal weakening of the edge occurs at low speeds. Thus, plastic deformation of the tool edge and nose radius emerges as a dominant mechanism. For the materials of lower strength, such as medium carbon steel and copper, the cutting edge region is much cooler than the rake face. As a result, flank wear mechanisms such as adhesion or abrasion, that are more associated with lower temperatures, are proposed to be dominant in these cases.

5. CONCLUDING REMARKS

A main emphasis of this paper is that the wear land formed on the clearance surface below the nose radius of a turning tool is the area of most commercial significance. Five main mechanisms of wear have been proposed for this surface, namely (i) adhesion, (ii) abrasion, (iii) oxidation and subsequent adhesion wear, (iv) diffusion and (v) plastic deformation of the cutting edge. A qualitative model, outlining the parameters which affect these individual wear mechanisms, has been presented as a basis for a future quantitative analysis. The discussions and supporting micrographs indicate that several of the mechanisms can operate simultaneously, and that the dominant wear mechanisms vary according to the tool-work combination. For example, in comparative machining tests on two grades of stainless steel, one containing hard TiC particles and the other without such particles, the evidence indicates that abrasion is the dominant form of wear in the former case whereas adhesive wear plays the major role in the latter case. In other experiments, on work-hardening materials such as austenitic manganese steel and on aerospace alloys, the elevated-temperature, high-strain-rate flow strength of the workmaterial has been found to remain high. This has resulted in high normal stresses on the tool face and the dominant form of wear being plastic collapse of the cutting edge - nose radius region.

ACKNOWLEDGMENT

We are grateful to the International Harvester Company for interest in this study, and their continuing support of work with transparent sapphire tools.

REFERENCES

(1) P.K. Wright and E.M. Trent, "Metallurgical Appraisal of Wear Mechanisms and Processes on High-Speed Steel Tools", Journal of Metals Technology, Vol. 1, 1974, p. 13.

(2) P.K. Wright, J.G. Horne and D. Tabor, "Boundary Conditions at the Chip-Tool Interface in Machining: Comparisons Between Seizure and Sliding Friction", Journal of Wear, Vol. 54, 1979, p. 371.

(3) E.M. Trent, Metal Cutting, Butterworths, London, 1977.

(4) S. Ramalingam and P.K. Wright, "Abrasive Wear in Machining: Experiments with Materials of Controlled Microstructure", submitted to Journal of Engineering Materials and Technology, 1980.

(5) E.D. Doyle, J.G. Horne and D. Tabor, "Frictional Interactions Between Chip and Rake Face in Continuous Chip Formation", Proc. R. Soc. London, Vol. A366, 1979, p. 173.

(6) P.K. Wright, S.P. McCormick and T.R. Miller, "Effect of Rake Face Design on Cutting Tool Temperature Distributions", J. of Engineering for Industry (ASME), Paper No. 79-WA PROD/3 to be published in Transactions, 1980.

DISCUSSION:

Bertil Colding (Sandvik)

The following are general comments applying to a great number of presentations made by researchers in conferences like this one. They are partially applicable to Dr. Wright's paper.

Why this emphasis on speed-feed regions where built-up-edge (b.u.e.) occurrs? Industry tries to operate with economic machining data where b.u.e. is not present. Work should be directed primarily at research pertaining to productive machining data. I think the main reason for this concentration on b.u.e. in Universities is due to the existance of poor lab equipment (old, shaky, low HP etc.), even today. While technical Universities in Germany and Scandinavia possess up-to-date machine tools most U.S. counterparts do not. In order to make the stated objectives in Dr. Wright's paper (on wear research) feasable in the 1980-1990 time period, it is essential that modern equipment be obtained. I suggest SME/ASM aid by providing funding for all colleges where Manufacturing education is given in order to modernize its research for the benefit of the student and industry.

Dr. Wright

I would like to reply to that by saying that if Professor Colding or the SME would like to give me a check for something like $150,000 I would be only too delighted and would rush off and buy a machine instantly.

I don't do much work on b.u.e. research for the reason you suggested. I showed the slides because they illustrate metallurgically whats going on there. You did metion drilling and ofcourse in a drill point one goes down to a speed of zero feet/minute at the center, so there are parts along a drill where a b.u.e. could form if its not pushed out of the way. However I still agree with you that at no point should we be hung up on b.u.e. research. I think that one can also get an accelerated wear on taps from the b.u.e.

In doing University research it is quite difficult to get the money for this high powered equipment together. In any kind of computer-aided manufacturing research the sums of money are even worse. One talks about $100,000 for a robot and 1/4 million dollars for a flexible machining center. One can put a question mark against University research in manufacturing unless the SME and other societies are willing to put considerable money into it.

Dr. VK Sarin (GTE Laboratories)

Talking about diffusion, have you considered when you were measuring temperature any diffusion effects that would change the structure and so give you a false temperature reading; and part "b" of that question is how accurate a temperature are you getting?

Dr. Wright

The accuracy depends on the range but without going into too much detail, the accuracy is ± 20 to 25 degrees. To come back to your first question, the point about diffusion wear affecting the surface of the tool is true for, perhaps, the first 10 to 20 microns of the surface. Below that I don't think the diffusion wear, if there is any, will have a substantial effect; so that the extrapolation to the surface is ok.

IN-PROCESS DETECTION OF TOOL BREAKAGE BY MONITORING ACOUSTIC EMISSION

Yoshiaki KAKINO
Associate Professor, Dept. of Precision Engg.,
Kyoto Univ., Yoshida, Sakyo-ku, Kyoto 606, JAPAN

Abstract

The in-process method for detecting the breakage of the cutting tool has been investigated by use of acoustic emission. The following results have been obtained. (1) For a case of normal cutting acoustic emission was detected mainly in the frequency range below 100kHz, whereas the tool breakage was associated with the acoustic emission detected mainly from 100 to 300kHz. (2) By monitoring the acoustice emission detected at the instance of the tool breakage and differentiate it from normal cutting even at interrupted cutting. (3) The amplitude of the acoustic emission detected at the instance of the tool breakage was approximately proportional to the second order root of the surface area generated by the fracture. (4) The appropriate locations for the detector to be mounted on the types of machine tools used in these experiments were the tool post for the lathe, the spindle head for the milling machine and the spindle quill for the drilling machine.

1. Introduction

The in-process detection of tool breakage is very important for the automation of machining. Hence, various experiments about endeavors to cope with this problem have been made in recent years. For example, such main endeavors include ones utilizing the changes in vibration[1)], cutting force[2)], motor power[3)], surface roughness of the machined part[4)] and so on.

These methods, however, have some advantages and disadvantages and have many practical problems. First of these is the problem of reliability, and second is the particular devices are required for the structures of machine or cutting tools which have extremely adverse effects on cutting conditions. The third is the cost problem and fourth, most methods are difficult to apply to intermittent cutting even if they are applicable to continuous cutting.

In this study, the detecting method of the cutting tool breakage has been investigated, by using Acoustic Emission (hereinafter abbreviated as AE) which has comparatively less problems. Possibility of the detection of the cutting tool by means of AE has already been reported by Iwata and others[5)].

In this study, in order to investigate the basic properties of AE generated at the cutting tool breakage and at normal cutting, various materials were turned with various cutting tools, and the frequencies of AE generated in such cases were analysed. To investigate the relation between the magnitude of cutting tool breakage and the amplitude of AE signals, the experiments by the use of a planer were carried out. Finally, locations appropriate for the AE detector disposition were examined for three different types of machine tools.

2. Experimental apparatus and method

2.1 Frequency analysis of AE generated at normal cutting and cutting tool breakage.

To investigate the basic properties of AE generated during machining, 10 kinds of materials for common industrial use were cut by using 11 kinds of cutting tools. Table 1 shows the composition and hardness of the various materials used in the experiments.

Figure 1 shows the outline of the experiment set-up and the signal processing employed in the AE frequency analysis. Signals detected by the AE detector "A" mounted on the rear end of the tool holder is 40dB amplified by a pre-amplifier and an amplifier inserted in the discriminator, and stored in a digital memory. At normal cutting, the memory updating is stopped by the manual operation of the external trigger of the digital memory. At the cutting tool breakage, signals detected by the AE detector "B" mounted on the side face of the tool holder are amplified likewise by 40dB. Thereafter components ranging from 100kHz to 1MHz are picked up through a band-pass filter, and the detected signals are used as the external trigger signals of the digital memory. In both cases, data sampling interval is 0.2μs and the record of 2048 in the 8-bit format is performed. These data are reached with an interval of 20ms, thereby the power spectra are depicted by using a real-time analyser.

The main measurement equipments used in the experiments are as follows.

o Equipments associated with AE

Detector: PZT piezoelectric pick up AE 905S (NF Circuit Design Block Co.)
Pre-amplifier: AE911
Discriminator: AE921

Table 1 Chemical composition and hardness of the materials used in the experiments.

(unit: %)

Material	C	Si	Mn	P	S	Cu	Ni	Cr	Mo	Co	Res.	Hardness (HB)
S48C	0.50	0.24	0.84	0.018	0.023	0.01	0.02	0.11	—	—	—	440
S15C	0.17	0.23	0.36	0.08	0.016	0.15	0.09	0.07	—	—	—	120
SK4	0.99	0.19	0.37	0.018	0.011	0.19	0.08	0.10	—	—	—	293
FC20	2.78	1.28	0.53	0.17	0.094	—	—	—	0.04	—	—	197
SNCM8	0.39	—	0.71	—	—	—	1.85	0.87	0.22	—	—	324
SUS50B	0.11	0.32	0.85	0.02	0.01	—	—	12.9	—	—	—	249
SUS 304	0.07	0.60	1.63	0.027	0.006	0.04	9.50	18.80	0.11	0.17	—	282
Nimonic	0.06	0.06	0.04	0.0004	0.0004	—	Res.	19.73	—	0.13	Al 1.41 Fe 1.79	331

Material	Mn	Cu	Ni	Fe	Al	Sn	Pb	Zn	Hardness (HB)
Brass	0.32	58.39	0.12	0.51	0.49	0.43	2.81	36.95	140

Material	H	O	N	Fe	C	Ti	Hardness (HB)
Pure Titanium	0.026	0.063	0.007	0.029	0.009	Res.	127

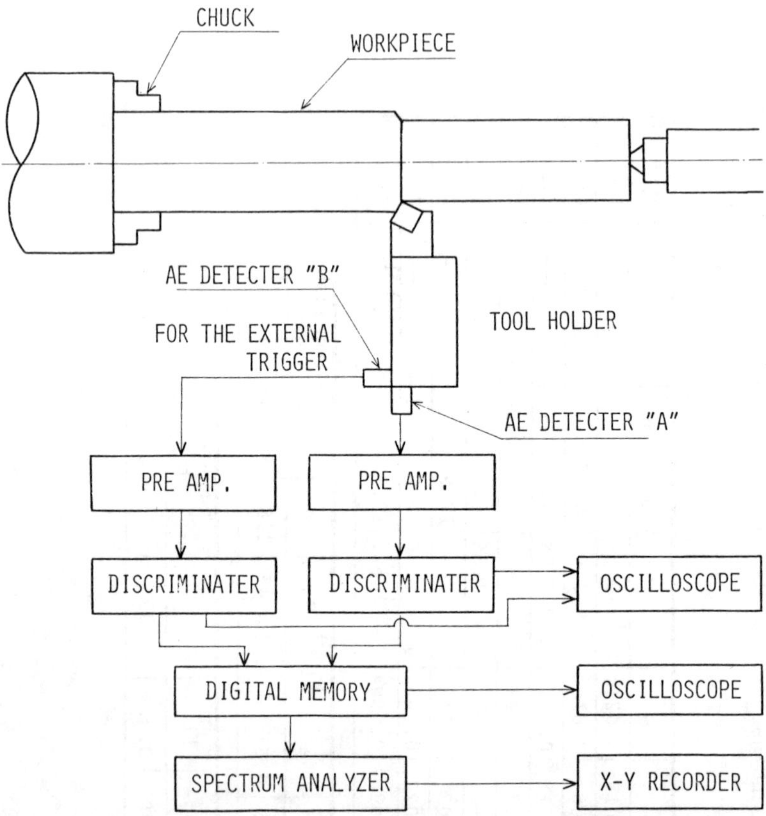

Fig.1 Measuring set-up for power spectrs of AE generated at normal cutting, chip clinging and cutting tool breakage.

- Digital memory: DM-901 (Iwasaki Tsushinki Co.)
- Real-time analyser: 440A (20kHz) (Nicolet Co.)

Table 2 shows the combinations of work and tool materials for the experiments, and the cutting conditions. For the measurement at tool breakage, conditions are selected where the tool breakage is most likely to take place corresponding to a given tool materials. Workpieces with a groove of 20mm width were cut when the tool material is of a difficult-to-break type.

Table 2 Cutting conditions used in the experiments.

(a) Where work materials are varied (normal cutting).

Material	Tool	Cutting speed (m/min)	Deoth of cut (mm)	Feed (mm/rev)
S48C	P20	120	2.0	0.4
S15C	P20	120	1.0	0.32
SK4	P20	100	1.0	0.315
FC20	P20	120	1.0	0.32
SNCM8	P20	120	1.0	0.32
SUS50B	P20	100	1.0	0.4
SUS304	P20	70	1.0	0.32
Brass	P20	120	1.0	0.315
Pure Aluminum	P20	120	1.0	0.315
Pure Titanium	K10	70	1.0	0.4
Nimonic	P30	40	1.0	0.2

(b) Where cutting tools are varied (normal cutting, cutting tool breakage)

Material	Tool	Cutting speed (m/min)	Depth of cut (mm)	Feed (mm/rev)
S48C	P10	120	1.5	0.4
	P20	120	2.0	0.4
	P30	100	2.0	0.4
	P40	80	2.0	0.4
	M10	120	1.5	0.4
	M20	120	2.0	0.4
	K10	120	1.5	0.4
	K20	100	2.0	0.4
	Cermet	160	2.0	0.4
	Ceramic	20	1.5	0.63
	SKH4	36	1.5	0.36

2.2 Relationship between magnitude of tool breakage and amplitude of AE signals.

In this experiment, the work material is limited to S48C (0.48% plain carbon steel), being cut with two kinds of sintered carbide tool, P10 and P20. These experiments were carried out by using a planer. By this a detailed examination is easily made on the relation between the tool breakage caused during machining and the AE signals.

That is, motion of the table was stopped at every cut to observe the state of the cutting tool. When an abnormal state appeared at the tip a microscopic photograph was taken. This procedure was repeated until the tool was completely broken. When a small chipping appeared in the tool, the fractured surface area was measured by using a microscope and when a great fracture occurred, the surface area was measured by copying it on a sheet of paper with the fractured surface painted. The AE signals detected by the detector mounted on the rear end of the tool holder were 40dB amplified. This passes through a band-pass filter with a pass band width of 100kHz to 1MHz.

Further after being envelope-demodulated it was finally recorded on the data recorder (TEAC R410, nominal frequency range; DC to 5kHz), data were reproduced on a storage oscilloscope to read their amplitudes.

The experiments were carried out under selected conditions where the tool is fractured as easily as possible. Furthermore, in order to cause different-sized tool breakages, the cutting was made with proper combinations of the following conditions of cutting speed, feed and depth of cut.

1. Cutting speed: 13, 22 and 28 m/min.
2. Depth of cut: 1, 1.5, 2, 2.5 and 3mm
3. Feed: 0.4, 0.65, 0.8 and 1.0mm/stroke

2.3 Mounting location of the detector

In order to investigate the difference between the AE signals depending upon the mounting location of the detector, the experiments were carried out by using the following three typical types of machine tools.

- Lathe (West Germany VDF high speed lathe S500)
  Swing over bed: 520 mm, Maximum center-to-center distance: 1,000mm, Spindle power: 15kW, Maximum spindle revolutions per minute: 3,030 rpm.
- Milling machine (Hitachi Seiki MK 3)
  Stroke: X850mm, Y350mm, Z500mm, Spindle power: 19kW, Maximum spindle revolutions per minute: 1,500 rpm.
- Drilling machine (Yoshida Tekkosho YUD-650)
  Table stroke: 200mm x 570mm, Spindle power: 3.7kW, Maximum spindle revolutions per minute: 1,500 rpm.

The experimental set-up and processing method are just the same as in Section 2.2 except for the mounting position of the detector and the employment of two sets of AE instruments. First, a detector to be standard is mounted on the standard location for each machine (lathe: rear end of tool holder, milling machine: spindle head, drilling machine: workpiece), and another detector is mounted on a characteristic location of each machine tool, (refer to the below-mentioned Figs. 7, 8 and 9). The AE signals at tool breakage are detected simultaneously at two locations, and the amplitude of the AE signals detected by the detector mounted on the standard location is compared with theat mounted on the other location.

In the experiment S48C (0.48% plain carbon stell) was used as the work material, and the other cutting conditions are as follows:

- o Turning
  Cutting tool: ceramics, depth of cut: 1.5mm, feed: 0.8mm/rev., cutting speed: 20m/min.
- o Milling (face milling)
  Diameter of milling cutter: 6" (single blade, cutting tool: sintered carbide tool P20, depth of cut: 1.5mm, feed: 1.76mm/tooth, cutting speed: 40.7m/min.
- o Drilling
  Cutting tool: high speed steel SKH2, diameter: 5mm, revolutions per minute: 210 rpm, feed: manual operation.

(The table is moved manually in the horizontal direction while cutting feed is applied, thereby drills are forcibly broken.)

3. Results of experiments and consideration

3.1 Frequency analysis of AE generated at normal cutting and cutting tool breakage.

Figure 2 shows a typical example obtained through the frequency analysis of AE measured at normal cutting and the cutting tool breakage when S48C (0.48% plain carbon steel) is cut with the sintered carbide tool P20. AE generated at plastic deformation is considered to be the major part at normal cutting and in this case, it is found that the high power components exist only within the range below 100kHz. At the chip-clinging, the AE level becomes high compared with that at normal cutting, while the distribution is nearly unchangeable. On the contrary, high power components exist within the range from 100 to 300kHz at tool breakage, and such components are considered principally as AE generated at the tool material breakage.

But the AE level at normal cutting as shown in Fig. 2 also becomes high when the cutting speed, feed and the depth of cut become large; that is, when the chip removal rate becomes high.

Almost the same results as those shown in Fig. 2 were obtained both at normal cutting and the tool breakage for all kinds of work materials and cutting tools in the scope of these experiments. The detecting method of tool breakage as shown in Fig. 3 has been devised on the basis of these results; that is, after the AE signals detected

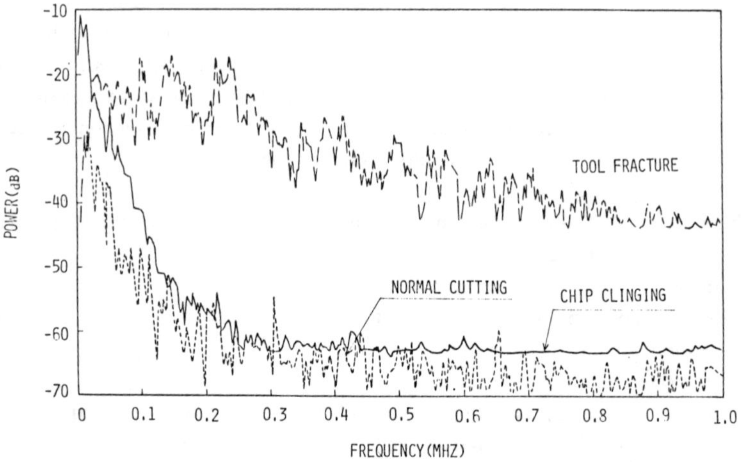

Fig.2 AE power spectra detected at normal cutting, chip-clinging and cutting tool breakage. (material used: S48C plain carbon steel, cutting tool:sintered carbide tool P20, gain:40dB, cutting speed:120m/min, depth of cut:2mm, feed:0.4mm/rev.)

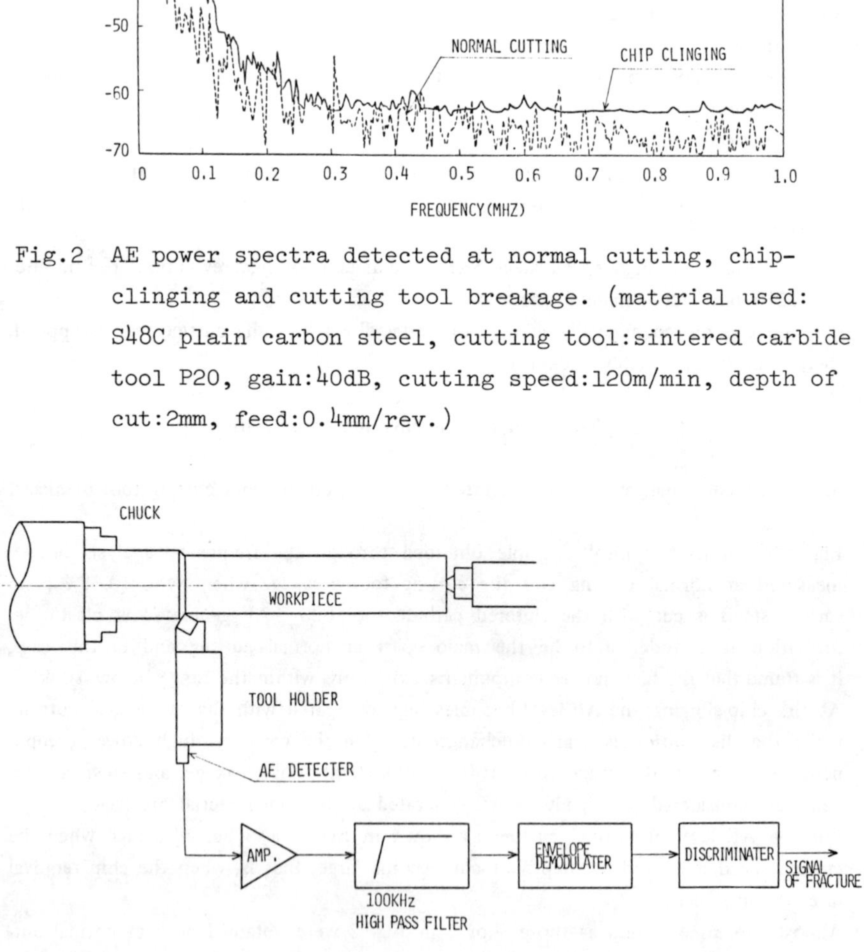

Fig.3 Detecting method of tool breakage by AE.

during machining are amplified, they are passed through a high-pass filter of 100kHz, being envelope-demodulated and entered into a discriminator. When a signal above the reference voltage is detected the discriminator outputs the signal of breakage. The experiments described in Section 2.2 and thereafter are based on these results. This method has the following advantage. Now a case is considered where a material with a groove is cut. When the entire frequency range of AE is de-tected, a nearly constant amplitude of AE is generated under a certain cutting condi-tion. Corres-ponding to such condition, therefore, some abnormality can be known by finding out the generation of AE greater than this as shown in Fig. 4 (a). However, in a case where the chip removal rate is varied such as the variable speed cutting and the

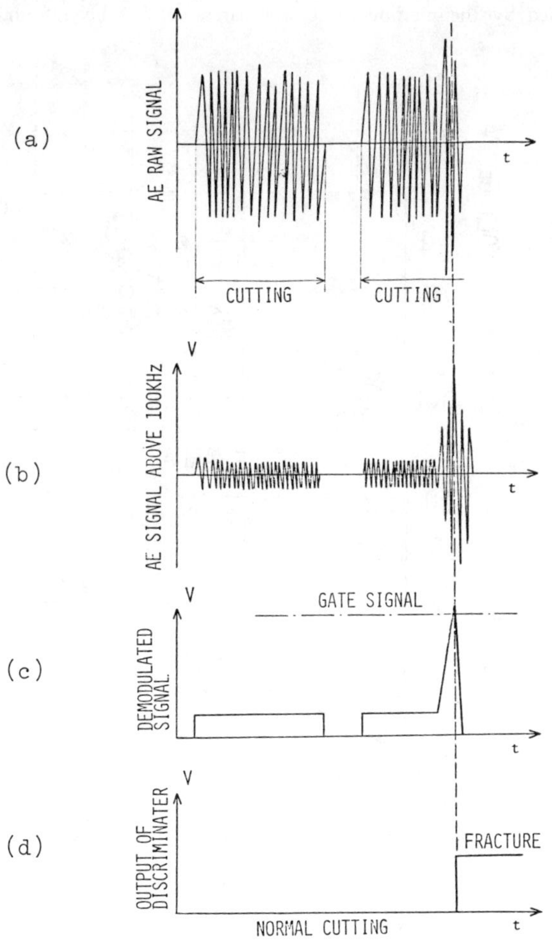

Fig.4 State of signals at each process in Fig.3.

variable depth cutting, the reference voltage must be varied corresponding to the case. In the case where only AE above 100kHz is detected, there is no need to vary the reference voltage even if the chip removal rate is varied, and also no reference voltage change is required to vary for different kinds of work materials.

3.2 Relationship between magnitude of tool breakage and amplitude of AE signals

Fig. 5 (a) and (b) show the relation between the surface area S(mm^2) of tool breakage occurring at the planing of S48C (0.48% plain carbon steel) with the respective sintered carbide tools P10 and P20, and the amplitude of the AE signals V(V) detected by the detector mounted on the rear end of the tool holder. The solid lines in the figures were obtained by the method of least squares. It has been found that con-

(a) P20

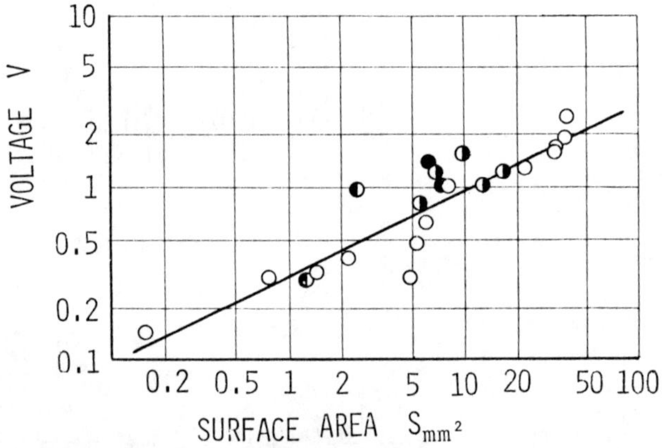

(b) P10

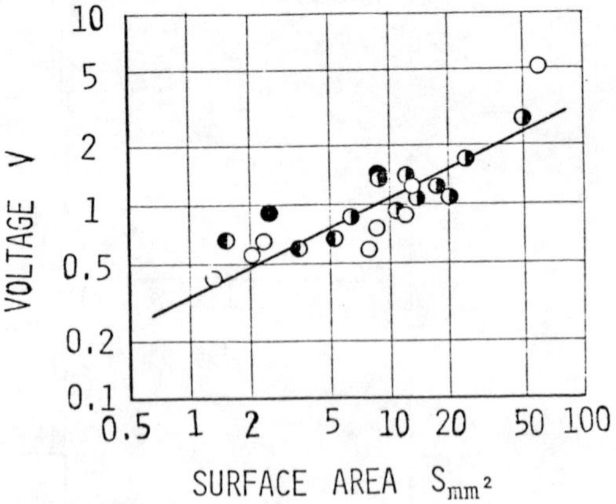

Fig.5 Relation between magnitude of tool breakage and amplitude of detected signals.

siderable amount of variation exists in these figures; but the slopes of the two straight lines are 0.49 and 0.47, both being close to 0.5, and the amplitude of the detected AE signals V is in proportion to the square root of the surface area S of tool breakage. The presence of such relation is practically of much convenience, that is, in the machining, workpieces are often machined in sequence in separate stages of rough machining, semi-finish machining and finish machining. The finest fracture is required to be detected according to this sequence. Conversely, some extent of chipping is often allowable in the rough machining. Consequently, the magnitude of fracture which must be detected in each machining stage is also different and the reference voltage corresponding to the magnitude of fracture which is desired to be detected can be preset from the relation as shown in Fig. 5.

Among the dots plotted in Fig. 5, the black ones do not show the voltages detected at final fracture. In this case, a large AE signal was detected at the cutting, one or two rotations before the final fracture. But no notable fracture or crack was seen on the surface of the cutting tool, so the cutting was continued. But while cutting, after one or two rotations, the tool was broken down; only the small AE signals being detected This is considered to be the cause of cracks existing inside the tool. The semi-black dots plotted in Fig. 5 show the cases where cracks were observed on the surface of the tool but no final fracture occurred. In all such cases, the greatest fracture was considered to take place when the maximum AE signal is generated, and then the plotting as made with the amplitude V of the AE signals detected at this time and with the surface area S at the final breakage.

3.3 Mounting location of detectors

3.3.1 Lathe

Figure 6 shows a sample of the AE signals at tool breakage detected by the two detectors mounted on the standard location (rear end of the tool holder) and on the tool post. At the rear end of the tool holder which is the standard location, the gain was 40dB, the signal of 5V being obtained. Since the sensitivities of the AE measuring instruments used for this are not the same, a comparative experiment was carried out with two detectors mounted in the same position. Then the detector mounted on the standard location showed 27% better sensitivity than the other. From these results, if the amplitude of signals obtained at the standard location is taken as 100, the amplitude of signals obtained at the tool post is taken as 38.

Figure 7 shows a comparison of amplitudes of signals obtained at tool breakage which were detected by the detectors mounted on each location by the same method, that is, signals of amplitude 3.6 were detected at the center (C), 1.3 at the tail stock (D), 2.1 at the head stock (E) and 4.9 at the carriage (F) respectively. However, in the case where the detectors are mounted on the rear end of the tool holder, tool post and the carriage, nearly same amplitude of signals were detected, while amplitudes of the detected signals varied in the case where the detectors are mounted on the tail stock, center and the head stock. This is because the distance between the cutting point and the detector varied with the movement of the cutting point. Fig. 7 shows that the amplitude of the detected signals is decreased as the distance from the cutting point, and is further greatly attenuated by passing through the contact surface.

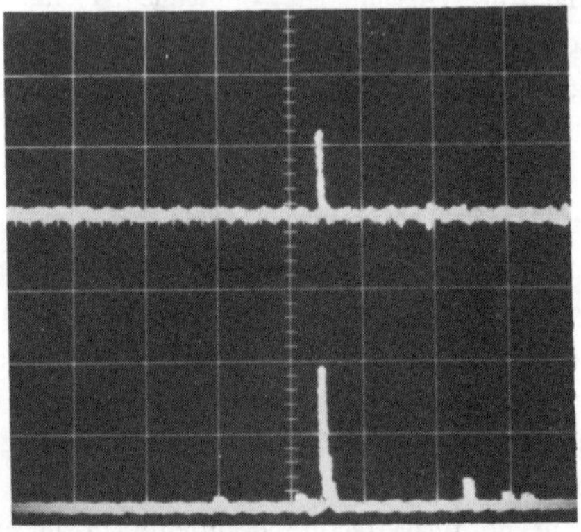

Fig.6 Comparison of amplitudes of tool breakage signals detected by two detectors. (Up: tool post, voltage 1V/div., gain 40dB, Down: tool holder, voltage 2V/div., gain 40dB)

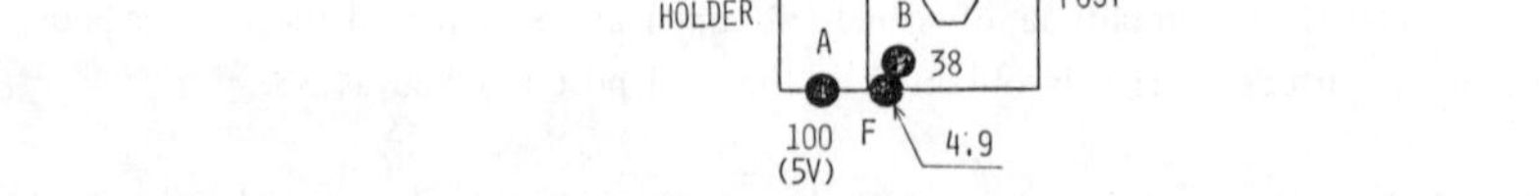

Fig.7 Relation between detector location in the lathe and amplitude of AE signals obtained at tool breakage.

Considering the above results, the maximum amplitude is obtained when the detector is mounted on the rear end of the tool holder. But this case has problems such as tool change, cable breakage due to chip-clining and so on. Therefore, it will be of practical advantage to mount the detector inside the tool post where a secondly large amplitude of signals is obtainable.

3.3.2 Milling machine

Figure 8 shows the relation between the amplitude of the AE signals at the tool breakage measured by the same method as in the case of the lathe and the mounting location of the detector. Taking the amplitude of the signals obtained at the spindle head which is the standard location as 100, signals of amplitude 600 were obtained at the workpiece, similarly 472 at the vise, and 32 at the table. In this case, it will be of practical advantage to mount the detector on the spindle head to which the chips and cutting fluid are difficult to access and at which the distance between the cutting point and the detector is unchangeable.

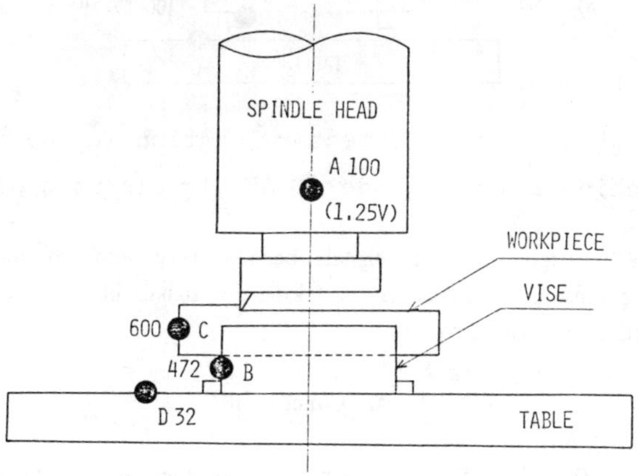

Fig.8 Relation between detector location in the milling machine and amplitude of AE signals at tool breakage.

3.3.3 Drilling machine

Figure 9 shows the relation between the amplitude of the AE signals at tool breakage measured by the same method as in the case of the lathe and the mounting location of the detector. Taking the amplitude of the signals obtained with the workpiece at the standard location as 100, signals of amplitude 27 were obtained at the vise, similarly 3.0 at the table, 4.9 at the spindle quill, and 1.1 at the spindle head. In this case, it is desirable to mount the detector on the spindle quill and the spindle head in view of the prevention of the chip-clinging, but the spindle quill will be

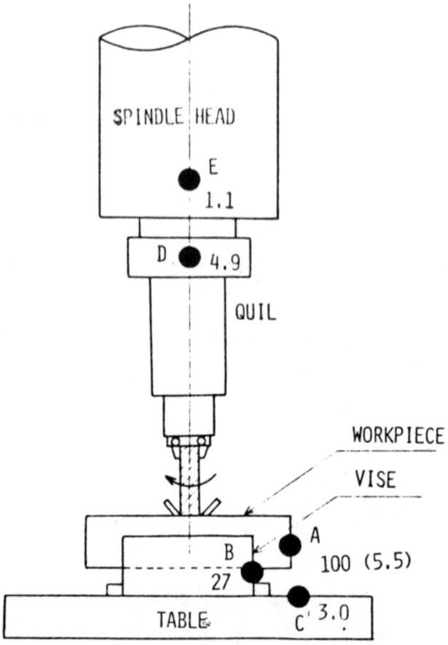

Fig.9 Relation between detector location in the drilling machine and amplitude of AE signals at tool breakage.

an appropriate location since the signals become very small at the spindle head. However, the extension of the cable for taking out signals in response to the amount of quill projection must be devised.

4. Conclusion

The in-process method for detecting the breakage of the cutting tool has been studied by using Acoustic Emission, and the following results have been obtained.

(1) Acoustic Emission (AE) detected at the normal cutting and at the chipclinging existed mainly in the frequency range below 100kHz, and Acoustic Emission (AE) detected at the instance of the breakage of the cutting tool existed mainly in the frequency range from 100 to 300kHz.

(2) Considering (1), the breakage of the cutting tool was able to be in-process-detected even at intermittent cutting by detecting only the AE components above 100kHz with the AE components below 100kHz cut off.

(3) The amplitude of the AE signals detected at the breakage of the cutting tool was approximately proportional to the square root of the fractured surface area.

(4) As for the machine with which these experiments were carried out, the appropriate locations of the detector disposition was the tool post for the lathe, the

spindle head for the milling machine and the spindle quill for the drilling machine.

Acknowledgement

This research work is supported by the Japan – U. S. Cooperative Science Program sponsored by National Science Foundation and Japan Sociey for Promotion of Science.

References

(1) A. Noh et al., "Cutting Tool Failure Detector for Turning," Toshiba Review, (1977–2), p. 120. (In Japanese).

(2) W. Koenig, "Automate Detection of Tool-breakage in Turning Operations", presented at the Meeting of STC "Cutting" of CIRP, (1978).

(3) H. Hirota et al., "Measurement of Main Cutting Force using the Load Characteristics of an Induction Motor", Journal of JSPE, 43, 8 (1977), p. 909. (In Japanese).

(4) H. Takeyama et al, "In-Process Detection of Surface Roughness in Machining", Ann. of CIRP, 25 (1976).

(5) K. Iwata & T. Moriwaki, "An Application of Acoustic Emission Measurement to In Process Sensing of Tool Wear", Ann. of CIRP, 26 (1977).

A NEW FACE MILLING TEST TO EVALUATE CARBIDE TOOLS

A. Ber, S. Kaldor,
E. Lenz

Faculty of Mechanical
Engineering, Technion, Haifa

SYNOPSIS

A short milling test is described for evaluating cemented carbide inserts for milling. A rating system is devised which takes into account both chippage and breakage as well as regular tool wear. The test is simple to perform and can normally be completed within only a few hours.

INTRODUCTION

In the past decade, metal cutting efficiency has improved mostly due to innovations in cemented carbide tool materials. Yet, besides their excellent wear resistance and their survival at relatively high cutting temperatures, these sintered materials also tend to chip and break when loaded by impacts. It has been previously established that both thermal and mechanical loads influence the life duration of a carbide tool [1,2]. The resulting mode of failure can be by chippage, breakage or wear. Chippage and breakage are caused by impact, thermal loading or fatigue. Milling, and especially face milling, is an interrupted cutting operation in which the carbide ultimately fails by one or a combination of these modes.

In this paper, a short milling test is described which was originally designed for the development of milling grades of carbides. The carbide performance rating is based on a combination of these possible failure modes. This test may also be used as a standard short time milling test.

TEST REQUIREMENTS

As a basis for designing the short tool life test for milling, the following requirements were specified:

1) The test should be performed in a machining process as close as possible to production conditions.
2) As fracture is a dominant factor in milling, suitable cutting conditions must be chosen in order to cause failure by fracture. (This is achieved by increasing the feed rate in steps.)
3) Cutting conditions should be selected for minimum tool life scatter.
4) Performance rating must be based on both fracture and wear since both contribute to the termination of tool life.
5) The rating scale should initially be based on the performance data of well known milling grades. The scale is subject to recalibration when necessary.

RATING BASE

Survival of a tested tool or standability "S" can be evaluated in terms of the total cutting work that was performed until failure occurred, while taking into account the influence of the final impact. Assuming a linear relationship, the following equations can be written.

$$\text{"S"} = A \cdot K \qquad (1)$$

$$\text{with } A = F_1(A_c \cdot I_c) \qquad (2)$$

$$\text{and } K = F_2(\Sigma VB) \qquad (3)$$

where A_c - the total milling work
I_c - the final cutting impact
ΣVB- the sum of the radial and axial flankwear [3]

With reference to Fig. 1, it can be shown that

$$A_c = a_{sp} B\, k_c \ell \qquad (4)$$

$$I_c = \frac{F}{t} \propto k_c a_{sp} f_z v_c \qquad (5)$$

$$\Sigma VB = VB_H + VB_N \qquad (6)$$

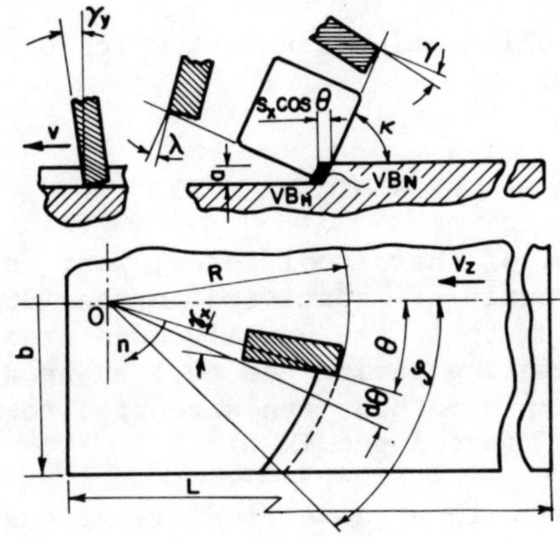

Fig. 1. Face Milling Geometry

where according to ISO standard 3002 III and IV

a_{sp} = back engagement
B = width of milled path
k_c = specific cutting force
ℓ = total length of cutting
F = force
t = time
f_z = feed per tooth
v_c = cutting velocity
VB_N = flank wear measure on primary cutting edges
VB_H = flank wear measure on secndary cutting edges (see Fig.1)

In a specified test some of the parameters do not change and may be considered constant. Therefore, Eqs. 4,5, and 3 may be rewritten

$$A_c \propto \ell \quad (7)$$

$$I_c \propto f_z \quad (8)$$

$$K = F_2\ (\Sigma VB\ , \ell\ ) \quad (9)$$

and consequently

$$"S" = C \cdot f_z \cdot \ell \cdot K \qquad (10)$$

where C is a calibration constant which depends on the workpiece material.

WEAR MEASUREMENT

In the case of sharp tool edges, wear measurement is carried out exactly as specified in the ISO standard [4,5]. However, milling tools are usually edge treated by either honing or chamfering giving the tool an apparent "initial wear." Two simple methods are suggested to account for this "initial wear." (see Appendix A):

1) Compensation with a geometrical relationship
2) Measurement of the "initial wear" on the cutting edge prior to any machining.

With both methods, the "initial wear" value (VB_o) is subtracted from the subsequent wear measurements after each pass. The corrected wear is (see appendix A):

$$\Sigma VB_c = VB_{NC} + VB_{HC} = \Sigma VB - 2VB_o$$

THE TESTS [6]

Tests were carried out according to this procedure and presented to the CIRP as a draft for a "Standard Short Milling Test."* This test was in use for some time in our laboratory and has been found to give accurate results with relatively little scatter. The general procedures of the test are as follows:

1. The Cutter

Only one tooth is in operation. Two types of cutters may be used: (a) a regular 100 mm diameter cutter with all inserts and seats removed but one; (b) a specifically designed single point cutter, well-balanced, 100 mm diameter; and 15° Lead Angle ($\kappa = 75^o$).

* Copies of the draft test procedure are available from the authors.

2. The Workpiece

Three workpiece materials are suggested: (a) a medium carbon steel (AISI 1060); (b) a Cr-Mo steel in the 4000 series (AISI 4140); (c) a stainless steel in the 300 series (AISI 316).

The workpiece dimensions:

Carbon steel and Cr-Mo steel: 60 mm width by 700 mm length with about 200 mm height.

Stainless steel: 350 mm length with the width and height as above.

The workpiece is solidly and rigidly clamped in a specially designed vice as shown in Fig.2.

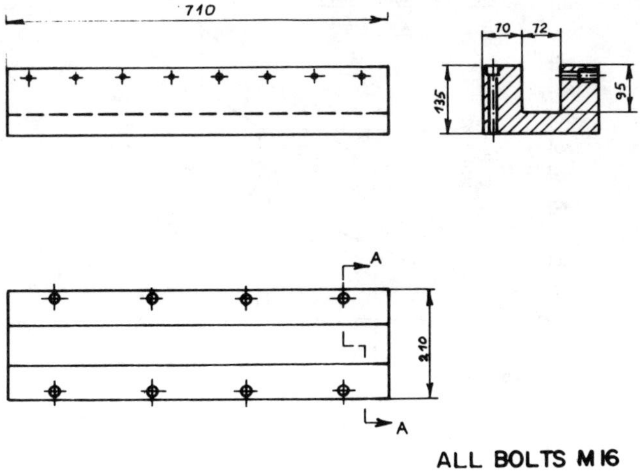

Fig. 2. Material Clamping Device

3. Cutting Conditions

(a) Speed. A cutting speed between 80 to 100 m/min is recommended: 88 m/min was chosen for our standard test. (b) Depth of Cut. A depth of cut between 3-6 mm is recommended: 3 mm was chosen for our standard test. (c) Feeds. The feeds are increased by one step every pass. The following feeds (mm/tooth) were used for our standard test:

(1) carbon steel and Cr-Mo steel: 0.3 0.4 0.5 0.6 0.7 0.8 0.9 1.0 1.1 1.2 mm/tooth

(2) for the stainless steel the feeds are much lower: (0.07) 0.08 0.12 0.16 0.20 0.24 0.28 0.32 mm/tooth

THE RESULTS

The results were normalized and presented as nomograms (see Figs. 3 and 4a-f).

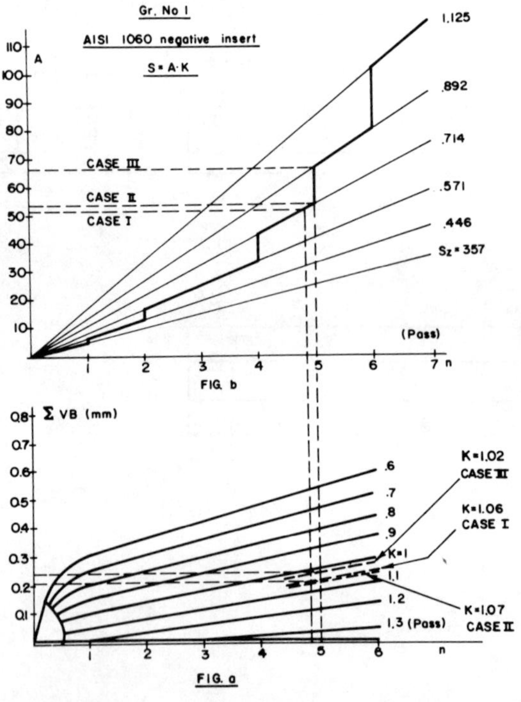

Fig. 3. Examples of rating technique with the nomogram

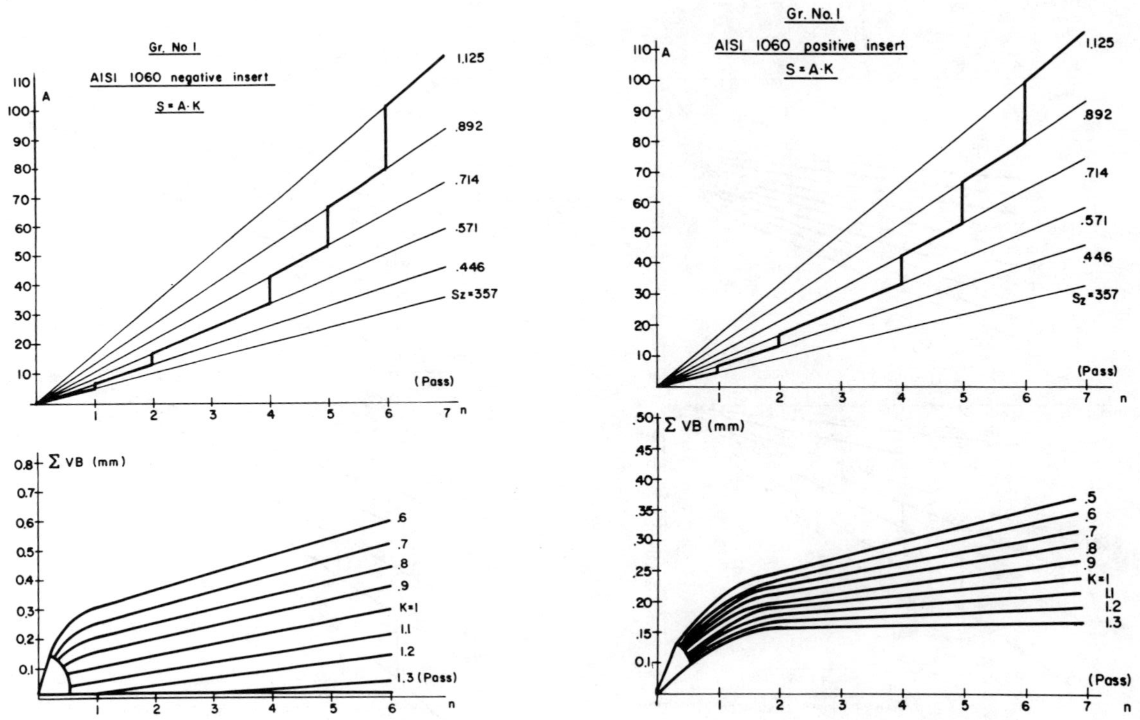

Fig. 4. Rating nomograms for: negative, positive tools and three work materials - (1) AISI 1010-1060

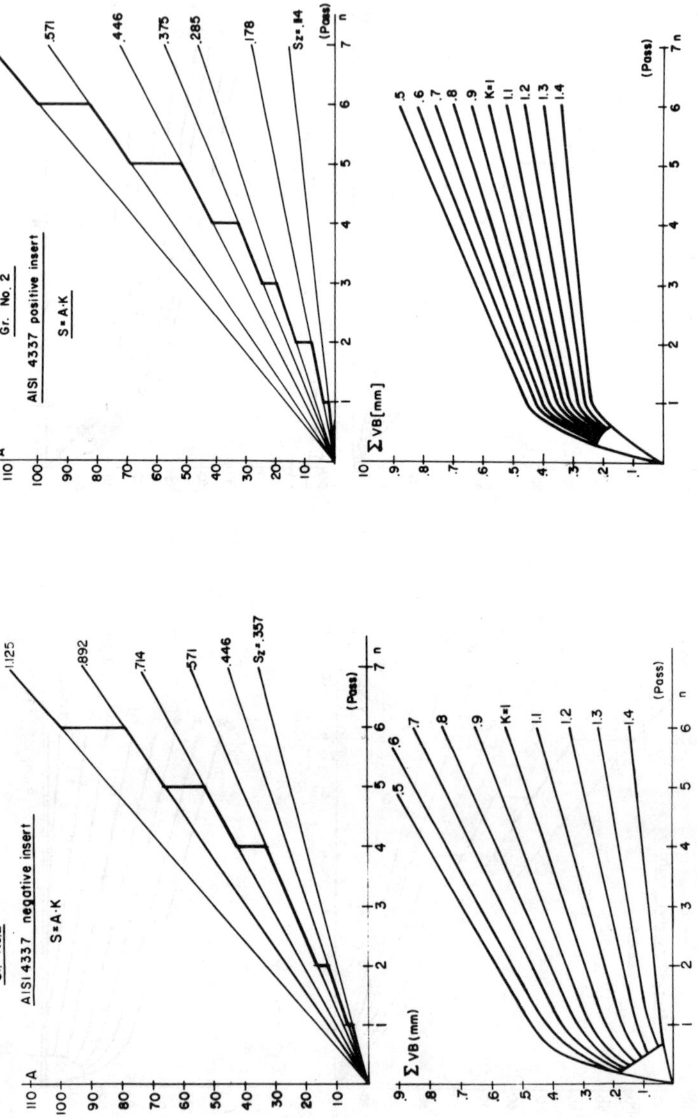

Fig. 4. (2) AISI 4140-4340

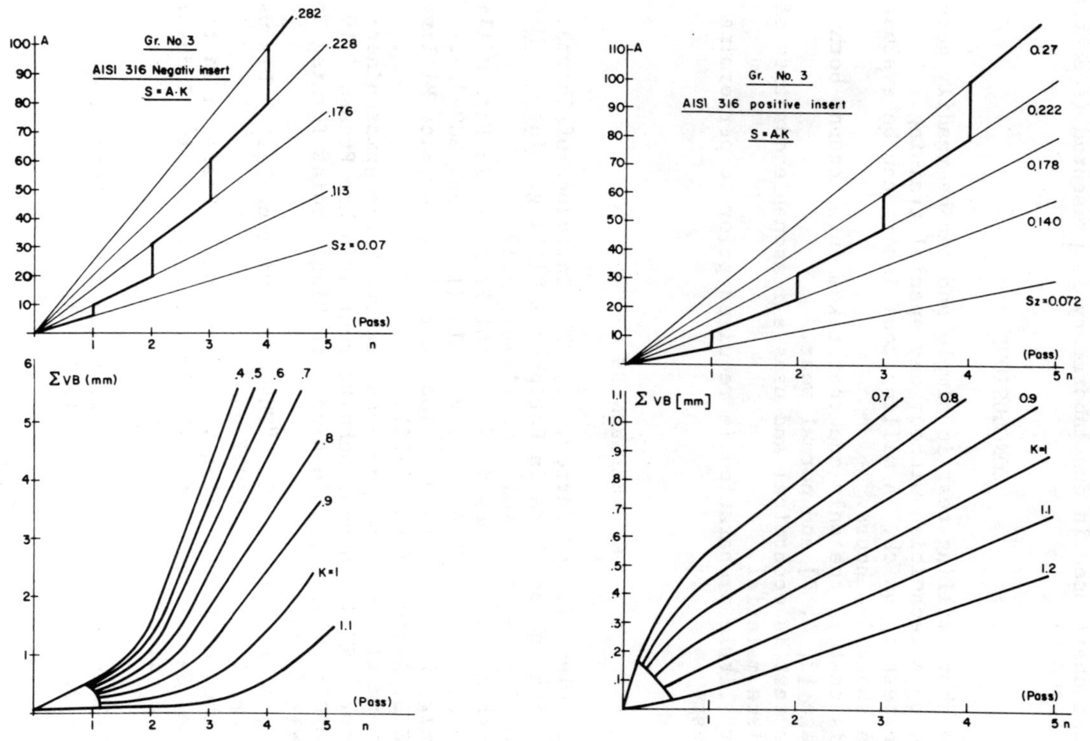

Fig. 4. (3) AISI 316

With these nomograms the parameters A and K are calculated and the "S" rating for any set of test data can be evaluated. The test chart used in our laboratory is presented (see Appendices B and C).

CONCLUSIONS

1. The short milling test is simple and can be readily performed by a technician without any special training.
2. The test is quick. A milling tool insert can be evaluated within about 3-4 hours.
3. This test is the only one that takes into account both breakage [3, 4, 7] and normal wear.
4. The test is economical and uses a reasonable quantity of workpiece material.
5. Very little preparation is required prior to performing the test.

REFERENCES

1. Yellowley, I. and Barrow, G., "The Influence of Thermal Cycling on Tool Life in Peripheral Milling," Int. J. Mach. Tool Des. & Res., Vol. 16, (1976), p.12.
2. Okushima, K. and Hoshi, T., "Tool Fracture in Face Milling Operations," CIRP Annals, Vol. 15 (1967) pp. 309-324.
3. Konig, W. and Lung, D., "Tool Life Test for Face Milling," W.Z.L. Aachen, (Jan. 1976).
4. "Informationszentrum fur Schnittwerte Werkzeugmaschinenlabor," RWTH, Aachen. Schnitterwertaufnahme Frasen, (1975).
5. ISO/Dis 3685. "Tool Testing with Single Point Turning Tools," Paris (1975).
6. Ber, A. and Kaldor, S., "The Short Milling Test," Discussion Paper, CIRP, Paris (1977 and 1978).
7. "Method of Face Milling Test," Technical Research Institute, Japan Society for the Promotion of Machine Industry, (June, 1975).
8. Pekelharing, A.J., "The Exit Failure in Interrupted Cutting," CIRP Annals, Vol. 27, No. 1, (1978).

APPENDIX A. COMPENSATION METHODS

The cutting edge cross-section in the plane P_n is shown in Fig. 5a for a honed edge and in Fig. 5b for a **chamfered** edge, together with the approximation for the "initial wear" VB_o in each case.

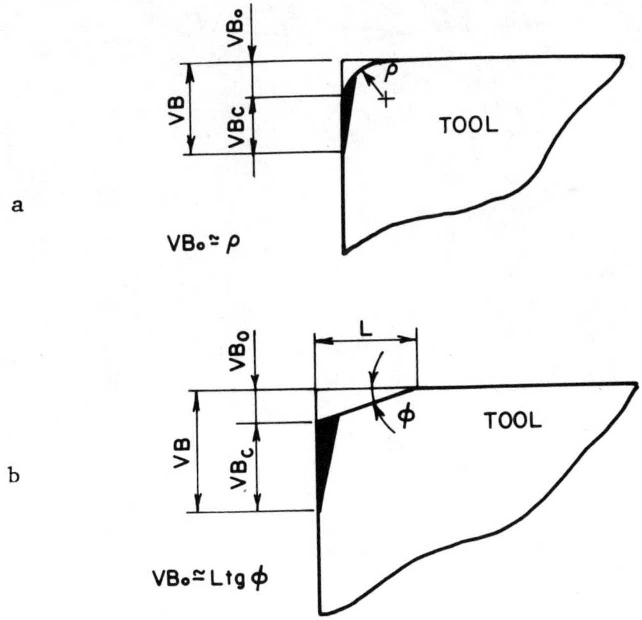

Fig. 5. Cross-section of cutting edge: (a) honed; (b) chamfered

VB_o can also be measured with a toolmaker's microscope. For either honed or chamfered edges, the method is to slightly polish and then directly measure the width of the dark area (not polished) as illustrated in Fig. 6.

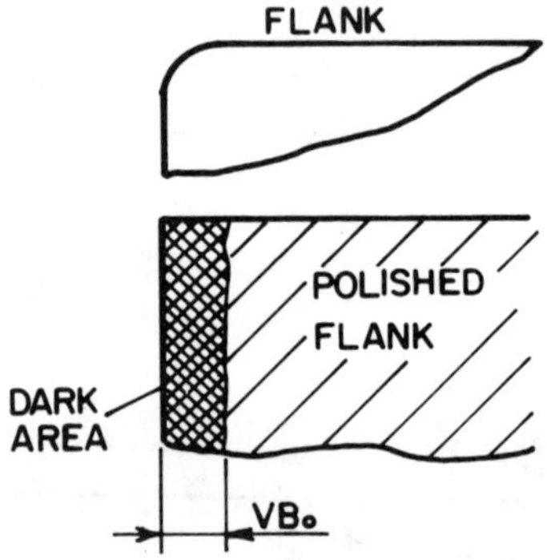

Fig. 6. Measurement of VB_o

APPENDIX B. ON THE RATING TECHNIQUE

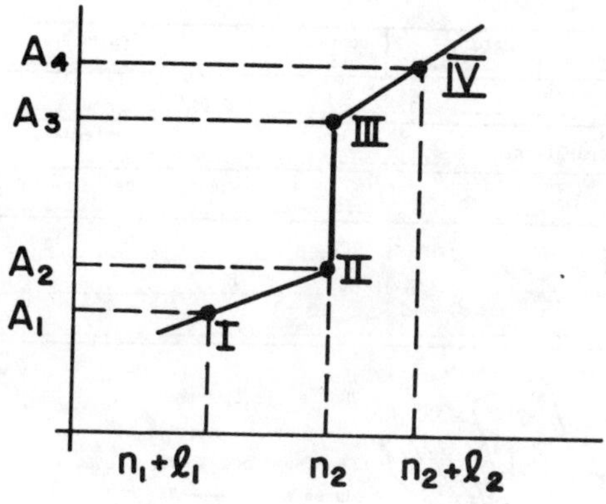

Fig. 7. Typical "step" on the rating nomogram

The stepped line of the feed's influence in the nomograms (see Fig. 3b and Fig 7) has to be used carefully in order to obtain the correct rating. As in almost all the tests, every further pass is being machined with a higher feed, a step in the rating is obtained on the nomogram according to Eq. 4 and 5. Examples I and IV (Fig. 7) cause no special difficulty in evaluation because for every number of passes only one value of A can be found. This example is typical for all the tests in which breakage, etc. occurs during machining somewhere before the pass is finished.

However, cases II and III are different evaluations of practically the same length of machining. This is explained by the fact that the exit from the cut of the tool is much more dangerous than the entrance [8]. In fact, most of the tools are expected to fail at the end of the pass, and this evaluation is marked by Case No. II (Fig. 7). Yet, some tools do not fail at the exit. Instead, the failure occurs at the next entrance. These rare cases usually represent better tools; consequently, the evaluation is presented as in case III, which results in a higher rating of the tool.

APPENDIX C. EXAMPLE OF A RATING CHART

SHORT MILLING TEST "S"

Test No.:	Date:	Operator:	Tested by:
----------	----------	------------------	------------------------------

Tool: Grade: ---------------- Type: -------------- Tip No. --

Edge Preparation: R.H. mm ---------------------- Chamfer ϕ , L mm ---------- ----------

Dimensions of cut: length ---------- x width -------- x depth of cut ------------ mm^3

Work material: Type: • AISI________ or __________

HB-------- $\frac{N}{mm^2}$ Condition:--

VBoN = ------- mm

VBoH = ------- mm

V_c = ------- $\frac{m}{min}$

	VBoN	VBoH
Chamfered edge → $VB_o = L \cdot tg\ \phi$ = · · ·	------	------
Honed edge —— VB_o = R.H. = · · ·	------	------
	mm	mm

Rating: $\frac{\text{nomogram no.}}{a \quad b \quad c \quad d \quad e \quad f}$ "S" = ---------------

Pass No.	mm ±	T.C.	chipping ≤ 0.2 mm	VB_N (mm)	VBC_N (mm)	VB_H (mm)	VBC_H (mm)	ΣVBC (mm)	Remarks
0									
1									
2									
3									
4									
5									
6									
7									
8									
9									

DISCUSSION:

Professor Tetsutaro Hoshi (Toyohashi Univ.)

How many teeth were on the test cutter and how often do you repeat a set of test conditions?

Dr. Abraham Ber

There was one tooth in the cutter. When evaluating a new tool material we run three tests and the scatter is very low but if we just check tools that we have a knowledge of the expected performance, one test is enough.

Frank Gorsler (General Electric Co., AEG)

I interpreted Professor Hoshi's question as referring to the effect of single VS multiple tooth milling.

Dr. Abraham Ber

There is a difference. It has been shown that once you get to four teeth there is little difference. Here we are comparing tools and whether there is one tooth or four the result will be the same, since its a relative comparison. We tried one, two and four teeth and documented the results in a 1977 CIRP paper.

George Wilson (General Electric Co., SMBD)

Why do you get more VB_n wear in one case and VB_h in another? Have you machined any nonferrous materials using this milling technique?

Dr. Abraham Ber

Once we thought we knew why. In a recent investigation we found we know nothing. The main differences are in 316 (stainless steel) and 4340. We are doing further work to understand this and it should appear in CIRP in a year or two.

We have intentions of working on Silicon Aluminum because this is a future material for the Automotive business. We are trying to convince somebody to finance the project.

Tom Hale (General Electric, Carboloy)

How sensitive is the breakage factor of this test to the hone level? If you double the hone on one grade of carbide, how much does this affect breakage?

Dr. Abraham Ber

We did investigate the hone level as well as different types of hone, and compared them with chamfering. Our results appeared in an ASME paper in Atlanta, Georgia in which we showed that chamfering is much better than honing. Further there is an influence of hone level but the wear is more rapid and breakage is not determining. We have a standard hone or chamfer for our tests.

THEORETICAL CONSIDERATIONS IN THE MACHINING OF NICKEL-BASED ALLOYS

Bruce M. Kramer and Paul D. Hartung
Massachusetts Institute of Technology
Cambridge, Massachusetts

ABSTRACT

The solution wear rates of five refractory carbides have been determined experimentally in the turning of nickel-based alloys. Test results show good agreement with the relative wear rates predicted from thermochemical data.

It is suggested that the solution wear mechanism limits the wear resistance of the carbides and that TiC and HfC are potentially the most wear resistant carbides in the machining of nickel-based alloys. However, coating delamination was a recurring problem with all of the carbide coatings that were employed; and plastic deformation limited the cutting life of the cemented TiC inserts that were tested. Improved coating adhesion might make HfC and TiC coatings attractive for the machining of nickel-based alloys. If the plastic deformation resistance can be improved, our tests indicate that cemented TiC tooling might be from four to six times more wear resistant than the WC-based tooling that is customarily used to machine nickel-based alloys.

INTRODUCTION

The high speed machining limit may be defined as the maximum speed of application of cemented carbide tooling. At higher speeds, in the very high speed range, oxide-based, polycrystalline diamond, and cubic boron nitride tooling are used. In the very high speed range, cemented carbide tooling fails rapidly due to thermal softening and gross plastic deformation of the cutting edge. Kramer and Suh [1] have shown that, in the turning of alloy steel near the high speed machining limit, the solution wear mechanism controls the wear rate of the refractory carbides. They have made accurate predictions of the relative wear rates of the carbides from theoretical considerations and have extended the analysis to other potentially useful cutting tool materials.

Hartung [2] extended this analysis to predict the relative wear rates of potential tool materials in the machining of nickel-based alloys, assuming that the chemical dissolution of the tool material in the material being cut is the primary wear mechanism. The solution wear theory explains the relatively small differences in the wear rates of the various carbides in the machining of nickel-based alloys, as compared to the large differences observed in the machining of steel.

THEORY

Theoretically, the relative wear rates of the tool materials can be calculated directly from the thermochemical properties of the materials involved in the cutting process. In practice, the wear of cemented carbide tooling in the machining of nickel-based alloys is complex and is affected by the plastic deformation of the tool material. With oxide-based tooling, it is suggested that another wear mechanism entirely may be involved and that the mutual solubility of the tool and workpiece is not the controlling factor in the wear of the oxides. In any case, however, the solution wear rate of a given tool material will determine the upper limit for the wear resistance of the material.

If the mechanics of chip flow are assumed to be similar for different tool materials cutting under identical cutting conditions, the relative wear rates of potential tool materials may be taken as the ratio of the products of the solubility and the molar volume for each material:

$$\text{Relative Wear Rate} = \frac{v_{\text{wear 1}}}{v_{\text{wear 2}}} = \frac{V_1 C_1}{V_2 C_2} \qquad (1)$$

where:

$v_{\text{wear i}}$ = the wear rate of tool material i (cm/sec)

V_i = the molar volume of tool material i (cm^3/mole)

C_i = the equilibrium concentration (solubility) of tool material i in the material being cut at the cutting temperature (mole/mole)

The molar volumes of the materials of interest are easily determined and are tabulated in [3]. There is currently very little experimental data concerning the solubility of refractory materials in nickel. Edwards and Raine [4] did what is probably the only systematic study of the solid solubilities of the carbides in the iron-group metals and measured the solubilities of various carbides in nickel, steel, and cobalt at 1,523°K. This data has been extrapolated to various temperatures in the range of interest and is presented in reference [2].

Because of the solubilities involved, it is felt that experimental determinations of solubilities may be unreliable. In this study, analytical techniques have been employed to estimate solubilities from thermochemical data. Briefly, for dissociation of the tool material to occur, chemical equilibrium requires that the free energy of formation of the tool material equals the sum of the relative partial molar free energies of solution of the constituents [1]:

$$\Delta G_{ij} = \Delta\overline{G}_i^M + \Delta\overline{G}_j^M \tag{2}$$

where:

ΔG_{ij} = the free energy of formation of the tool material

ΔG_i^M = the relative partial molar free energy of solution of component i

ΔG_j^M = the relative partial molar free energy of solution of component j

At a given temperature, T, if the solution is assumed to be regular, the relative partial molar free energy of solution of component i can be expressed in terms of the relative partial molar excess free energy of solution of that component, G_i, a term that indicates the departure of the behavior of the actual solution from that of an ideal solution, and the concentration of the component, c_i:

$$\Delta\overline{G}_i^M = \Delta\overline{G}_i^{xs} + RT\ln c_i \qquad (3)$$

For the carbides of composition XC, where X is a metal atom, Equations (2) and (3) may be combined to give for the carbides at equilibrium:

$$\Delta G_{XC} = \Delta\overline{G}_X^{xs} + \Delta\overline{G}_C^{xs} + 2RT\ln C_{XC} \qquad (4)$$

Solving for the solubility of the tool material in Equation (4), the solubility can be expressed in terms of free energy of formation of the tool material and the excess free energies of solution of the tool constituents. For a carbide, C_{XC}, the solubility of the tool material is given by:

$$C_{XC} = \exp\left[\frac{\Delta G_{XC} - \Delta\overline{G}_X^{xs} - \Delta\overline{G}_C^{xs}}{2RT}\right] \qquad (5)$$

The estimated solubilities of potential tool materials are listed in Table 1 with their free energies of formation at 1,473°K. Substitution of the estimated solubilities of Table 1 into Equation (1) yields the predicted solution wear rates that are given in Table 2.

The excess free energies of the tool constituents of interest are contained in Table 3. The methods used in calculating the excess free energies are discussed in detail in reference [4]. Essentially, no data is available concerning the thermodynamic properties of solid solutions between nickel and the elements of interest. The excess free energies have been estimated from the scant data that is available for intermetallic compounds of nickel. At the solubility limit of an element of interest in nickel, the relative partial molar free energy of formation of the saturated solution must equal the free energy of formation per mole of solute atom of the intermetallic compound with which the solution is in equilibrium. Using solubility limits obtained from phase diagrams and relative partial molar free energies of solution estimated from available data, Equation (3) was used to estimate the excess free energies. This technique is far from ideal, and experimental determinations of the thermochemical properties of interest would greatly improve the accuracy of the analysis.

Table 1. Estimated Solubilities of Potential Tool Materials in Nickel at 1,473°K

Tool Material	Free Energy of Formation (Cal/Mole)	Estimated Equilibrium Concentration (Solubility) in Mole %
HfC	- 49,407 [5]	1.38
NbC	- 32,299 [5]	3.58
TaC	- 34,616 [5]	1.90
TiC	- 39,977 [5]	1.72
VC	- 23,639*[5]	1.65
$V_{1.14}C_{0.86}$ (V_4C_3)	- 23,639*[5]	2.79
WC	- 8,209 [5]	1.83
ZrC	- 43,127 [5]	7.62
Al_2O_3	-288,187 [6]	3.80×10^{-5}
HfO_2	-198,000 [7]	8.13×10^{-6}

*Both values are for a carbide of composition $V_{1.06}C_{0.94}$. Therefore, the solubility of VC is probably overestimated and that of $V_{1.14}C_{0.86}$ is probably underestimated.

Table 2: Predicted Solution Wear Rates of Various Tool Materials Relative to TiC at Various Temperatures [v_{wear} (TiC) = 1]

Carbide	Temperature 1,473°K	1,373°K	1,273°K
HfC	0.99	0.97	0.96
NbC	2.30	2.42	2.58
TaC	1.21	1.21	1.23
TiC	1.00	1.00	1.00
VC	0.86	0.85	0.85
V_4C_3	1.45	1.50	1.56
WC	1.08	1.09	1.09
ZrC	5.69	6.34	7.20
HfO_2	8.43×10^{-6}	3.45×10^{-6}	1.23×10^{-6}
Al_2O_3	4.66×10^{-5}	2.13×10^{-5}	8.64×10^{-6}

Table 3. Estimated Excess Free Energies of Solution of Tool Constituents in Nickel, $\Delta\overline{G}_i^{xs}$, at 1,473°K

Tool Constituent, i	$\Delta\overline{G}_i^{xs}$ (Cal/Mole)
Hf	- 35,782
Nb	- 24,274
Ta	- 22,862
Ti	- 27,622
V	- 11,061
W	+ 3,592
Zr	- 39,559
C	+ 11,600
Al	- 36,552
O	- 3,703

The free energies of formation of Ni_3Ti, Ni_5Hf, Ni_5Zr, Ni_3V, and Ni_4W have been estimated from the known heat of formation of Ni_3Ti by assuming that the heat of formation per unit volume change of the constituent elements is identical for all of the compounds [8]. The hexagonal structure of Ni_5Hf and Ni_5Zr has similar packing to the face-centered cubic structure of Ni_3Ti, and all three elements are members of Group IVB of the Periodic Table, so the estimates for these two compounds should be reasonably good.

There is some uncertainty as to the actual structures of the intermetallic compounds of tungsten and vanadium at 1,473°K [9,10,11], so a significant amount of error may be present in the estimated values for the free energy of solution of these Group VIB elements.

The free energy of formation of Ni_3Ta was calculated from the known free energy of formation of Ni_3Nb since the compounds have identical structures and both elements are in Group VB.

EXPERIMENTAL PROCEDURE

To determine the relative wear rates of the various carbides in the machining of nickel, cutting tests were performed on two different nickel-based alloys, Inconel X and Inconel 718, with an assortment of inserts. Table 4 lists the cutting tools used and the number of tests performed on each.

Table 4: Cutting Tools Tested and Number of Tests Performed on Each

Tool Material	Supplier	Coating Thickness (mm)	#Tests on Inconel X	#Tests on Inconel 718
HfC coated	Surftech	5	7	5
TaC coated	Kramer*	2	3	3
TiC coated	Carboloy, Kramer*	5	5	3
Cemented TiC	Ex-Cell-O, Teledyne	-	6	1
Cemented WC	Surftech	-	3	4
ZrC coated	Surftech	3	3	4

*These tools were coated using a deposition process covered by U.S. Patent Number 4,066,821 [12].

The Inconel X (HT9243 X) was in the form of a 546 mm (21.5 inch) long, 94 mm (3.7 inch) in diameter bar. It had a Rockwell C hardness of 29 (282 Brinell) at a diameter of 81.5 mm (3.21 inches).

The Inconel 718 (HT96378) used in the experiments was provided by the Hamilton Standard Company of Windsor Locks, Connecticut. It was in the form of a 120 mm (4.7 inch) in diameter, 25 mm (1 inch) thick, disk with a 51 mm (2 inch) hole in the center. It had a Rockwell C hardness of 25 (255 Brinell) at a diameter of 74.2 mm (2.92 inches).

The cutting tests were performed on an American Pacemaker 14 x 30 lathe. The lathe is powered by a 18.6 kW (25 horsepower) D.C. motor, and the speed is continuously variable.

The depth of cut was standardized to 1.27 mm (.050 inch) and the feed was standardized to 0.127 mm (.005 inch) per revolution. All of the tests on Inconel X were one minute long while several 15, 30, and 45 second tests were done in addition to those on Inconel 718.

Except for the solid titanium carbide and tungsten carbide inserts, all of the cutting tools used were coated inserts. The coatings were from 2 to 5 microns thick. The Teledyne Firth Sterling insert is a grade SD3, developed by

Rudy [13] and composed of titanium carbonitrides, titanium molybdenum transition phases, and a nickel alloy binder. This grade is known to have good plastic deformation resistance.

Cutting speed for the tests was chosen so as to minimize flank wear on tungsten carbide while maintaining a crater wear rate of approximately 5 microns per minute. While it was desired to go as fast as possible, it was necessary to choose a speed at which the coatings would not wear through in one minute and at which gross plastic deformation of the substrate material would not occur.

Preliminary tests were performed at speeds ranging from 0.508 m/sec (100 sfpm) to 1.500 m/sec (295 sfpm), using tools with a 1.19 mm (3/64 inch) nose radius, at 2.54 mm (.100 inch) depth of cut. The effects of vibration seemed to be too great at 2.54 mm (.100 inch) so 1.27 mm (.050 inch) was chosen for the depth of cut, and tools with 0.79 mm (1/32 inch) nose radius were used.

A cutting speed of 0.686 m/sec (135 sfpm) was chosen for the final tests since there seemed to be a transition in wear mechanisms occurring between 0.584 m/sec (115 sfpm) and 0.660 m/sec (130 sfpm). The wear rate increases with speed until somewhere in that speed range where wear begins to decrease with increasing speed. Such a phenomenon is known to occur in the flank wear of tungsten carbide inserts in the machining of steel [14]. In this case, the increase in wear rate with decreasing speed is associated with the formation of the built-up edge on the tool face. With increase in speed, the built-up edge disappears and solution wear predominates. It is expected that this may also be the case in the wear of nickel-based alloys. At 0.686 m/sec (135 sfpm), we are out of the transition range and are in a speed range where solution wear should be the primary wear mechanism.

The surface speed of the workpiece was accurately measured with a tachometer before every cut, and the cutting time was measured with a stopwatch. After a cut was made, the flank wear was measured; and the crater wear was observed with a toolmaker's microscope.

Crater traces were taken with a Taylor-Hobson Model 3 surface measuring instrument. Crater traces were produced by moving the stylus across the tool crater at three different locations.

To eliminate adherent nickel in the crater area, the tools were etched for approximately 24 hours in hydrochloric acid; and crater traces were taken on the etched tools.

EXPERIMENTAL RESULTS

It can be inferred from the crater traces that some plastic deformation had occurred in all of the carbide tools after one minute of cutting. It can be observed microscopically that, in general, there is no wear on the leading edge of the rake face before the crater starts (Figure 1). The maximum crater depths were measured as the maximum distance perpendicular to the unworn rate surface of the tool, from the bottom of the crater to the line connecting the leading and trailing margins of the crater (see Figure 2).

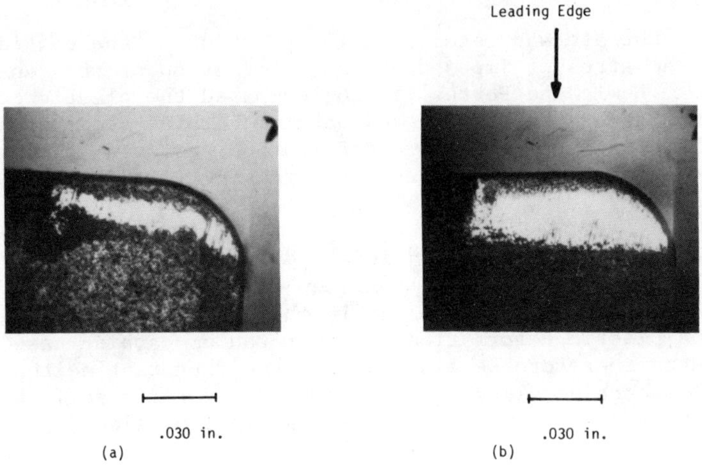

Figure 1: (a) HAFNIUM CARBIDE AND (b) TUNGSTEN CARBIDE CRATERS SHOWING NO WEAR ON THE LEADING EDGE OF THE RAKE FACE

Many of the coated tools had problems with coating adhesion. It is believed that plastic deformation of the substrate is the cause of much of the cracking and spalling in the coatings, due to the differences in mechanical properties between the coating and the substrate. The low fracture resistance of the brittle coatings makes it impossible for them to conform to the deformation of the tungsten carbide

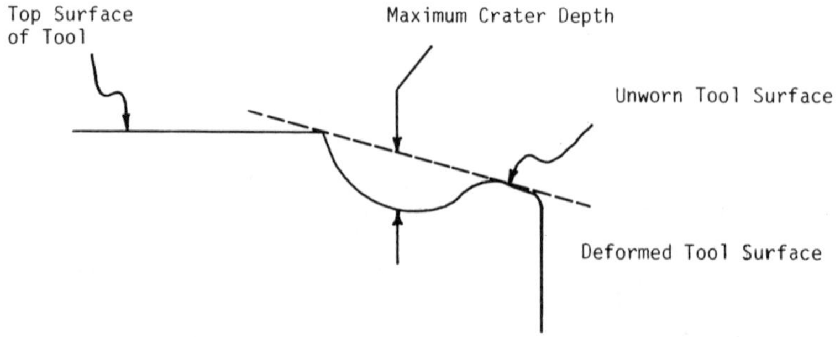

Figure 2: SCHEMATIC ILLUSTRATION OF CRATER DEPTH MEASUREMENT ON A DEFORMED TOOL CRATER

based substrate and results in the cracking of the coating under the stresses imposed by the deformation of the substrate. Lewis and Porter [15,16] compared the plastic deformation of tungsten carbide and hafnium carbide, concluding that hafnium carbide is more deformation resistant than tungsten carbide and is more prone to cracking along certain slip planes.

Dawihl and Frisch [17] inferred from hardness tests that the plastic deformation of tungsten carbide-cobalt hard metals proceeds mainly by slip in the cobalt phase, The increased plastic deformation of the titanium carbide based tools can therefore be explained by the fact that neither titanium carbide, like tantalum carbide, nor its solid solution in tungsten carbide is wetted as well as liquid binder metals as is tungsten carbide [18]. Poor wetting requires a greater binder volume fraction to get acceptable toughness, and it is possible that the solid titanium carbide tools have a higher binder fraction than the tungsten carbide tools. Solid titanium carbide tools use a binder that consists of nickel and molybdenum added.

The results of the tests were inconsistent mainly because some of the coatings remained intact, and some were removed immediately upon contact of the tool with the workpiece. Figure 3 shows two traces taken on two different cutting edges of the same hafnium carbide coated insert under identical cutting conditions. The difference between the two is that on one the coating remained intact while on the other the coating was removed. In some cases, the chip appeared to

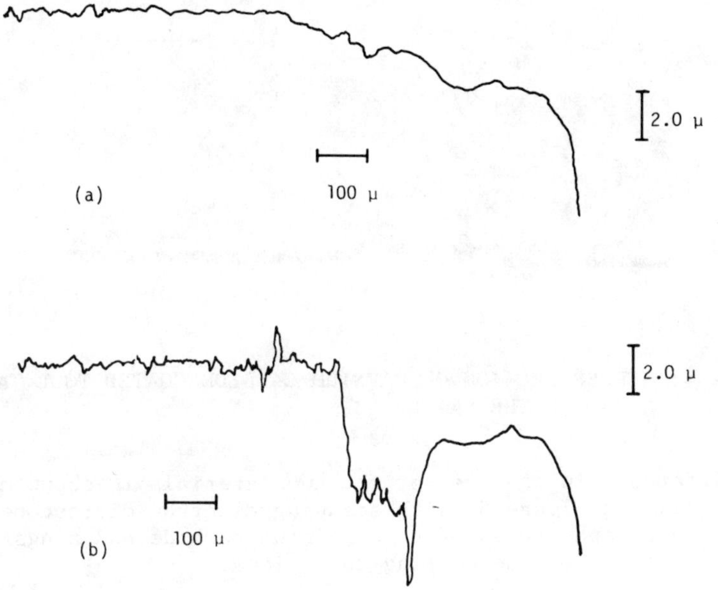

Figure 3: (a) CRATER TRACE OF AN INTACT COATING AND
(b) CRATER TRACE OF A DESTROYED COATING OF HAFNIUM CARBIDE

have welded to the coating and to have removed a large portion of the coating in the crater area when the cut was stopped. The coating failure of Figure 3b is characteristic and provides indirect evidence that a transition between sliding and sticking of the chip occurs on the tool rake face near the maximum wear point, as is suggested in [3] on diferent grounds.

Figure 4 shows a cross-section of a hafnium carbide tool which was etched to remove the adherent nickel from the crater surface. As can be seen from the photograph, the coating remained intact in this case; and the tool showed approximately 1.5 microns of wear after one minute of cutting time at 0.686 m/sec (135 sfpm). This measurement confirms the results of crater wear rates determined from surface traces taken on tools that were believed to have intact coatings and shows that hafnium carbide has only one-third to one-sixth the wear rate of tungsten carbide in the machining of Inconel.

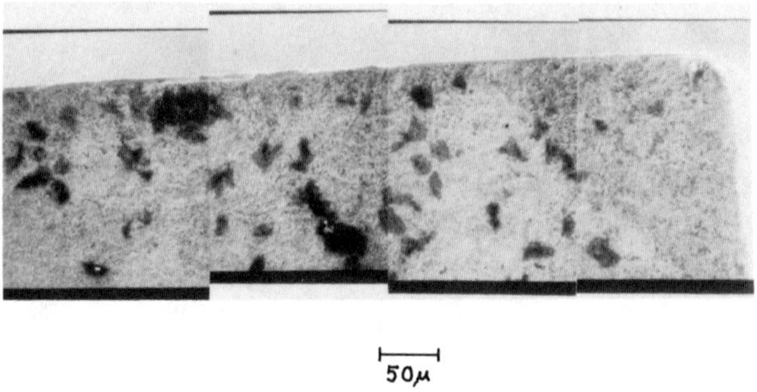

Figure 4: CROSS SECTION OF HAFNIUM CARBIDE COATED TOOL TAKEN THROUGH THE CRATER

The difference in the wear of the two materials is shown more graphically in Figure 5, with scanning electron microscope (SEM) photographs taken of both hafnium carbide and tungsten carbide after the same cutting conditions.

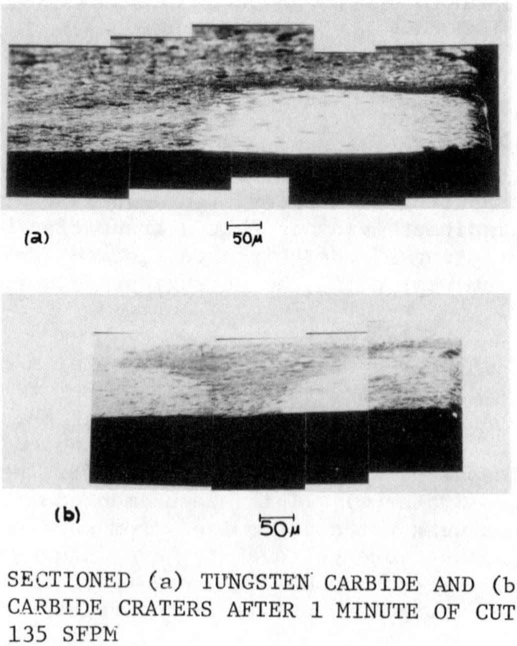

Figure 5 : SECTIONED (a) TUNGSTEN CARBIDE AND (b) HAFNIUM CARBIDE CRATERS AFTER 1 MINUTE OF CUTTING AT 135 SFPM

The crater depth data for the tools on which the coating remained intact was compiled and is presented in Table 6. Because the tantalum carbide coatings were only 2 microns thick, all of the coatings were penetrated during the cutting interval. In this case, the crater depth through the coating was taken as a measure of the crater depth. Since the uncoated K7H inserts proved to have a lower wear rate than that predicted for tantalum carbide, the wear rate through the coating should give a lower bound estimate of the actual wear rate of tantalum carbide.

Table 6. Averages and Standard Deviations of Crater Depth Data Where the Coating Remained Intact

Tool Material	Work Material: Inconel X — Average 1 Min Crater Depth (μ)	Inconel X — Standard Deviation	Inconel 718 — Average 1 Min Crater Depth (μ)	Inconel 718 — Standard Deviation
HfC	1.3	0.4	2.2*	0.8
TaC	1.7	-	10.8*†	3.7
TiC	1.4	1.2	1.2	0.4
WC	5.3	0.6	6.6	2.2
ZrC	4.8	0.4	6.0	2.3
Uncoated K7H	1.6	0.4	-	-

*Problems were encountered with coating adhesion in these tests.

†This value is extrapolated from 15 and 30 second test results.

Table 7 is a comparison of experimental with theoretical results. The wear rates of the carbides are normalized relative to the wear rate of titanium carbide. In each case, the wear rate of titanium carbide is set equal to unity; and theoretical predictions for a cutting temperature of 1,473°K are used for the comparison. As is discussed elsewhere, uncertainties with regard to the composition of the phase diagrams of the Ni-W and Ni-V binary systems may produce large errors in our estimates of thermochemical properties. This may account for the discrepancy in the case of WC.

Table 7. Comparison of Experimental with Theoretical Results for the Wear Rates of the Carbides Relative to the Wear Rate of Titanium Carbide [V_{wear} (TiC) = 1]

Carbide	Predicted at 1473°K	Relative Wear Rates Test Results on Inconel X	on Inconel 718
HfC	1.0	1.0	1.9*
TaC	1.2	1.2	9.2*
TiC	1.0	1.0	1.0
WC	1.1	3.8	5.6
ZrC	5.7	3.4	5.1

*Severe problems were encountered with coating adhesion in these tests

Preliminary cutting tests were performed with aluminum oxide and hafnium oxide on Inconel 718. The tests were performed for one minute at a cutting speed of 135 sfpm. On the basis of the limited test data, it was not possible to determine quantitative wear rates.

Crater traces of the aluminum oxide tooling showed what appeared to be slight cratering, with pronounced rounding of the cutting edge, and substantial flank wear.

The hafnium oxide tooling had the form of a hafnium oxide coating on an aluminum oxide substrate. In this case, it appeared as if the hafnium oxide layer was quickly removed from the surface.

Since for these two oxides the wear rates predicted from their solubilities are on the order of 10^{-6} times the wear rate of titanium carbide, if solution wear predominates, the crater wear of these two oxides should be so small as to be unmeasurable after one minute of cutting Inconel.

The experimental results indicate that at least some cratering did occur in the oxides. This suggests that oxide tooling may wear by a mechanically activated mechanism. This possiblity has been suggested in the case of the machining of steel [2] and is attributed to the relatively low flow stress of oxide materials and, particularly, to the low hardness of hafnium oxide.

CONCLUSIONS

It is clear from the agreement of theoretical predictions with the experimental results that solution wear is the primary wear mechanism in the cratering of the carbides in the high speed machining of nickel-based alloys. However, it is suggested that unless more adherent coatings can be produced, coated tools may offer little benefit over solid tungsten carbide tools.

Titanium carbide and hafnium carbide proved to have substantially greater wear resistance than tungsten carbide, and it is suggested that further work be done on finding suitable binder metals for these carbides. The high cost of hafnium carbide will probably be the limiting factor in the development of cemented hafnium carbide compositions. At present, the wear of solid titanium carbide tooling is dominated by the plastic deformation of the binder metal.

Except for titanium carbide and hafnium carbide, none of the Group IVB and VB carbides seem to offer any potential advantage over tungsten carbide in the machining of nickel-based alloys. This is particularly true in the case of zirconium carbide.

It is suggested that the carbides are limited by their solubilities in nickel and that further work be done on the development of binder materials for the more stable oxides.

ACKNOWLEDGEMENTS

The authors would like to gratefully acknowledge the research initiation support of the Procter and Gamble Company. It is hoped that the results of this research justify the support. In addition, the Boeing Aircraft Corporation has provided financial support to the project and the Hamilton Standard Company, through Mr. Ron Mador, has provided valuable advice, encouragement, and raw materials. Mr. Ralph Whittemore of MIT was of invaluable assistance in the conduct of the experimental aspects of the work.

REFERENCES

[1] B.M. Kramer and N.P. Suh, "Tool Wear by Solution: A Quantitative Understanding," Trans. of the ASME, J. of Engineering for Industry, Vol. 102, No. 3, 1980.

[2] P.D. Hartung, B.S.Thesis, Department of Mechanical Engineering, M.I.T., 1980.
[3] B.M. Kramer, Ph.D. Thesis, Department of Mechanical Engineering, M.I.T., 1979.
[4] R. Edwards and T. Raine, "The Solid Solubilities of Some Stable Carbides in Cobalt, Nickel and Iron at 1,250°C," Plansee Proceedings, 1952, F. Benesovsky, ed., Metallwerk Plansee Ges. M.B.H. Reutte/Tyrol, 1953, p. 232.
[5] R. Hultgren et al., Selected Values of the Thermodynamic Properties of Binary Alloys, American Society for Metals, Metals Park, Ohio, 1973.
[6] D. Stull and H. Prophet, JANAF Thermochemical Tables, Second Edition, U.S. Superintendent of Documents Number C13.48:37, 1971.
[7] T.B. Reed, An Atlas of Charts for High-Temperature Chemical Calculations, M.I.T. Press, Cambridge, Massachusetts, 1971, p. 9.
[8] O. Kubaschewski and E. Evans, Metallurgical Thermochemistry, Pergamon Press, New York, 1958.
[9] M. Hansen, Constitution of the Binary Alloys, McGraw-Hill, New York, 1958.
[10] R. Elliott, Constitution of the Binary Alloys, First Supplement, McGraw-Hill, New York, 1965.
[11] F. Shunk, Constitution of the Binary Alloys, Second Supplement, McGraw-Hill, New York, 1969.
[12] N.H. Cook and B.M. Kramer "Tungsten Carbide Tools Treated with Group IVB and VB Metals," U.S. Patent Number 4,066,821, 1978.
[13] E. Rudy, NSF Hard Materials Research, Volume 2, MIT Conference, NITS Number PB-221908, 1973, p. 349.
[14] G. Vieregge, "Der Werkzeugvershleiss beider spanabherbenden Bearbeitung im Speigel der Vershleiss-Schnittgeschwindigkertskurven," Stahl Eisen 77, 1957, No. 18, Vol. 5, 1957.
[15] D. Lewis and L.J. Porter, "Plastic Deformation in Tungsten Carbide," J. Appl Cryst., Vol. 2, 1969, p. 249.
[16] D. Lewis and L.J. Porter, "Plastic Deformation in Hafnium Carbide," J. Less-Common Metals, Vol. 22, 1970, pp. 439-444.
[17] W. Dawihl and B. Frisch, "The Mechanical Characteristics of the Cobalt Phase in Tungsten Carbide-Cobalt Hard Metals, Cobalt, Vol. 22, March, 1964, p. 22.
[18] W. Dawihl and M.K. Mal, "Contribution to the Study of the Deformation Behavior and Structure of WC-TiC-TaC-Co Alloys," Cobalt, Vol. 26, March, 1965, p. 25.

DISCUSSION

VK Sarin (GTE Labs)

Talking about solubility wear from a thermodynamic point of view, isn't the wear process controlled by the kinetics of diffusion?

Dr. Bruce Kramer

Very surprisingly that is not the case. It is an equilibrium process.

VK Sarin (GTE Labs)

When cutting a 1020 steel, how is the carbon forming on the tool? Do you get carbon diffusion out of the 1020?

Dr. Bruce Kramer

If you machine 1020 steel at high speeds and produce temperatures in the range of 1000-1300°C there is no question it would be the chemical dissolution of the tool material that will control wear. If diffusion controls, the wear rate of the tool material should correlate with the diffusion of something in something else. Looking at the self diffusivity of the tool material in order to support a kinetic argument in which the transport of tool atoms into the chip controls, you find there is no correlation between the predicted and measured values. If you look at the diffusivity of the tool constituents in the chip and do a limited analysis to estimate the upper bound of the wear rate, again you find there is no correlation. The only analysis that correlates with experimental results is one based on chemical dissolution and the maintenance of equilibrium solubility of the tool material at the tool/chip interface.

Dr. V.A. Tipnis (Tipnis Associates)

What can you do to apply this approach to the cutting of Titanium? Can you suggest some method of looking at coatings to put on carbide of other tool materials which will be good for turning and milling of Titanium?

Dr. Bruce Kramer

We are studing this at this time and I cannot divulge

too much because it's all preliminary data. It is a very difficult problem. We are predicting certain materials will show some improvement over existing materials but, perhaps, less than we would have hoped. We may have some results in the near future but it's all theoretical at this point, without experimental verification.

Dr. Marcus P. Borom (General Electric, CRD)

It appears that we are dealing with a semantic problem as to whether diffusion is involved in your "solution-wear mechanism". Thermodynamic data certainly define the propensity of reaction, and solubility limits will influence the kinetics. If interfacial reactions are not rate limiting, equilibrium concentrations will be established at the tool/work-piece interface, but we are not dealing with an equilibrium system. The boundary conditions appear to be best fitted by dissolution of a solid by a strongly stirred, infinite liquid in which the tool is the solid and the chip is the stirred liquid. Such solutions are found in "Mathematics of Diffussion" by J. Crank. It would be interesting to determine if wear rate as a function of cutting speed follows the same relationship as dissolution in a stirred solution as function of stirring velocity.

Bruce Kramer

It is interesting that the "diffusion" theory has been around for some 30 years without the appearance of any quantitative predictions of the wear rates of materials based on the diffusivities of the relevant species. There is no quantitative correlation between diffusivities and observed wear rates. On the other hand, excellent agreement is obtained between test results and predictions based on the assumption that the convection of dissolved tool material from the interface controls the wear rate.

SESSION II

HIGH SPEED STEEL TOOL MATERIALS

METALLURGICAL INVESTIGATION OF SEVERAL HIGH SPEED TOOL STEEL FAILURES

by

G. K. Bouse, General Electric Company
Large Steam Turbine's Materials & Processes Laboratory
Schenectady, New York 12345

INTRODUCTION

Material acceptance of a finished tool may be limited to dimensional checks, to actual machining of parts, or to hardness measurements supplied by the vendor, so the tool user may never get much information about the tool until it fails. Since many of the high speed steel tools discussed in this paper are large, up to 15 cm (6 in) in diameter, and available from few tool shops, therefore expensive, many failures can be ill afforded.

If the tools can be non-destructively tested (NDT) before use, the tool user stands a chance to confirm the original order, because once used, the tool usually becomes the property and responsibility of the purchaser. With NDT the material grade, hardness, microstructure, retained austenite, cracks, etc., can be checked, affording the tool user an opportunity to decide how to best use the tool (before it causes a failure) or to return it to the tool shop for rework or credit. Such complete protection, however, is expensive, and not always convenient, therefore, the tool user must use discretion. Utilizing the results of failure analyses like the ones presented here can then be aimed at convincing purchasing agents, tool shops, and ultimately steel mills that higher quality tools are in demand and that they will pay for themselves in the long run. Since tool material costs are low compared to the overall machining costs, better quality tools will accomodate higher feeds and speeds with less down time, producing an overall increase in productivity.

The tool failures presented below occurred during the manufacture of two types of steam turbine components, namely, rotors and buckets. The rotor material can be described as a 3 Wt% Ni, 1.5% Cr, 0.5%Mo, 0.1%V, bal. Fe (ASTM A470 class 5-7 equivalent), and the buckets are plain 12% Cr-bal. Fe (ASTM A193 Grade B6-AISI Type 410).

POWDER METAL (PM) TOOL FAILURES

AISI T-15 (Modified) Ball End Mill

Background. This PM ball end mill is used in a vertical two spindel NC controlled milling machine to obtain complex contours on buckets. The cutter is held in the milling machine by a single setscrew. The ball end mills began breaking at unacceptably frequent intervals, much like those shown in Figure 1a and 1b. The failures were primarily in the shank where the cutter often broke into two or more pieces. However, flute failures also occurred and sometimes caused the ejection of large chunks from the cutting edges near the ball end. The composition of the tool is approximately 1.5 Wt%C, 4% Cr, 3% V, 5.3% Mo, 9% Co, and 10% W. This PM tool is believed to be manufactured by hot isostatic pressing (HIP), followed by hot deformation. Hardening is near 1204°C (2200°F) and tempering for 2+2+2 hours near 585°C (1085°F) produces a hardness in the flute between 67-68 Rc. The shank, however, is tempered back to 45-56 Rc to promote toughness.

Experimental. Rockwell C hardness measurements were taken both transverse and longitudinal to the mill axis on a number of failed tools. For rigidity while taking the hardness measurements, the sections were mounted in epoxy, and surface ground to a 32 RMS or better finish. To reduce hardness error introduced by the epoxy, the mounts were hardened four hours at 66°C (150°F).

Several of the mills were prepared for optical and transmission electron microscopy (TEM) carbon replicas using automated procedures documented elsewhere (1,2). Retained austenite was also measured on several mills using the x-ray diffraction (XRD) direct comparison method (3).

To evaluate toughness, 3/4 size Charpy "C" notch specimens were removed from the flute and tested at room temperature. In an attempt to induce toughness, one mill was also tempered at 649°C (1200°F)/1 hr.

Results. Transverse hardness measurements indicated that center-to-edge variations were typically 1 point Rc. Longitudinal variations, however, were 23 points Rc, which were expected due to the tempered shank. A plot of these hardnesses, Figure 1c, indicates that if the hardness at the bottom of the setscrew flat (in the shank) was 50 Rc or less, failure did not occur, or occurred only at the ball end. If the hardness at this setscrew flat location was 52 Rc or above, then failure always occurred at the setscrew flat location. Once this became known, cutters with 50 Rc hardness or above at the setscrew flat location were tempered back, which significantly reduced the occurrence of failures at this location.

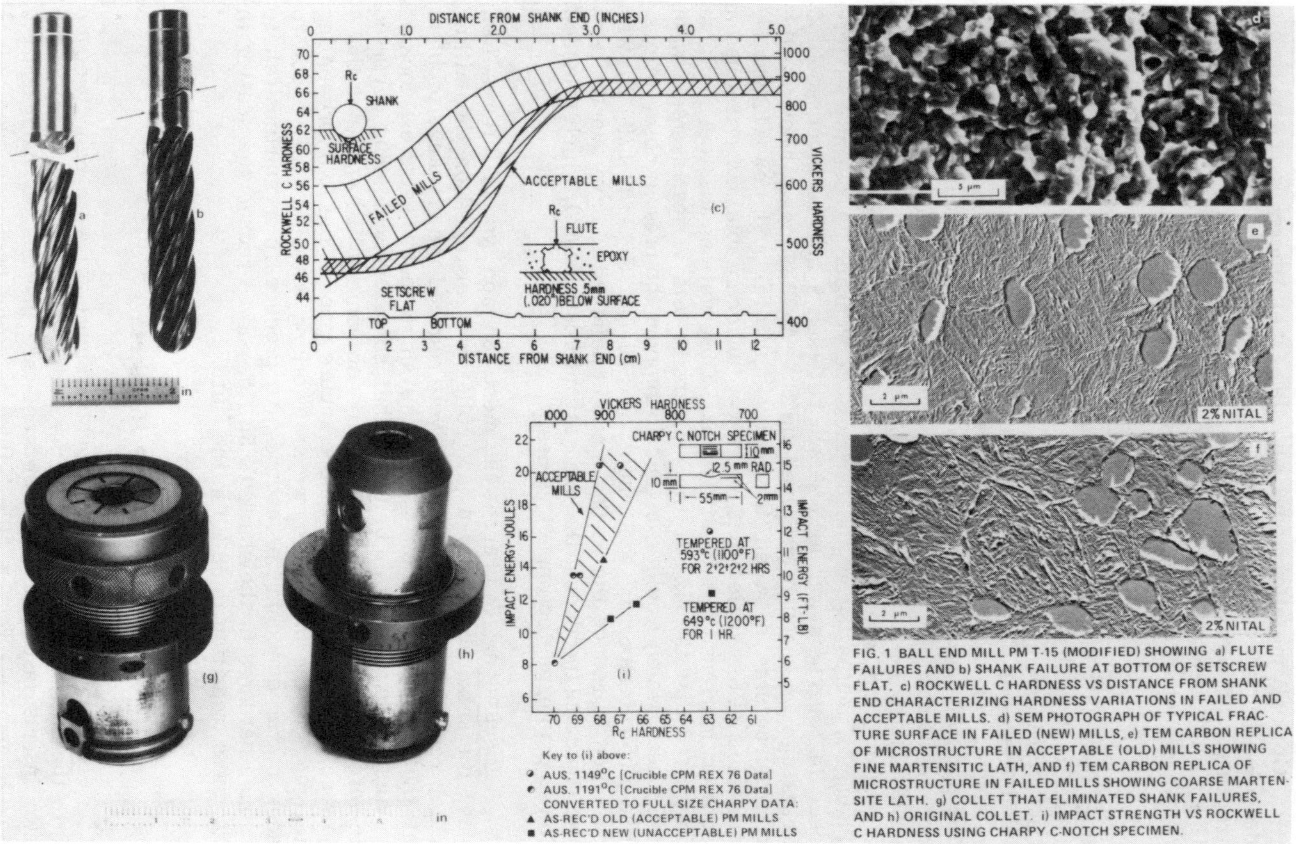

FIG. 1 BALL END MILL PM T-15 (MODIFIED) SHOWING a) FLUTE FAILURES AND b) SHANK FAILURE AT BOTTOM OF SETSCREW FLAT. c) ROCKWELL C HARDNESS VS DISTANCE FROM SHANK END CHARACTERIZING HARDNESS VARIATIONS IN FAILED AND ACCEPTABLE MILLS. d) SEM PHOTOGRAPH OF TYPICAL FRACTURE SURFACE IN FAILED (NEW) MILLS, e) TEM CARBON REPLICA OF MICROSTRUCTURE IN ACCEPTABLE (OLD) MILLS SHOWING FINE MARTENSITIC LATH, AND f) TEM CARBON REPLICA OF MICROSTRUCTURE IN FAILED MILLS SHOWING COARSE MARTENSITE LATH. g) COLLET THAT ELIMINATED SHANK FAILURES, AND h) ORIGINAL COLLET. i) IMPACT STRENGTH VS ROCKWELL C HARDNESS USING CHARPY C-NOTCH SPECIMEN.

Optical metallography revealed that there were no gross defects which contributed to failure. Observing several of the fracture surfaces with the scanning electron microscope (SEM) (Figure 1d) also revealed no defects. There was no evidence that retained austenite was present in about five of the mills tested.

Still, the newly received cutters (those experiencing unacceptably frequent failures) were fracturing near the ball nosed end much more so than the same cutters purchased earlier. A comparison of base metal chemistry from one old and one new mill revealed them to be similar, but that the new cutter had 340 ppm oxygen and 220 ppm nitrogen, while the older cutter had only 260 ppm oxygen and 180 ppm nitrogen, respectively. While these measurements alone are not significant, it suggests that the new material contains more residuals than the older material, possibly due to a change in the manufacturing process. Toughness was shown to vary with hardness, Figure 1i, increasing as the hardness decreased. A definite difference in toughness was also shown between a mill received about 2 years ago vs recently received mills. A TEM comparison of the microstructures between the old (Figure 1e) and new (Figure 1f) materials also revealed that the newer (unacceptable) materials had a coarser martensitic lath structure. Observations in steel (4) have suggested that the size of the martensite lath varies inversely with toughness, which would also tend to support the present observations. Use of a temper in the range of 593-649°C (1100-1200°F) was observed to reduce the hardness and is believed to have overtemperd the structure (producing the coarsest microstructure), which provided lower than expected impact values.

Discussion. To prevent the most common type of failures, i.e., near the setscrew, a new 4 jaw collet was installed on the milling machines, Figure 1g. This effectively reduced the breakage problem at the setscrew location. With the single point setscrew locking device (Figure 1h), it is believed that some cyclic loading occurred (macro fatigue striations were noted on some mills) and that the 4 jaw collet provided a much more secure grip on the tool, preventing fatigue. In the meantime, we are working with the tool vendor to seek better quenching procedures (to produce finer martensite), to reduce hardness in the shank to 50 Rc, and to reduce the scatter in hardness transition to 68 Rc in the flute.

AISI T-15 Insert

Background. As these inserts (Figure 2a) were being wedged into the cutter body (20 inserts per body), one would break for every two inserts installed. The wedges were installed by hand using a small hammer. Thus, a 50% failure rate was occurring even before the tool piece met the work material!

Experimental and Results. The broken pieces were examined with the SEM, and no irregularities were revealed. After mounting and polishing, the insert was etched (with some difficulty), indicating that an undertempered condition might exist. The grain size for all inserts studied was ASTM 13 (3.5 microns). X-ray diffraction was used to check the retained austenite in the failed vendor supplied inserts and none was found. However, the (211) martensite peak on the diffractometer scan was quite broad, suggesting that the failed

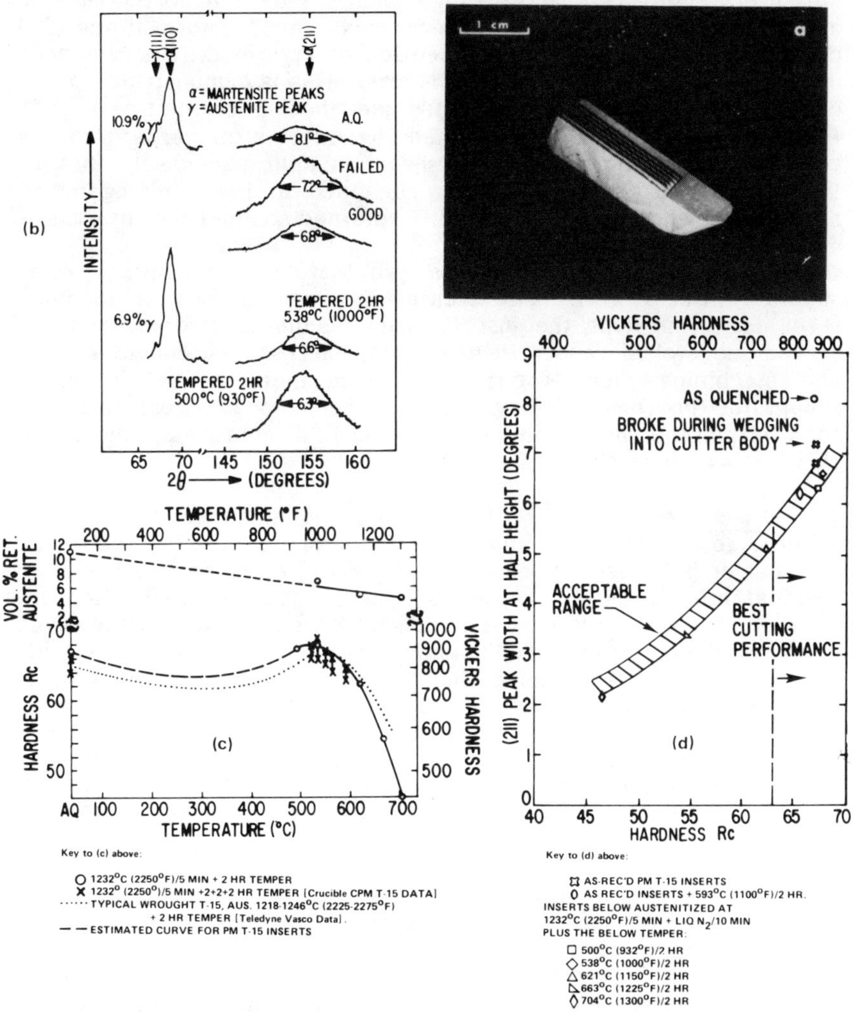

FIG. 2 a) PM T-15 INSERT. b) X-RAY DIFFRACTION SCAN SHOWING SOME RETAINED AUSTENITE PEAKS AND (211) LINE BROADENING FOR SEVERAL CONDITIONS NEAR 67 RC HARDNESS. c) RETAINED AUSTENITE AND RC HARDNESS VS TEMPERING TEMPERATURE FOR PM AND WROUGHT T-15 STEEL. d) PLOT OF (211) PEAK WIDTH AT HALF HEIGHT VS RC HARDNESS FOR THE T-15 INSERT.

material contained a residual stress. To test the hypothesis that the structure was undertempered, several unused inserts were oil quenched after being austenitized at 1232°C (2250°F) for 5 minutes. The tools were quenched to room temperature then cooled in liquid nitrogen and held 10 minutes, warmed back to room temperature, and tempered at temperatures in the 500-704°C (932-1300°F) range for 2 hours. The inserts were then mounted and polished, the Rc hardness was recorded and the (211) peak width at half height measured, Figure 2d. These results clearly show that as the hardness decreases, the peak width also decreases. Furthermore, if the (211) peak width is above the shaded section of Figure 2d, it is considered undertempered and prone to brittleness. This is demonstrated by not only the laboratory heat treated (as quenched) insert, but also by the failed insert. Hardness and retained austenite vs tempering temperature for the inserts studied are shown in Figure 2c. Note that the laboratory treated (austenitized, quenched in liquid nitrogen, and single 2 hr tempers) left 4-7.5 vol % retained austenite in the inserts.

To remedy the brittleness problem, the remaining vendor supplied inserts were heat treated in vacuum 2 hours at 593°C (1100°F). This kept the inserts from becoming oxidized and preserved the etched vendor identification marks. Subsequent wedging and machining then proceeded without incident, indicating the success of this remedy. Properly tempered, the (211) peak half width falls within the shaded portion of Figure 2d, confirming the use of this test as a material control tool.

Discussion. The XRD technique is a quantitative tool that can be used to measure not only the amount of retained austenite, but whether or not a tool is properly tempered, i.e., some measure of the residual microstresses. Although the concept of diffraction line broadening and microstresses are discussed elsewhere (5), the reader can get a "feel" of the phenomenon by equating the excessive width of the peak to the distance that individual atoms are out of their equilibrium position. With a wide (211) peak (e.g., as-quenched specimen), the material contains many internal stresses set up by the non-equilibrium cooling conditions. As the material is tempered, the stresses (characterized by peak width) are reduced because the structure begins to attain equilibrium, due to the formation of stable carbides (6,7).

By referring to the Peak Width vs Hardness graph in Figure 2d, the as-quenched insert has a diffraction peak width well above the acceptable range, a condition that we know by experience is very brittle. For the inserts that failed while being wedged into the cutter body, they too are above the acceptable range and were also brittle. By using this method then, the user can see why with two tools having the same hardness, one may break under seemingly low loads, while the other lasts and lasts. Thus, the user can gain an appreciation of hardness and toughness.

It is not within the scope of this paper to explain hardness vs toughness relationship in tool steels, but to demonstrate how they can be differentiated by non-destructive tests. With this is mind, XRD has much potential as a material control tool.

WROUGHT TOOL FAILURES

AISI M2 (1.25%C) Dovetail Slotter

Background. The dovetail slotter, Figure 3a, is a form cutter used to machine longitudinal slots in turbine rotors for the smaller turbine buckets. This was a finishing operation, with extremely tight

a)

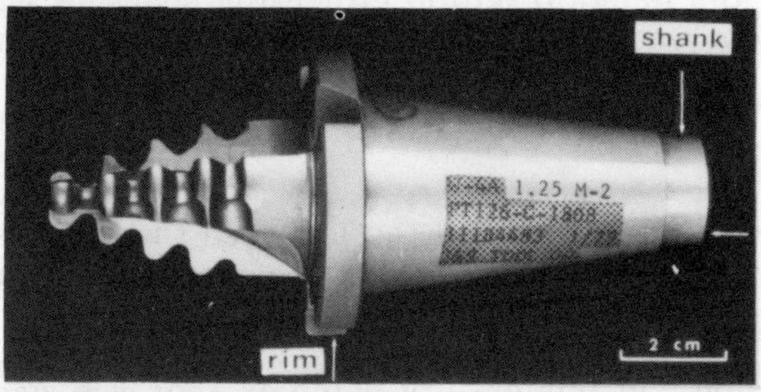

b)

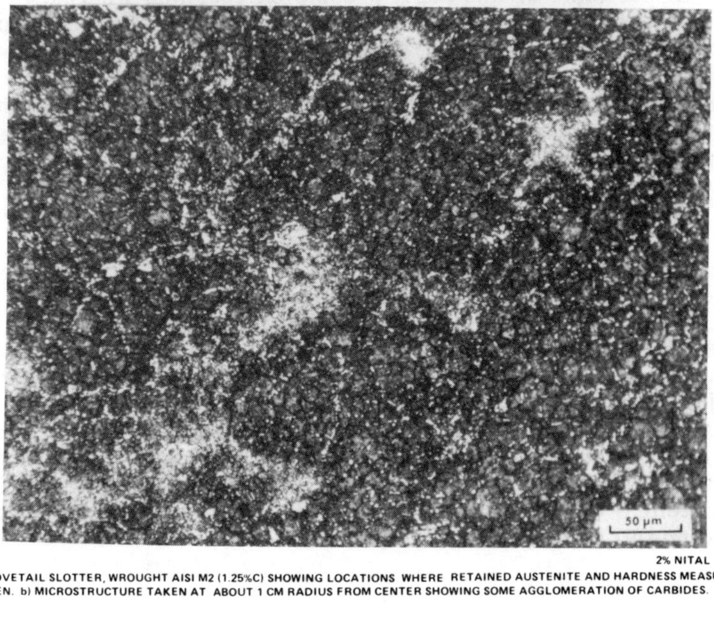

2% NITAL

FIG. 3 a) DOVETAIL SLOTTER, WROUGHT AISI M2 (1.25%C) SHOWING LOCATIONS WHERE RETAINED AUSTENITE AND HARDNESS MEASUREMENTS WERE TAKEN. b) MICROSTRUCTURE TAKEN AT ABOUT 1 CM RADIUS FROM CENTER SHOWING SOME AGGLOMERATION OF CARBIDES.

tolerances and the cutters were wearing at a faster than normal rate. With a limited supply on hand, action was needed to insure that they would last for the required amount of machining.

Experimental and Results. Hardness, retained austenite, and dimensional checks were taken on the tool. It was found that 10.2 vol% retained austenite was contained in the shank. The tool was then cooled in alcohol to -85°C (-121°F) at a rate of 1°C (1.8°F)/min and held for 20 minutes and heated to room temperature in a like manner. The heat treatment then consisted of a 1 hour stress relief at 150°C (302°F), followed by a 1 hour temper at 540°C (1004°F) in vacuum. The tool was then slow cooled to room temperature and re-examined. The most noticeable change was that the retained austenite reduced to 4.2 vol%, the hardness in the shank increased .5 points Rc to 50.4 Rc, and the hardness in the rim increased 1.5 points Rc to 64.2 Rc. No dimensional changes were recorded (to the fourth place) in either the shank or rim. The shank end of the tool was then polished and etched, and the microstructure is shown in Figure 3b. The grain size was revealed to be approximately ASTM 10 (10 microns) with some agglomeration of carbides evident.

The remaining cutters in the lot were then processed similarly, and all were returned to service, performing much better than previously.

Discussion. In this case, a complex tool was wearing fast, and retained austenite was identified as the microconstituent responsible for the poor performance. Checking the retained austenite in any tool is a non-standard practice, and it is not the intention of this author to suggest that such a practice be standardized. But for complicated, expensive tools such as these, it would be advantageous to spot check samples on a per-lot basis to insure maximum performance.

Large AISI M3-2 Milling Cutters

Background. This study was initiated when ten slot mills, Figure 4a, failed by chipping (Figure 4b) and one shell mill, Figure 4d, failed as one tooth broke. Both of these 15 cm (6 in.) diameter milling cutters were made by vendor "A" and were received within two weeks of each other, suggesting that the same material was used for both cutters. The mills were used in roughing operations to machine large 12 Cr turbine buckets. A shell mill (Figure 4g) from a consistently reliable vendor "B" which did not fail, was used as a reference to the other failures.

Experimental and Results. At a radius of approximately 4 cm (1.6 in) from the center of each tool, a small area was polished using "field" metallography techniques and the microstructures were characterized (Figures 4c,e,f,h,i). Portable microscopes were necessary to photograph the microstructures of the shell mills, due to their excessive weight. The microstructures of all three cutters were fairly equivalent, consisting of extensive carbide networks and a grain diameter near ASTM 10 (10 microns). Hardnesses were taken in these same areas, those from vendor "A" being 66 Rc, and the one from vendor "B" being 60 Rc. Although these microstructures are considered undesireable from a metallurgy viewpoint, they are apparently typical for wrought tools of this size and cannot be blamed for failure. To salvage the slot mills, the axial rake angle was increased from 0 to 6 degrees, which eliminated the chipping

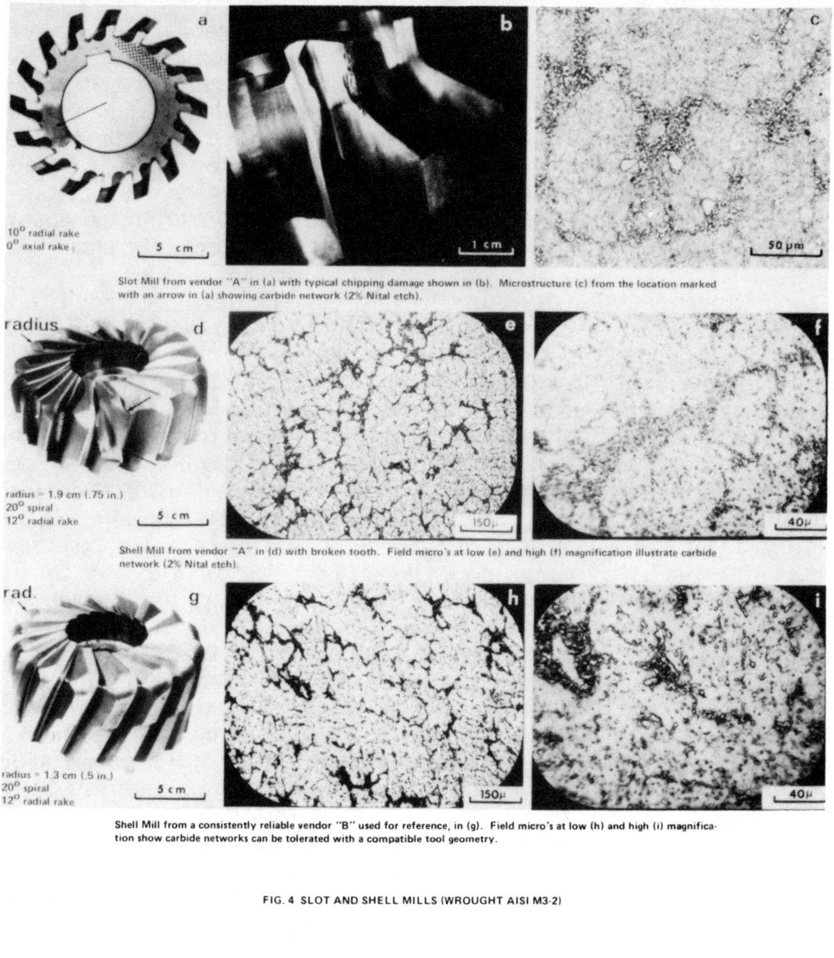

Slot Mill from vendor "A" in (a) with typical chipping damage shown in (b). Microstructure (c) from the location marked with an arrow in (a) showing carbide network (2% Nital etch).

Shell Mill from vendor "A" in (d) with broken tooth. Field micro's at low (e) and high (f) magnification illustrate carbide network (2% Nital etch).

Shell Mill from a consistently reliable vendor "B" used for reference, in (g). Field micro's at low (h) and high (i) magnification show carbide networks can be tolerated with a compatible tool geometry.

FIG. 4 SLOT AND SHELL MILLS (WROUGHT AISI M3-2)

problem. For the shell mills, changing the tool geometry was considered too complicated so they were tempered back to 60 Rc, which eliminated the tooth breakage problems.

Discussion. The "fix" for these cutters was an additional heat treatment for the shell mill, and the addition of an axial rake angle on the slot mill. Even though the microstructures were far from what we know as being optimum, they were not determined as the main cause of failure. While the tool designs seemed to be limiting in these cases, sooner or later the designs will become "optimum". Once the tools have reached this point, improvements in performance will come only by improvements in the cutter material, which could include changing grades or going to PM.

AISI M3 Turbine Bucket Vane Cutters

Background. The two cutters shown in Figure 5a and 5b are used interchangeably on multi-spindel tracing milling machines. They are similar with the exception that vendor "C" used a .95 cm (.375 in) step before turning down the inside diameter, while vendor "D" used a 1.43 cm (.56 in) step. Vendor "C's" cutter, however, was outlasting its companion by a ratio of 3 to 1 in this continuous cutting operation. Typical chipping damage is shown in Figures 5c and 5d.

Experimental and Results. Both cutters were pretended to be of AISI M3-2 composition (1.2wt%C, 6%W, 5%Mo, 4%Cr, and 3%V), however, chemical analysis determined cutter "C" to be of AISI M3-1 composition (1.05%C, 6%W, 5%Mo, 4%Cr, and 2.4%V). Where chipping damage is predominant, this information would be consistent with experience in that M3-1 compositions are tougher than M3-2 compositions. To evaluate the cutters metallurgically, they were mounted in epoxy and ground down to the centerline of the tooth. After the mounts were polished and etched (Figure 5e and 5f), the microstructures revealed a difference in the carbide size, showing cutter D to have many more large carbides than cutter C. A quantitative plot of the carbide population using the Bausch and Lomb Omnicon Image Analysis System FASII confirmed this (Figure 5g - 50 fields counted at 2000X), and it was found that for cutter C, the mean and maximum carbide size is 1.65 microns (.000065 in) and 22.6 microns (.00089 in) respectively, and for cutter D, the mean and maximum are 2.06 microns (.000081 in) and 27.4 microns (.00108 in) respectively. Both cutters contained about 10 vol% carbide, but cutter C had a grain size of ASTM 11 (7 microns) and cutter D had a grain size of ASTM 10 (10 microns). To complete the investigation, the hardness was taken near the cutting edge and found to be 65.4 Rc in cutter C and 65.8 in cutter D. X-ray diffraction showed that there was not an appreciable amount of retained austenite (less than 2%) in the general microstructure of either cutter.

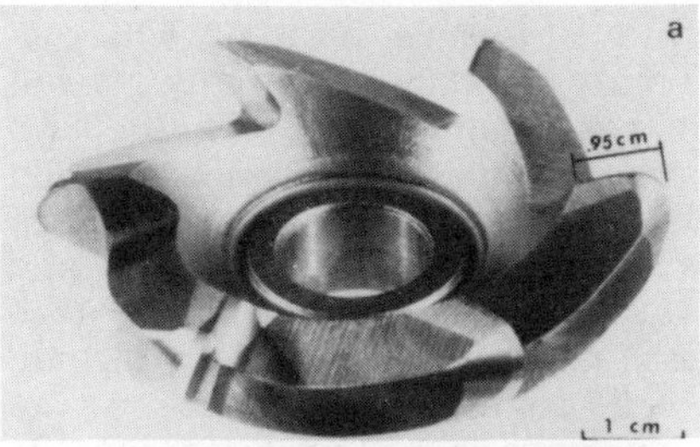

Fig. 5a Best performance from this cutter, vendor "C", 6.7 cm (2.625 in.) in diameter.

Fig. 5b Vendor "D" cutter, also 6.7 cm (2.625 in.) in diameter.

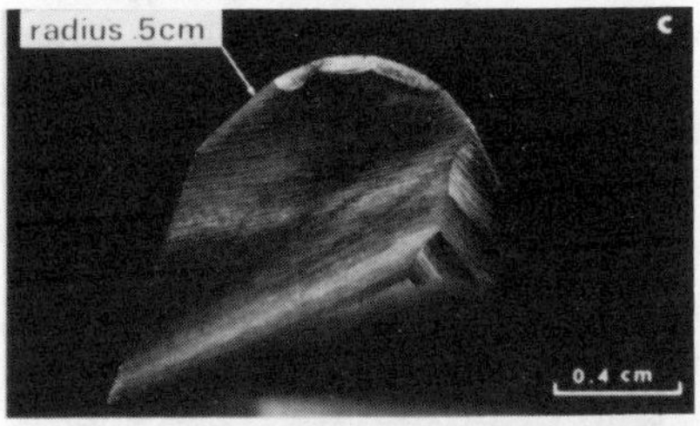

Fig. 5c Typical damage from cutter "C" after machining about 3 buckets.

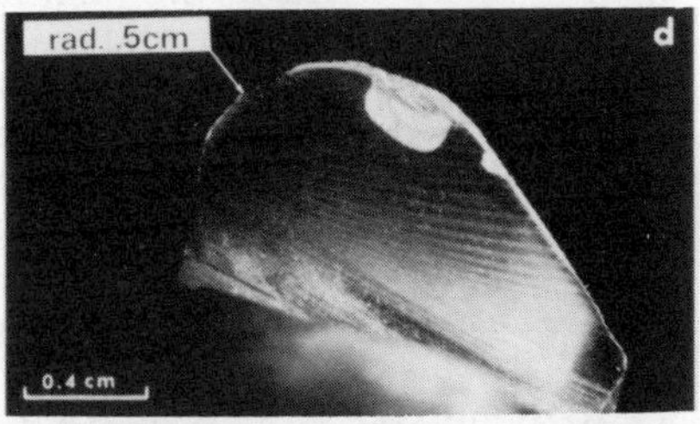

Fig. 5d Typical damage from cutter "D" after machining about 1 bucket.

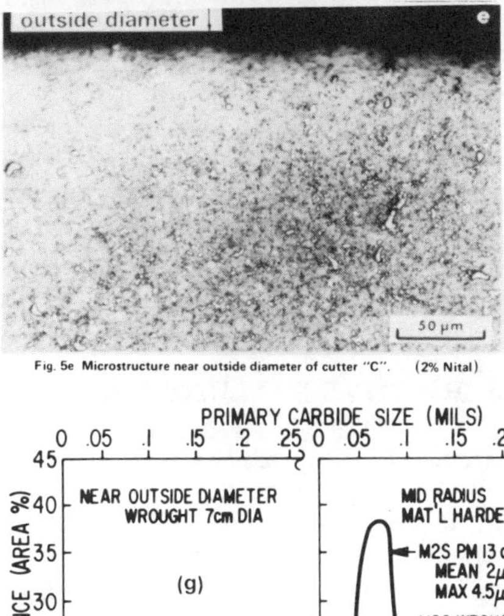

Fig. 5e Microstructure near outside diameter of cutter "C". (2% Nital)

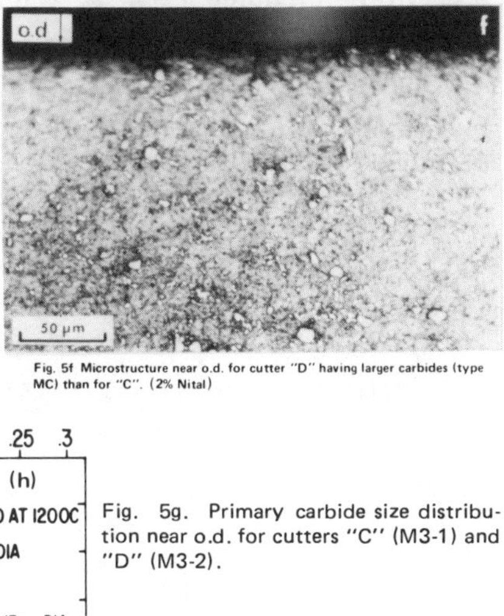

Fig. 5f Microstructure near o.d. for cutter "D" having larger carbides (type MC) than for "C". (2% Nital)

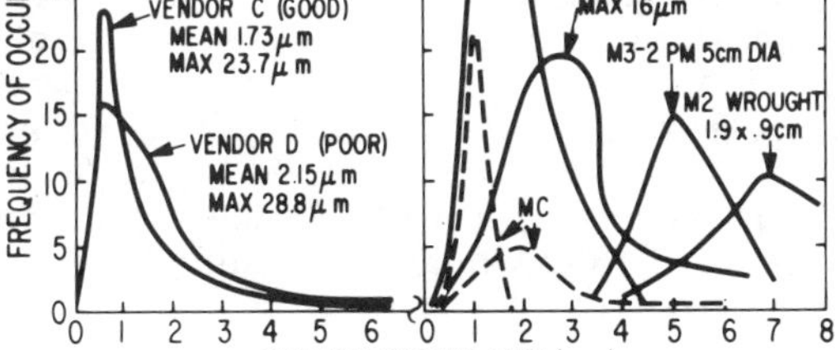

Fig. 5g. Primary carbide size distribution near o.d. for cutters "C" (M3-1) and "D" (M3-2).

Fig. 5h. Primary carbide size distributions at mid radius for M2S (PM and wrought ref. 8) and M3-2PM and M2 wrought (both from ref. 13) shown for comparison.

Fig. 5. TURBINE BUCKET VANE CUTTERS (AISI M3-1, M3-2).

Discussion. The above example is a rare case of where a deviation (M3-1) from the requested material (M3-2) actually proved to be beneficial. The difference in performance between the two cutters can be directly related to the structural differences, which include the grain size, matrix, and carbides. Since grain size has been determined to play a minor role in continuous cutting (9,10), and the matrix's of the two materials are assumed to be similar (the matrix of both having a hardness of 832 KHN), the carbide differences between M3-1 and M3-2 compositions must be the important feature in the cutting performance. The difference is directly observable in the microstructure, with cutter D having more carbon and vanadium and also many larger MC carbides than cutter C. The above observation becomes significant when one considers that hardness, geometry, and volume percent carbide are essentially identical, which is in agreement with some (11) but not all (12) wear studies.

The difference in carbide size between the two steels can also be attributed to the steel mill, where solidification rate (related to ingot diameter) or deformation ratio's were different. Figure 5h is presented for comparison of primary carbide sizes for M2S (PM and wrought-ref. 8) and M2 (wrought-ref. 13) and M3-2 (PM-ref. 13), and although the distributions are not directly applicable, they do put the information obtained in this study into perspective. At the steel mill in an attempt to reduce primary carbide size, the ingot diameter is reduced, but this means that the reduction ratio is also lower. A lower reduction ratio means that carbides will not be broken up as well as if a high reduction ratio were used. It is for this reason that wrought products present a limit on the smallest attainable carbide sizes.

ONE LAST NOTE ON NONDESTRUCTIVE EVALUATIONS

This discussion applies to the largest of tools that one may wish to evaluate. Large tools are heavy and hard to handle, especially for metallography. A simple technique is available that will produce large "in-focus" areas, and allow observations with normal optical metallographic equipment at magnifications up to about 1000X. Higher magnifications are available with TEM, but this is more expensive. The procedure involves polishing as one normally would for a "field" micro, and then heavy etching with the desired solutions. An acetate replica is then applied to the area of interest, which can be a flat or curved surface. After the replica is dry, it is stripped off of the tool and attached to a glass slide with double stick tape. Using a hard rubber roller (a solid rubber stopper will do) the replica is rolled to 1) flatten it and 2) secure it to the slide. The replica is then

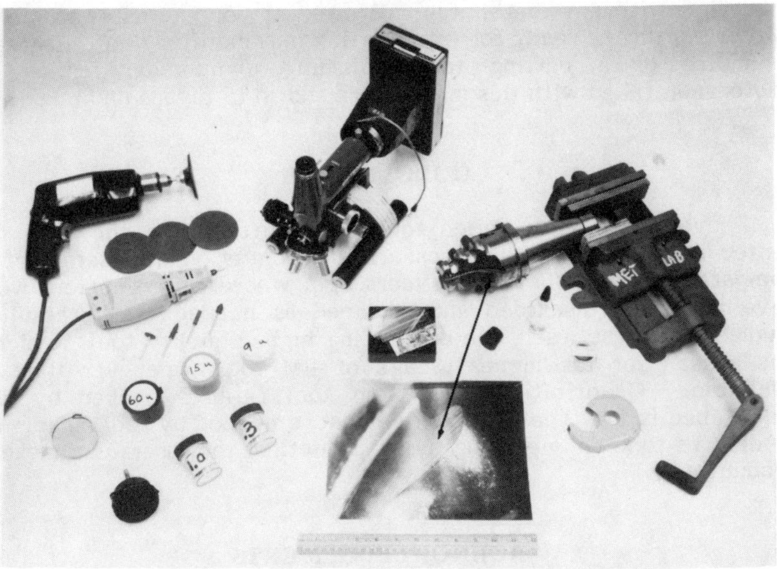

Fig. 6a Typical polishing, replicating, and photo equipment for "field" metallography. Arrow in picture indicates location from which micro below was taken.

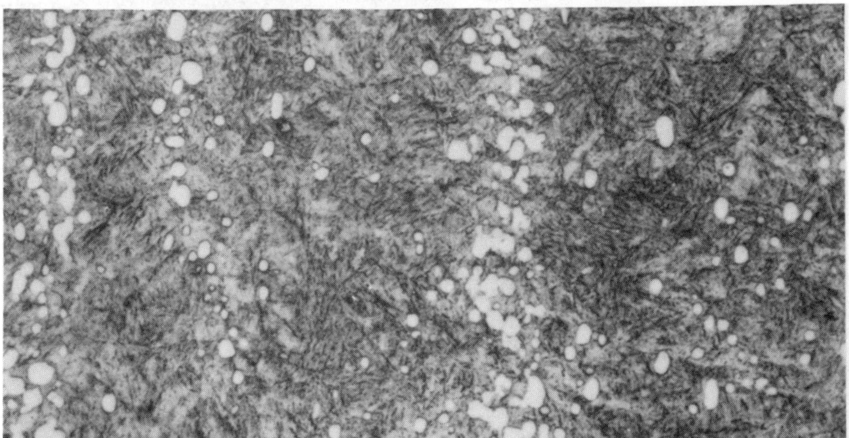

Fig. 6b Microstructure from cutter tooth shown above using an acetate replica, 5% Nital etch, silver shadowing, and optical metallography.

shadowed with any metal, typically Au, Ag, or Cr, after which the microstructure is ready for examination on laboratory metallographs. Examples of a cutting tool, polishing equipment, replica and photograph taken with this method are shown in Figure 6a and 6b.

CONCLUSIONS

Investigations of failed tool steel cutters like the ones presented here result in an increase in awareness on the users part, as compared to the situation 10 years ago, where failed cutters would have just been discarded and replaced as needed. As a result of studies like these, users are demanding higher quality tools, which is one reason for the higher useage of PM tools whenever they are available. Quantitative data from tool failures backed by shop experience is the type of incentive that is needed by tool makers and steel mills to keep improving their product, so tool users can increase productivity.

ACKNOWLEDGEMENTS

The author would like to thank the Large Steam Turbine Department for permission to publish this work. J.D. Martin and W.H. Pattison are also acknowledged for providing the failed tools and much encouragement. The following Materials and Processes Laboratory personnel who contributed technical expertise and whose help I could not have done without include R. Ling; R. Betlejeski, R. Carpenter, E. Drebitko, and N. Gendron (Metallography); R. Matteson (TEM); and H. Beattie (XRD).

REFERENCES

1. N.J. Gendron, Metals Engineering Quarterly, 13, (1), 1973 p13.
2. N.J. Gendron, "Mounting, Lapping and Polishing Large Sizes and Quantities of Metallographic Specimens," Presented at Int'l. Metallographic Society/ASM Joint Meeting, Beverly Hills, Ca., Sept. 23-24, 1973.
3. B.D. Cullity, Elements of X-Ray Diffraction, Addison Wesley, 1956, p391.
4. A.M. Turkalo, Trans AIME, 218, 1960, p24.
5. A. Taylor, X-Ray Metallography, John Wiley and Sons, Inc., 1961, p724.
6. J.H. Woodhead and A.G. Quarrell, "The Role of Carbides in Low Alloy Creep Resisting Steels," University of Sheffield, Available thru Climax Molybdenum Co., Publication M111, 1965, p23.

7. R. Wilson, Metallurgy and Heat Treatment of Tool Steels, McGraw Hill Book Co., (UK) Limited, 1975.

8. A. Kasak, G. Steven, and T.A. Neumeyer, SAE Publication No. 720182, Presented at Automotive Eng. Congress, Detroit, MI, Jan. 10-14, 1972.

9. T.A. Neumeyer and A. Kasak, Met. Trans., 3, 1972, p2281.

10. D. Bhattacharya et.at., Wear, 42, 1977, p63.

11. A. Kasak and T.A. Neumeyer, Wear 14, 1969, p445.

12. P.R. Borneman, Met. Eng. Q., 13, (2), 1973, p50.

13. G. Berry and M.J. Kadhim Al-Tornachi, Metals Technology, June 1977, p289.

[Discussion is presented on the next page.]

DISCUSSION

Richard Bratt (Crucible)

Do you go beyond surface examination in your non-destructive metallographic techniques?

Gregory Bouse

To insure that the microstructures represent bulk material, we take micro's from the surfaces that have been finished machined only, which is about .005-.01" below the depth of the original heat treated surfaces. If the structure is not immediately obvious, we will hand grind to levels where normal structures are observed. If hand grinding is necessary, there are usually places where this can be done that will not permanently damage the tool. Using the acetate replica, curved surfaces can also be used to obtain microstructures if micro's on flat surfaces are not allowable.

THE PROPERTIES OF HIGH SPEED TOOL STEEL PRODUCED BY HORIZONTAL CONTINUOUS CASTING

David C. Briggs and Robert Thomson
Canada Centre for Mineral and Energy Technology
Energy, Mines and Resources, Canada

ABSTRACT

Equipment developed for the closed-head horizontal continuous casting of ferrous alloys was used to cast 1-in. dia. bar of high speed tool steels. The apparatus consists of a water-cooled, unlubricated, copper mould attached to a stationary tundish by a connecting nozzle of boron nitride. The cast bar is extracted intermittently from the mould by means of pinch rolls. The single strand operation is capable of producing ~1000 lb (500 kg) of cast bar per hour.

Continuously cast M2, M7, and M42 tool steels have been evaluated as cutting tool material in both the as-cast and hot-swaged conditions. The hardening response of the test materials was equivalent to that of commercial steels of the same grade. In single-point turning of annealed 4140 steel, tool life for all three steels, whether cast or worked, was also found to be comparable to that of commercially produced tools. Unnotched Charpy impact data showed a lack of toughness in the as-cast product but confirmed that the impact behaviour of the worked and commercial materials were comparable. The cast microstructure showed no coring and contained a discontinuous network of primary carbides. The ASTM grain size was 9-11.

The paper concludes with a description of the conceptual design and economics of a plant for the horizontal continuous casting and rolling of high speed tool steels.

INTRODUCTION

Conventional techniques of producing high speed tool steel by ingot casting, forging and rolling give low yields of ~50% finished product from liquid metal. A potentially more efficient production method, that of horizontal continuous casting of small sections, has been developed by Thomson at CANMET (1). This paper briefly describes the CANMET process, examines the physical metallurgy and performance of concast high-speed steels and sketches a conceptual design and the economics of a small plant dedicated to continuous casting of tool steel.

The world-wide development of horizontal continuous steel casting has not been well documented in the literature because of the proprietary nature of the various processes. Information which is available on the productivities of both commercial and experimental processes is shown in Fig. 1 (2).

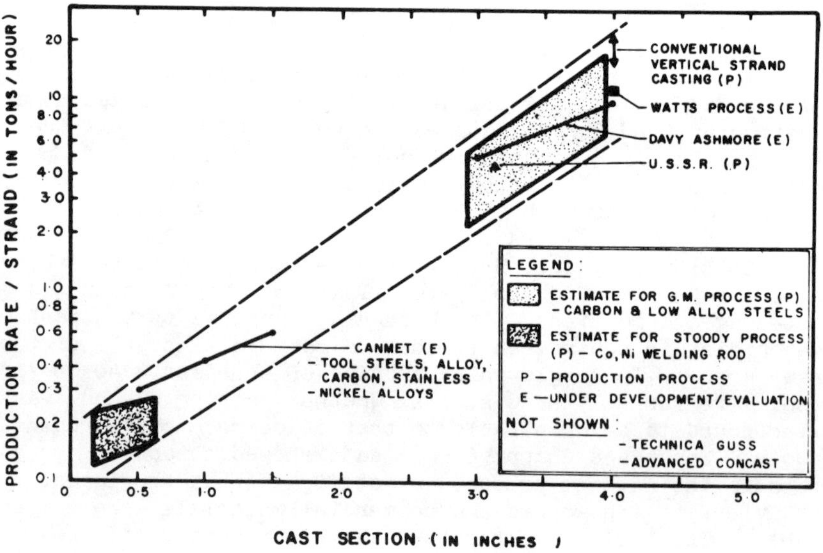

Fig. 1. Production rates of horizontal continuous steel-casting process.

Four of the most significant designs were originated by General Motors (3), Davy-Ashmore (4), the U.S.S.R. (5) and the Technicon Corporation (the Watts Process) (6). They all produce round or square billets of 3-6 in. (75-150 mm) thickness at estimated production rates slightly less than that of the conventional vertical strand casting. At the low extreme of production rate lies the Stoody process which produces wire rod in the range 3/16 to 5/8 in. dia. (7). Of these designs, both the horizontal Watts process and conventional vertical casting have been used in the production of high speed steel. From reports about these applications it appears that coarse carbide networks persist in the concast billet structure causing problems during subsequent fabrication. This severely reduces the potential economies of concast production at these sizes. With the CANMET process, which gives section sizes of the order of 1 in., the rate of solidification precludes the generation of networks of large carbides. Consequently, cracking problems in hot working are greatly reduced.

CANMET HORIZONTAL CONTINUOUS STEEL-CASTING PROCESS

The principle of the CANMET technique is shown schematically in Fig. 2. The tundish, made of exothermic moulding sand, is self-heating to facilitate start-up, and is attached to a water-cooled copper mould. The connecting sleeve or nozzle is machined from boron nitride rod. The casting is withdrawn from the stationary mould/tundish assembly by the

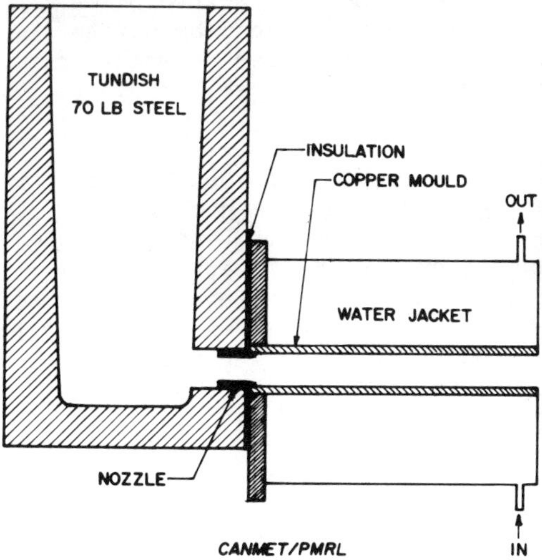

Fig. 2
Tundish/mould assembly

Fig.3
Casting tool steel

The CANMET closed-head continuous casting process

intermittent-extraction technique common to most horizontal casting systems. Pinch rolls gripping the casting are cyclically accelerated to ~200 in./min (5000 mm/min) producing short rapid extraction strokes of ~0.5 in. (13 mm) at the rate of 40 to 120 per min, yielding net casting rates of 20 to 60 in./min (500-1500 mm/min). The casting unit is shown in operation in Fig. 3. At these small section sizes spray cooling of the extracted casting is not employed.

The surface condition of tool-steel bars is illustrated in Fig. 4. Cyclic lap marks are minimized by very rapid movement of the bar during extraction. This action develops turbulence in the liquid steel entering the mould and, by melting and erosion, reduces surface discontinuities between newly solidifying metal and that solidified during the preceding stroke. The lap marks shown in Fig. 4 are the most severe that were encountered. Many of them are of sufficient depth that conditioning of the bar would be advisable before hot working.

The type of internal defect that results from thermal-stress cracking in the bar after casting and during air cooling is shown in the radiograph in Fig. 5. The frequency of cracks appears to increase with casting speed, and in these examples the cracking is too severe for manufacture of cast tool bits. Material such as this, however, can be satisfactorily swaged. The internal crack surfaces re-weld during fabrication to yield an homogenous, defect-free, wrought product.

A major potential benefit of continuous casting for tool-steel production is the elimination of a number of costly steps, e.g. soaking, forging, reheating and rolling, and the accompanying savings in energy, manpower, and plant. The advantages of closed-head continuous casting as described here, in which the tundish is attached directly to the mould, include prevention of reoxidation during casting and the ability to cast small sections with the accompanying elimination of coring.

EVALUATION OF HIGH SPEED TOOL STEEL PRODUCED BY HORIZONTAL CONTINUOUS CASTING

Following a favourable preliminary assessment (8), concast tool steels were characterized with respect to their response to heat treatment, performance as cutting tools, impact properties and microstructure. Type M2 high-speed steel was examined in detail in both the cast and hot-worked conditions, while types M7 and M42 were assessed in the cast condition only, using smaller numbers of replicates. Hot working of the M2 grade was achieved by swaging at 1100-925°C (2010-1700°F) from 1.0 in. dia. to 0.56 in. dia. (25 to 14 mm dia.), i.e. 70% reduction. All materials were fully annealed after casting; swaged materials were annealed again after working.

Fig. 4 The surface of CANMET concast tool steel bars, 1.0 in. dia. top, M2; centre, M7; bottom, M42.

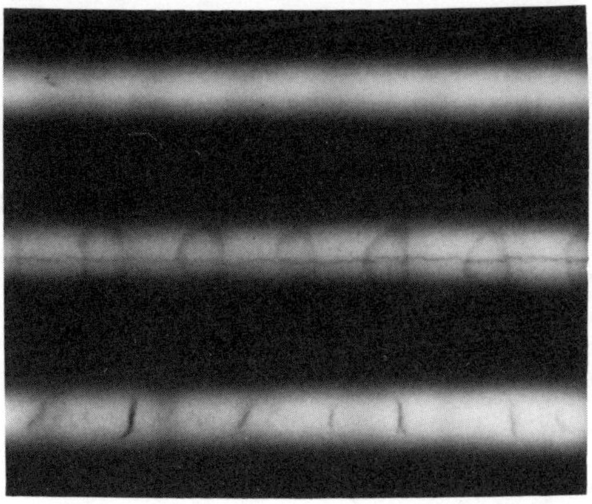

Fig. 5 Radiograph of 1.0 in. dia. Type M7 tool steel bars, concast at CANMET, showing transverse and longitudinal cracking and sound bar.

Hardness

Despite the microstructural differences between concast and commercially produced tool steel, it was found that standard heat-treating schedules produced the optimum hardening response in the concast alloys (8). The austenitizing temperatures employed were as follows: for concast M2, both the cast and the worked products, 1225°C (2240°F); for concast M7, 1210°C (2210°F); and for concast M42, 1190°C (2175°F). All three materials were hardened by quenching in oil at room temperature and were tempered at 550°C (1025°F). M2 and M7 were tempered twice and M42 three times, for one hour each temper.

The hardness of the three steels was monitored at each stage of the manufacturing process from the concast bar through to the finished lathe-tool bit. Data for all of the materials are summarized in Table 1. The commercial alloy, Atlas Sixix, mentioned in the table, is a Type M2 tool steel produced by Atlas Steels Co. It was used throughout this evaluation as a commercial reference material.

Table 1. Rockwell C hardness of concast and commercial high speed tool steels.

Condition	Worked		Cast		
	Atlas Sixix	Concast M2	Concast M2	Concast M7	Concast M42
Initial condition[a]	-	53.1	55.8	57.5	54.7
+ annealed[a]	-	24.3	29.0	31.7	29.0
+ hardened[a]	-	64.0	65.2	66.0	63.1
+ tempered	66.0[b]	67.0[b]	66.2[c]	66.2[d]	68.9[d]

a - average of 6 readings on 1 sample
b - average of 81 readings on 27 samples
c - average of 87 readings on 29 samples
d - average of 30 readings on 10 samples

The hardness values for annealed material, particularly M2 and M7, were high compared with typical values for commercial products. However, values for the as-worked, as-hardened and as-tempered materials agreed well with the commercial values. Clearly, concast tool steels are capable of achieving the hardness levels required for metal cutting service.

Tool Life

As an indication of tool-steel performance, the life of a tool bit in a lathe turning trial was measured. Catastrophic failure of the cutting edge was taken as the end point of the test. Trials were carried out on a Monarch Monamatic 13 in. lathe with continuously variable spindle speed and feed rate. The depth of cut was 0.0625 in. (1.6 mm), the feed rate 0.010 in/rev (0.25 mm/rev). The tool geometry was as follows:

Back rake	0°	End cutting edge angle	5°
Side rake	10°	Side cutting edge angle	15°
End and side relief	5°	Nose radius 0.30 in (0.76 mm)	

Esso Cutwell was used as cutting fluid. The machining stock was AISI 4140 steel, hot rolled and fully annealed. It was scalped at least 1/8 in. (0.32 mm) to the test log size of 3.00 in. dia. (76.2 mm) x 19 in. long (480 mm). At least three cutting speeds, in the range 120 - 210 ft/min (36.6 - 64.0 m/min), were used to evaluate each material. When failure had not occurred by the end of a pass, a new pass was started with the spindle speed and feed rate adjusted for the reduced diameter of the test log. The tool life recorded was the cumulative time to failure.

The test specimens, 5/8 in. (15.9 mm) square tool bits, were machined from 1.0 in. (25 mm) annealed bars of the three concast steels, then hardened, tempered and ground to the prescribed tool geometry. Similarly, 3/8 in. (9.5 mm) square tool bits were prepared from the swaged concast bars of Type M2 steel. Identical tests were performed on tool bits of the commercial reference material, Atlas Sixix.

The tool life results were interpreted in terms of the Taylor equation,

$$VT^n = C,$$

where V is the surface cutting speed in ft/min or m/min, T is the tool life in min, C is a constant depending on the test conditions and equal to the cutting speed for a tool life of one minute, and n = -m where m is the slope of a straight line fitted to the test data on a log-log plot. The graphs of tool life vs cutting speed obtained by regression analysis of the data are presented in Figs. 6 to 8. The quantity "r" in the Figures is the correlation coefficient. To illustrate test reproducibility, Fig. 6 shows the data points, 95% confidence limits, and regression line for the 18 tool bits of concast M2 tested in the cast and heat-treated condition.

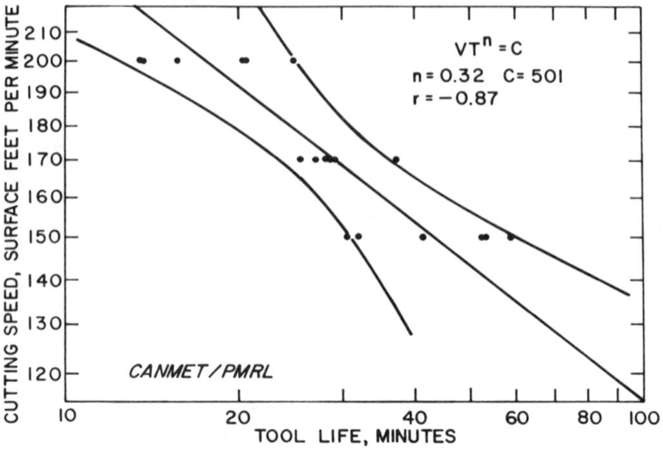

Fig. 6 Tool life - cutting speed relationship for concast Type M2 tool steel, as cast and heat treated, with experimental data and 95% confidence limits.

Figure 7 shows the results of all tests on concast material heat treated from the as-cast condition, with the corresponding reference data for Atlas Sixix. Note the comparable results with all four steels particularly near the centre of the range of observations where the reproducibility of test data is highest. Speculation about the superiority of one steel over another at any cutting speed is unwarranted because of the divergence of the confidence limits.

Figure 8 shows the test results for concast M2 which was hot worked before heat treatment. The wrought concast material is also seen to be comparable with the commercial reference material. In addition, there is little difference in behaviour between the worked M2 in Fig. 8 and the cast M2 shown in Fig. 7.

Impact Properties

Table 2 lists the Charpy impact energies for tests on unnotched specimens of each of the steels, including the reference material Atlas Sixix. It can be seen that the impact strength is low for all of the as-cast steels with M7 marginally better than M2 or M42. Hot working makes a great

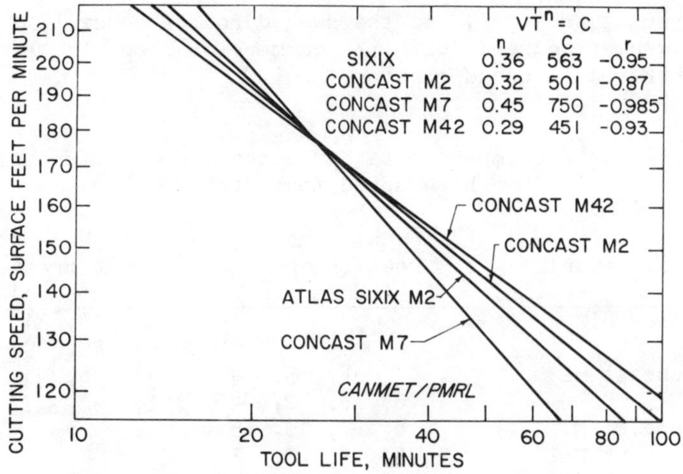

Fig. 7 Tool life - cutting speed relationship for concast-and-heat-treated and commercial high speed tool steels.

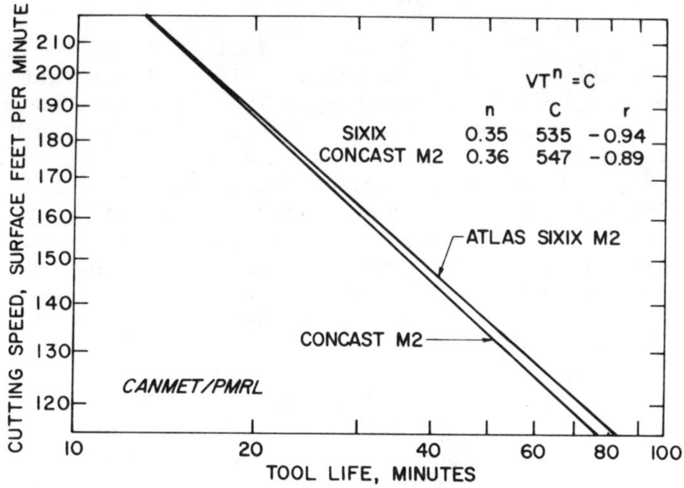

Fig. 8 Tool life - cutting speed relationship for concast-swaged-and-heat-treated and commercial high speed tool steels.

improvement, however, and the swaged rod of concast M2 compares very favourably with the commercial product; the data for both show a broad scatter.

Table 2 Impact data for concast and commercial high speed tool steels

Tool material	Cast and heat treated (joules)	Worked and heat treated (joules)
Atlas Sixix		35, 13, 31, 31
		1, 29, 29, 14
Concast M2	5, 1, 4, 3, 4	47, 16, 41, 37
	6, 6, 2, 4	16, 42, 17, 15
Concast M7	6, 7, 15, 5	
Concast M42	4, 5, 3, 3	

Unnotched Charpy impact bars
1 joule = 0.74 foot-pounds

Microstructure

The microstructure of concast tool steel is characteristic of the small section size and the rapid cooling in effect during casting. The grain size is small, ASTM 9 - 11. The grain boundary carbides are thin and discontinuous; they do not occur in massive form and seldom appear in broad eutectic arrays. Full annealing followed by austenitizing can effect some agglomeration and spheroidizing of the primary carbides but is incapable of reducing their prominence in the grain boundaries or of redistributing them significantly to the grain interiors. As an example, the microstructure of concast M2 after heat treatment is shown in Fig. 9 and comprises tempered martensite with a network of primary carbides. By comparison, Atlas Sixix also contains tempered martensite but has primary carbides that are spheroidized and well dispersed (Fig. 10). The microstructure of concast M2 after hot working and heat treatment is shown in Fig. 11. The grain-boundary carbides have become more spheroidized and dispersed. Evidence of the original carbide distribution still remains after 70% deformation by swaging, but, as has been shown in Table 2, the embrittling effect of the carbide network has been greatly reduced.

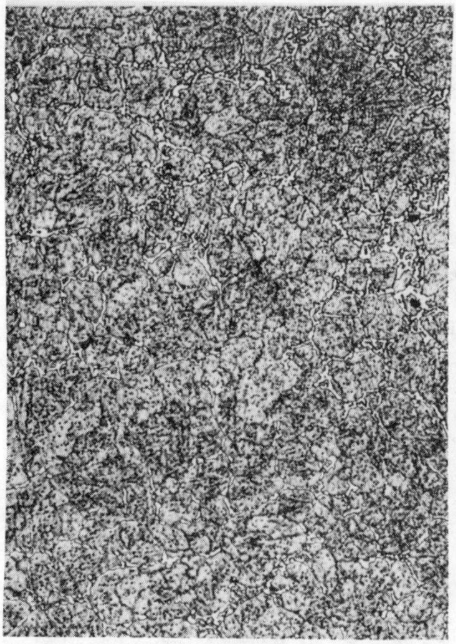

Fig. 9 Concast M2 tool steel, hardened and tempered from cast and annealed condition.

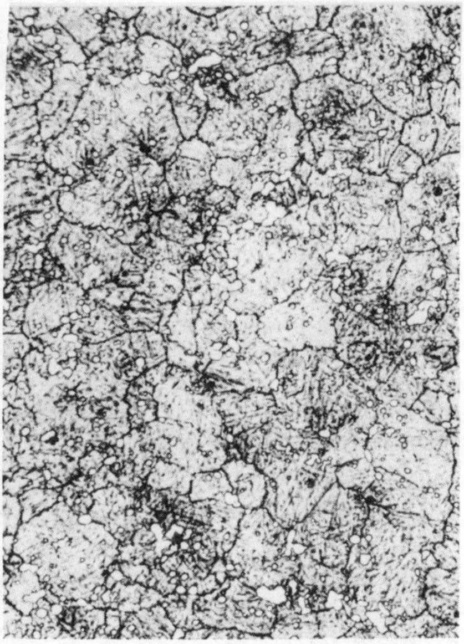

Fig. 10 Commercial M2 Tool steel, as received in hardened and tempered conditon.

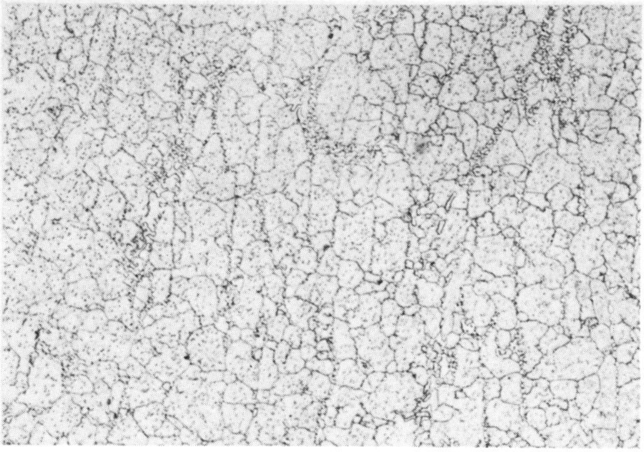

Fig. 11 Concast Type M2 tool steel, hardened and tempered after hot swaging.

A fractured impact specimen of each steel was examined by scanning electron microscopy to determine the mode of fracture and the rôle played by the grain-boundary carbides. The fractographs proved difficult to interpret and even the carbides were not readily distinguishable from the matrix in appearance. The fracture shown in Fig. 12 is from concast M2 which was heat treated from the cast condition. The dark areas outlined by white are similar in size to the grains in the preceding optical micrographs. Areas such as these may approximate grain surfaces exposed by the passage of the crack and thus may indicate intergranular fracture. On the other hand, grain separation was ragged, with the crack diverging occasionally into a grain. The fracture surface of Atlas Sixix in Fig. 13 shows much less roughness and is believed to be transgranular. The grain-boundary carbides of cast material, then, likely affect the impact properties by affecting the fracture mode.

Fig. 12 SEM fractograph of CANMET concast M2 tool steel, fractured after hardening and tempering from the cast and annealed condition.

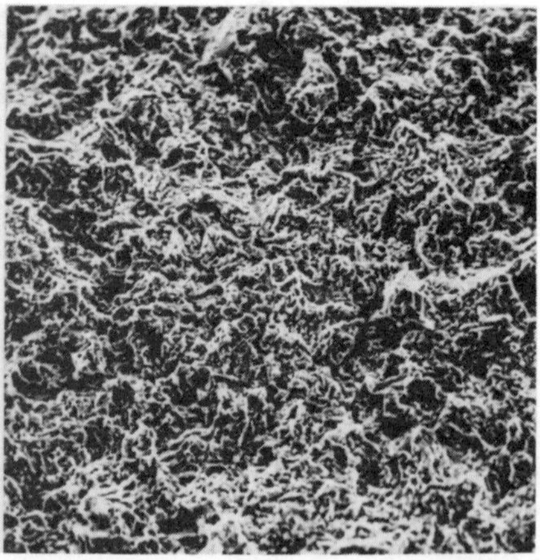

Fig. 13 SEM fractograph of Atlas Sixis commercial M2 tool steel, fractured in the hardened and tempered condition.

Discussion

The properties of horizontally concast high-speed tool steel can be summarized as follows:

1. The materials show a hardening response to heat treatment equivalent to that of the same tool steels produced commercially,
2. Tool life is comparable to that of the equivalent commercial material under the test conditions employed,
3. The impact properties are low for material heat treated from the cast and annealed condition, but can be raised to commercial levels by deformation prior to heat treatment.

The possibility of using these concast tool steels in the cast and heat treated condition, is clearly limited by their low impact properties. They can not, for example, be used in machining operations involving interrupted cutting. It may be possible, however, to tailor tool-steel compositions to suit the production process. For example, for the few applications in which good toughness is not required, tool steel grades that lack workability could be continuously cast as stock of near final size and shape. Alternatively, by lowering alloy and carbon content to reduce the volume of intergranular carbide, it may be feasible to improve impact properties while maintaining high matrix hardness and compressive yield strength (9). This would utilize the observed direct correlation of tool life with compressive strength (10, 11). In this way, tough, as-cast tool bits may be producible by continuous casting.

Hot swaging the concast tool steels produced tools with properties comparable to those of conventionally fabricated materials. This effect occurred with only 70% hot deformation as compared to the minimum 94% normally required in conventional commercial practice, and may be ascribed to the smaller size and more uniform distribution of carbide phases in the concast microstructure.

A MICRO MILL FOR TOOL STEEL PRODUCTION

The commercial viability of tool-steel manufacture by horizontal continuous casting was examined by Ferrco Engineering Ltd., Whitby, Ontario, Canada, in a study commissioned by CANMET (12). Their proposed plant design is shown in Fig. 14. A four-ton electric-arc furnace supplies liquid steel, via a ladle, to a 1½-ton tundish. The latter is mounted on a horizontal casting machine producing bar of 1.5 - 2.5 in. (38 - 64 mm) dia. A holding/reheat furnace provides a buffer zone for cast bar prior to rolling. The layout includes four reversing mills, two vertical and two horizontal, with facilities downstream for heat treatment, surface grinding and shipping. This scheme might be improved by including a forge rolling mill ahead of the reversing mill line to provide the initial 10 - 15% breakdown of the cast structure and for conditioning of the surface of the bar.

A plant such as this, shipping 4700 tons per annum of a mix of M2, M7 and T1 grades and assuming a yield of 75% at the mill, would have the following financial and operating parameters:

Capital requirement	Cdn $27,000,000
Cost of finished product	Cdn $2.14/lb
Average selling price	Cdn $4.00/lb
Employment	76 people
Productivity	62 tons/man yr.
Return of equity	47%

This output is about one half of the present market for all tool steels in Canada and ~5% of that in the United States.

Because of the difficulty of a new venture in penetrating a tool-steel market which is already adequately supplied, such a production technique may interest only existing tool-steel manufacturers. The Ferrco report also suggests that significant advantage would be obtained by a combination of products such as tool-steel stock, forging-bar stock and grinding balls.

REFERENCES

(1) R.Thomson, unpublished work, CANMET, Energy, Mines and Resources, Ottawa.

(2) R.Thomson, Closed-Head Continuous Casting Workshop, March 1979, CANMET, Energy, Mines and Resources, Ottawa.

(3) "G.M.Casts In-plant Scrap into In-plant Steel", Iron Age, Vol.208, No.24, Dec.9, 1971, p.53.

(4) J.Marsh, P.C.D.Gamble, and R.J.Dain, "Development of a new Horizontal Continuous Casting Process for Steel", Metals and Materials, Vol.5, July 1971, p.229.

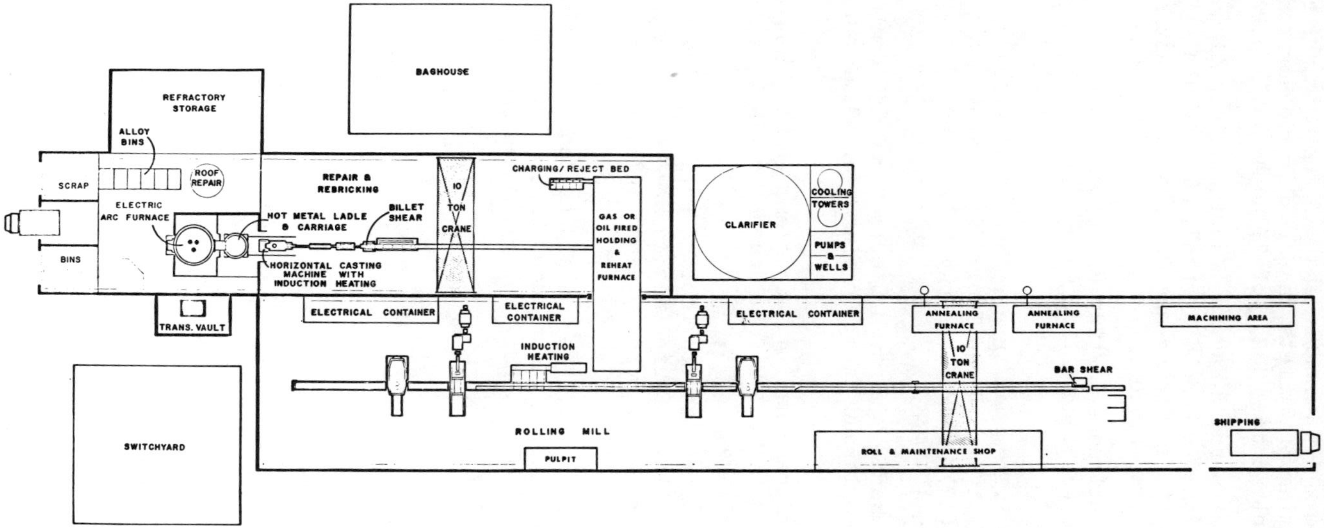

Fig. 14 Design of micro mill for production of tool steel by horizontal casting.

(5) V.T.Sladkoshteev, A.I.Manokhin and O.A.Shatagin, "The Continuous Casting of Billets on Multi-strand Horizontal Machines", Steel in the USSR, Vol.7, No.4, 1977, p.208.

(6) H.E.Allen, L.Watts and R.Hadden, "Horizontal Continuous Casting in a Closed-end Mold System", The Metals Society International Conference on Continuous Casting, May 1976, Biarritz, France.

(7) J.E.Gullon, "Continuous Casting Process for Co Hardfacing Weld Rod", Metal Progress, Vol.112, No.6, 1977, p.51.

(8) D.C.Briggs and R.Thomson, "Preliminary Evaluation of Continuously Cast High Speed Tool Steel", Rept. MRP/PMRL 78-1 (IR), CANMET, Energy, Mines and Resources, Ottawa, 1978.

(9) M.E.Bush and R.Ward, "Improved Steels for Cold Forming Tools", Metallurgia and Metal Forming, Vol.43, No.10, 1976, p.346.

(10) W.E.Henderer, "Strengthening Mechanisms in High Speed Steel as Related to Tool Life", Journal of Engineering for Industry, Vol.101, No.2, 1979, p. 217.

(11) E.M.Trent, "Requirements for Tool Materials for Cutting Metals", International Conference on New Tool Materials and Cutting Techniques, London, March 1977, Engineers Digest, 1977.

(12) Ferrco Engineering Ltd., "Technical and Economic Assessment of Horizontal Continuous Casting of Steel", Contract OSQ78-00197, CANMET, Energy, Mines and Resources, Ottawa, 1979.

DISCUSSION

Fred Kanaan (Kennametal)

Is there a problem with inclusions and/or porosity with this casting process of High-Speed tool steels; and if so, how do you control or reduce these defects?

David Briggs

We have seen no inclusions in our concast product and, indeed, this is one of the great advantages of the closed-head continuous casting, over conventional vertical strand casting. Attachment of the tundish directly to the mould precludes reoxidation of the liquid steel as it flows into the mould. With suitable design of tundish and the use of baffles, the propelling of inclusion materials into the mould by the pouring stream from the ladle or furnace can be avoided.

Macroporosity has been observed in bar cast under conditions that give a solidification front extending far down the bar. In such a case pools of liquid steel can become sealed off from the ferrostatic head and develop shrinkage voids upon solidification. Adjustment of casting rate and cooling conditions in order to establish a blunter cone-shaped solidification front cures this problem. Similarly, microporosity can result when liquid is trapped by growing dendrite arms but this has not been a noticeable problem in our concast bar.

John Knox (Vermont American Corp.)

Are the transverse defects shown in the radiograph associated with the intermittent withdrawal of the bar from the mold by the pinch rolls?

David Briggs

Certainly the spacing between the internal transverse defects matches the stroke length of the intermittent withdrawal and for this reason must be related to it. However, we have not been able to identify the relationship. We think the defects are cracks resulting from differential contraction between outer shell and interior core during cooling, and as such they could be avoided by annealing or controlled cooling immediately after casting.

Mohan Chaudhari (Gray Iron Research Institute)

What would you use as an inoculant to refine the as-cast structure?

David Briggs

We have not taken any steps to investigate the effect of inoculants on the structure of our concast material and, therefore, have not considered the choice of a specific inoculant. I can say only that those mentioned for high speed steel in the literature include oxides, carbides and nitrides of Al, Mg, Ti, Zr, Ca, V and Ce. As well, manganese should be carefully controlled for, above 0.4%, it is reported to contribute to grain growth.

Gregory Bouse (General Electric, LST)

Compared to some of the cast tools I've seen, yours look very good. We need 6" diameter tools with microstructures equivalent to or better than conventional wrought products. If you could cast a 3" diameter ingot, then upset it to 6" diameter, your process may be able to complete with larger diameter wrought products. Would you comment on upsetting instead of swagging your cast bar?

David Briggs

In the present work we have evaluated 1-in. dia. bar. Billets of 3-in. dia. would necessarily have experienced slower solidification rates and thus possess coarser grains and more massive carbide networks. The extent to which this structure survived the upsetting would influence the properties of the forged piece. However, the structure should not affect the ability of the piece to be upset, especially since the outer layers would be chill-cast and therefore fine-grained. More important to the success of the operation would be the surface condition; grinding of the lap marks and any other surface defects would be essential in order to prevent these from initiating surface cracking during upsetting. Press forging rather than hammer forging probably would be most successful.

FRACTURE TOUGHNESS AND RUPTURE STRENGTH OF HIGH SPEED STEELS

H. F. Fischmeister
L. R. Olsson

Austrian School of Mines (Montanuniversität), Leoben, Austria
Chalmers University of Technology, Gothenburg, Sweden

INTRODUCTION

Wear resistance, hot hardness and toughness are customarily regarded as the three most essential characteristics of high speed steels (HSS). Materials science offers two well-defined and universally accepted tests for toughness: rupture strength (usually tested in bending) and fracture toughness. The first indicates the probability for cracks to be nucleated at internal stress concentrators, the other measures the material's resistance to their propagation. Together, they provide today a satisfactory description of flaw-induced fracture in large specimens of high-speed steels (and cemented carbides as well).

It must be conceded that a description which satisfies the materials scientist cannot yet fully match the many facets of the tool engineer's toughness concept which today can only be grasped through various aspects of cutting performance.

In the last decade, powder metallurgy (PM) tool steels have appeared (1-4) whose microstructure is entirely different from conventionally produced (cast and wrought, CW) materials. Their carbide distribution is much finer and virtually free from stringers and other directionality. The finer carbide dispersion also allows finer austenite grain size to be developed during proper heat treatment, which should lead to increased toughness (5) by virtue of the classical Stroh-Petch grain size dependence (6).

This paper reviews work on the fracture toughness and bend rupture strength of both PM and CW tool steels. Our own interest was initially slanted towards fracture toughness as a new approach, which was expected (7) to bear some relation to wear behaviour, and because it would gage the average quality of particle bonding in PM materials (which could be impaired by oxide films on the powder particles, or improper processing). As processing methods were developed which gave virtually perfect bonding, the interest of materials scientists everywhere turned back to the classical property - rupture strength - and it turned out that this is a fine indicator of the singular sites of structural weakness which are present even in the most perfect materials.

MATERIALS, PROCESSES, TEST METHODS

The structural problems to be discussed will be better understood if prefaced by a description of the processes used for the manufacture of the newer types of material (PM). This will also serve as background documentation for the experimental data to be presented later.

HSS powders are produced by atomization (4,8) of fully prealloyed melts either with inert gas (usually nitrogen) or with water. Gas atomization produces spherical particles by solidification of the melt droplets in free fall; water atomization gives irregular shapes (fig. 1).

Typical oxygen contents are 50 ppm for gas atomized and several thousands of ppm for water atomized powders, corresponding to oxide film thicknesses of the order of $8 \cdot 10^{-9}$ m for gas atomized and $200 \cdot 10^{-9}$ m for water atomized powders (if all oxygen is assumed to be surface bound).

The powders are filled in metal cans and hot compacted either by hot isostatic pressing (HIP), as in the two processes

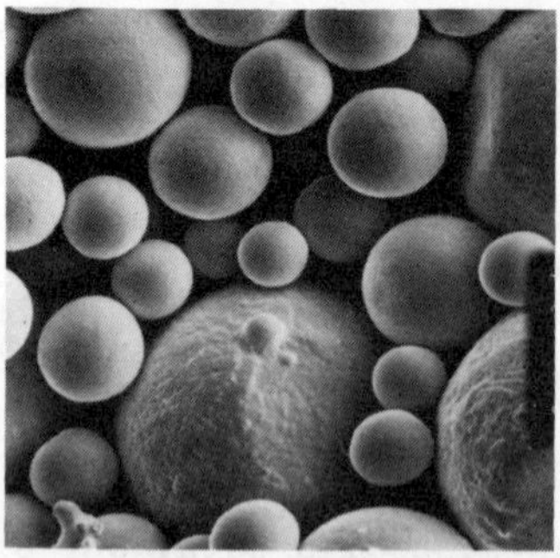

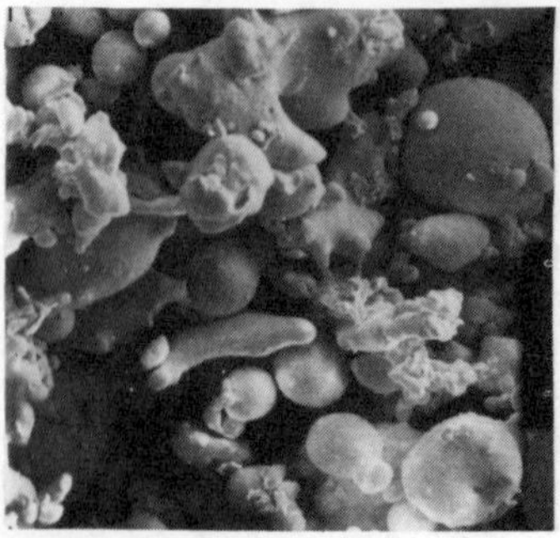

Fig. 1. High speed steel powders ——— 100 μm.
(a) gas atomized, (b) water atomized.

that have been commercially operative during the past decade (2,3), or by hot extrusion (9) or forging (10) followed by rolling. Initially, it was felt that high-oxygen powders could not be processed at all, but it was found that three factors can cooperate to give satisfactory particle bonding and toughness:

(i) the oxide can be reduced during heat treatment before compaction, by carbon contained in or added to the powder (9-12);

(ii) strong hot deformation of the powder mass, as in extrusion or forging-and-rolling, will break down oxide films and produce good metal-to-metal contact;

(iii) a given content of oxides is less deleterious in the finely dispersed form which results from processing of irregular powders than in the lumpy form found in CW material (9,11).

Toughness measurements reflecting the effect of oxygen and hot work in PM materials will be reported in this paper.

Recently, several versions of a process have been published which consolidates water-atomized, ground HSS powder by cold pressing and sintering in the region of incipient melting (13). Although the fine carbide structure of the powder is considerably coarsened during the process, fracture toughness values in the range of ordinary CW and PM materials have been reported (14). We have not yet studied any such material.

Included in the present paper are toughness measurements made on electroslag-remelted material (ESR). This process produces a finer and more homogeneous carbide distribution than conventional ingot practice, but the carbides are still coarser than in PM material. Our ESR material had been hot worked to an area reduction of only 75 pct., which left the carbides comparatively large.

The CW material included in this study was of commercial origin. We studied material of 45 and 25 mm ⌀, which - with a customary ingot size in Swedish HSS practice of ca. 300 mm ⌀ - would correspond to a probable area reduction of 97.8 and 99.3 pct. respectively.

Representative microstructures of the PM, CW and ESR materials are shown in fig. 2, which illustrates the large differences both in carbide size and in directionality.

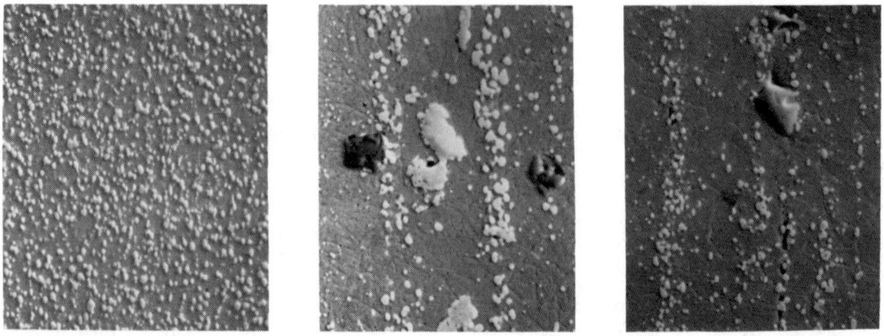

Fig. 2. Microstructures of M2 (SEM, nital etch) ⟝⟞ 20 μm (a) water atomized, extruded (in vertical direction); (b) ESR, rolled; (c) cast and wrought.

In the main, we have studied steels of type M2 processed by different routes. They are referred to by the following code symbols:

- CW - cast and wrought (with dimension in mm)
- ESR - electro slag remelted
- WE - water atomized, extruded
- WF - water atomized, compacted by forging
- WFR - water atomized, forged, and further reduced (60 pct) by hot rolling
- RF - atomized by the Rotating Electrode Process (15) (which gives spherical low-oxygen powder similar to gas atomization) and consolidated by forging.

Heat treatment followed conventional lines, with austenitization at 1050, 1150, and 1200°C for CW and PM materials and at 1200°C for the ESR grade, followed by double and triple tempering for CW and ESR and triple tempering for the PM materials whose easily dissolvable carbides form stabler austenite.

Fracture toughness tests were carried out on compact tension (CT) specimens of the type shown in fig. 3, mostly with W = 16, B = 4 mm and with the following orientations:

	crack plane	crack propagation
LL	longitudinal	longitudinal
LT	longitudinal	transverse
TT	transverse	transverse

The fracture toughness test determines the critical stress intensity, K_{Ic}, at which an existing, stationary, sharp crack will begin to propagate. The required sharp crack was

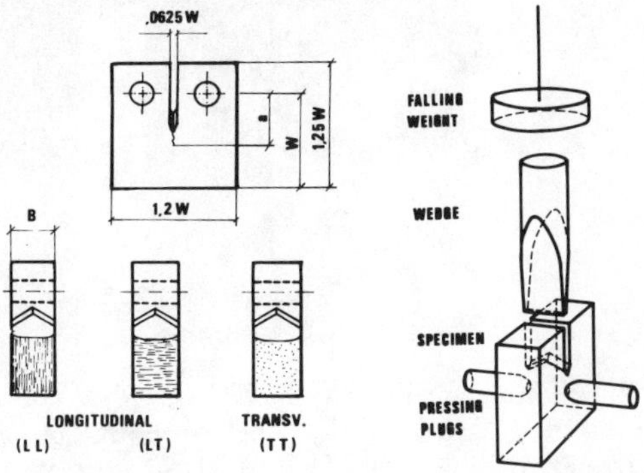

Fig. 3. Crack orientation and method of pre-cracking for fracture toughness specimens.

introduced by a method developed by Harris and Dunegan (15) and validated for HSS by Ericsson (17) and by ourselves (18), in which a brittle crack is started in a chevron-notched bar (cf. fig. 3) by a wedge under the impact of a guided drop weight. The crack comes to a standstill at the desired depth as it runs into an opposing stress field generated by local compression. Its actual depth is visible on the eventual fracture surface because of slight tarnishing. - The measurement and evaluation of K_{Ic} followed ASTM specification E 399.

Rupture strength was measured on prismatic test bars of 4×4 mm^2 section with a HSS load ball and HSS supports 40 mm apart, at a loading rate of 200 MPa/s.

FRACTURE PATHS AND FRACTURE SURFACES

Intuitively, one would expect the crack to associate with the stringers, and there are many examples of this behaviour (18), especially toward the ends of cracks initiated according to fig. 3, where the stress intensity is barely sufficient for propagation (fig. 4).

The degree to which the crack follows the stringers depends on the hardness of the material. This is because the crack can react to carbides in its vicinity only as far as it can 'feel' their presence, i.e. as far as the plastic zone created around the crack tip extends. Only at low hardness levels is the plastic zone large enough to span the distance between stringers. This has an important consequence for the

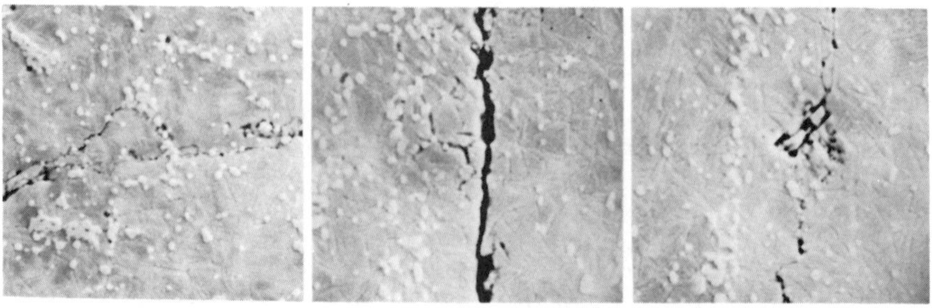

Fig. 4. Crack paths in precracked fracture toughness specimens. Nital etch, SEM ⟼ 20 μm. (a) LT fracture, (b) and (c) LL fractures.

fracture topography, which had escaped us in our earlier work. At hardness around the normal level for cutting tools (65 R_c), we had observed comparatively little directionality in the appearance of the fracture surface and in the values of fracture toughness (18). At low hardness (see fig. 5) there is pronounced directionality - except for the TT direction, where the fracture plane cuts right across the stringers. In the LL orientation the crack is corrugated with valleys and ridges running parallel to the stringers. In the LT orientation, the crack is no longer guided by continuous stringers. Instead, it is 'attracted' by stringers running at different height levels parallel to and ahead of its front. Once the front has come to coincide with such a stringer, it stops and starts looking for the next. The result is a random upwards or downwards deflection of adjacent parts of the crack to different levels, with abrupt steps between them, and occasional branching-off of secondary cracks.

A warning is necessary at this point: it is absolutely impossible to gain a true impression of the fracture topography without stereo viewing. Figs. 5 and 6 are stereo pairs, and a suitable viewer is found in every issue of Metals Handbook (19).

With stereo viewing, fig. 5 clearly reveals that LL and LT fractures are more serrated than TT fractures. All of them are more serrated than the fractures of PM materials. PM materials show no directionality, neither at low hardness nor at high. Their fracture surfaces invariably look like the TT fractures of CW materials, cf. fig. 6.

We have measured the surface fraction of carbides in the fracture surface. Even after correction for surface roughness (20), this clearly exceeds the volume fraction of carbides in

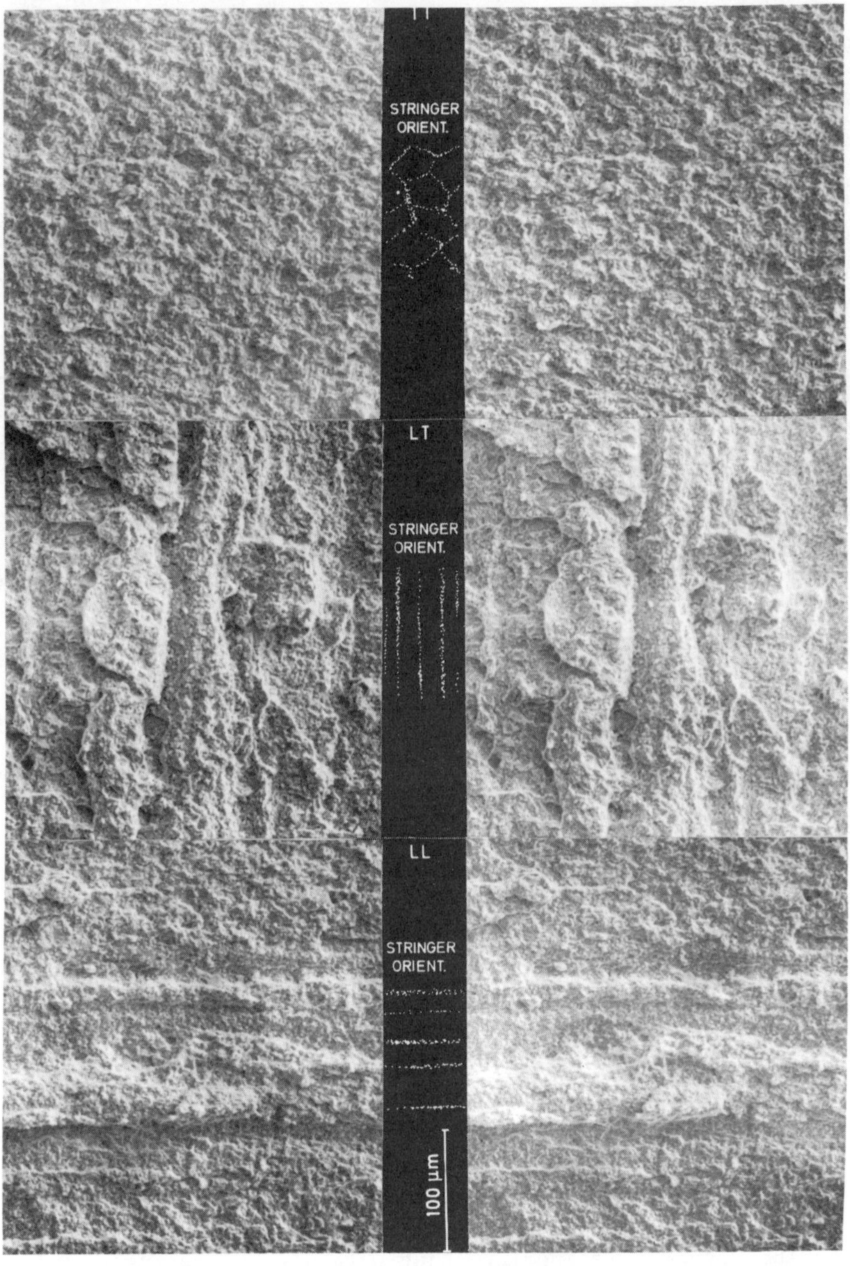

Fig. 5. Fracture surfaces of ESR material (58 R_c) 0.5 mm beyond end of precrack. Crack propagation → SEM, stereo pairs.

the material. Very few carbides on the fracture surface seem to be cracked, in agreement with observations by others (14, 21). These two observations together indicate a decided preference of the crack path for the carbide/matrix interface.

Fig. 6. Fracture surfaces of powder metallurgy material (EW). Ca. 0.5 mm beyond end of precrack. Crack propagation → SEM.

FRACTURE TOUGHNESS

Fracture toughness turns out to be so strongly related to the degree of tempering, that data can be meaningfully compared only when related to the material's hardness. Fig. 7 displays a summary of published fracture toughness data

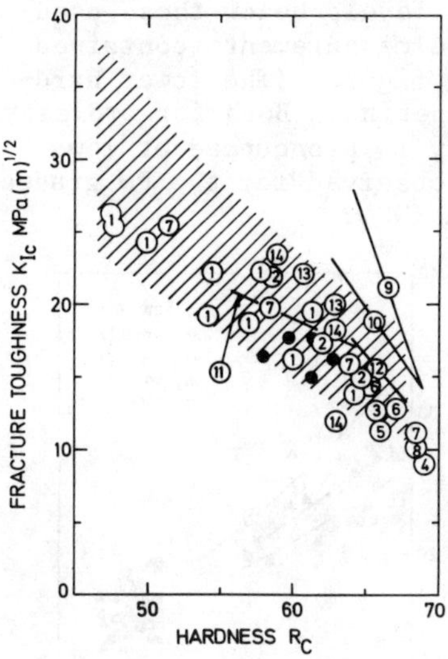

Data points Reference

PM M2, Olsson, Fischmeister (18)
CW M2, " "
ESR M2, this work
• *PM T1, Olsson, Fischmeister (18)*
1 *CW M2, Johnson (22)*
2 *CW M2, Hellman, Wisell (23)*
3 *CW M2, Eriksson (24)*
4 *PM ASP 30, Eriksson (24)*
5 *PM ASP 23, Hellman, Wisell (22)*
6 *PM ASP 23, Ingelström (25)*
7 *CW M7, Eriksson (24)*
8 *PM STORA 30, Eriksson (24)*
9 *Sintered T6, Wronski et al (14)*
10 *CW T6, -" -*
11 *CW M2, Berry, Al Tornachi (26)*
12 *PM M2/3, " "*
13 *CW T1, Rescalvo, Averbach (21)*
14 *CW M1, Queeney, Heffelfinger (27)*

Fig. 7. Compilation of fracture toughness values.

against a backdrop of the range observed in our own work. The figure contains data for both CW and PM variants of M1, M2, M7, T1, T6 and a PM super-HSS grade (ASP 30). With the possible exception of one set of measurements, the data all arrange themselves in a common band. Even our earlier fracture toughness values for CW and PM versions of a hot work die steel of composition 0.3 C, 11.5 Cr, 7.5 W, 0.5 V, 9.5 Co (18) could be fitted into this common band.

Fig. 7 demonstrates a surprising insensitivity of fracture toughness to alloy composition or even alloy type. This indicates that fracture toughness should be regarded mainly as a matrix property (18,22,28). If that is so, only small scope would seem to remain for the effect of carbide distribution which constitutes the microstructural difference between PM and CW grades. This holds especially at high hardness levels, where the range of observed fracture toughness values is clearly narrowed. Carbide size variations induced by soaking in CW M1 had no significant effect on fracture toughness in a study by Queeney and Heffelfinger (27). Neither was there a significant difference in fracture toughness between the gages 25 ⌀ and 45 ⌀ studied in our earlier work (18).

Yet, the range at a given hardness is not negligible in relation to the scatter of measurements. Some structural influence exists, although its importance may be largely academic because it emerges only at hardness levels below those used in cutting tools. Fig. 8 details the measurements contained in the shaded band in the preceding figure. (The lower hardnesses were all obtained by overtempering.) Both for ordinary CW and for ESR grades directionality is pronounced at low hardness, but no directionality is observed for the PM grades, not even after strong hot reduction (WE).

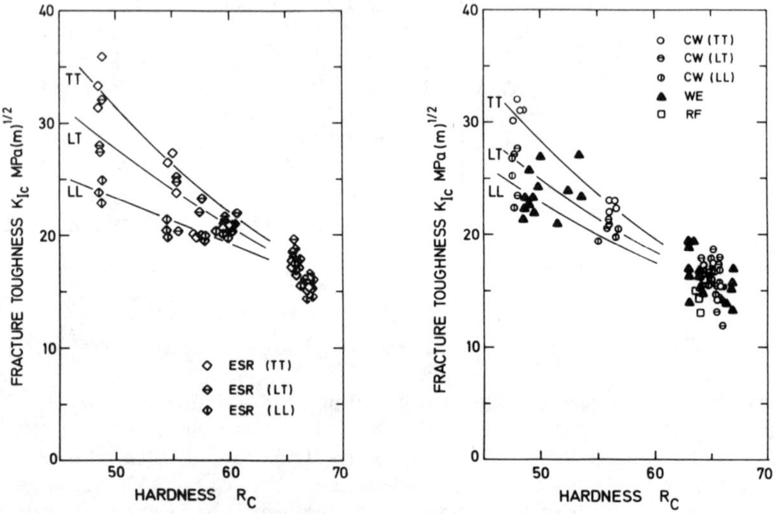

Fig. 8. Fracture toughness vs. hardness for M2 steels made by (a) electroslag remelting, (b) CW and PM processing.

As discussed in the preceding section, the fracture path seems to have strong preference for the carbide/matrix boundary. Also in cemented carbides, debonding at the carbide/matrix boundary becomes the dominant mode of crack propagation when the volume fraction of binder exceeds ca. 30 pct. (29,30). Thus the boundary has lower crack propagation resistance than the matrix. This would explain that LL fractures - which permit the least disturbed association of the crack with closely spaced carbides in stringers - exhibit the lowest fracture toughness, followed by LT fractures, where the association is less perfect, and has to be paid for in terms of frequent deflection and even crack branching - in other words, a longer fracture path.

That the influence of directionality virtually vanishes between 55 and 65 R_C could be explained by the observation (18,40) that in this region the plastic zone has shrunk to the size of the average carbide spacings (fig. 9). A contributory cause will be the diminishing advantage of phase boundary propagation over propagation through a matrix whose retarding power (K_{Ic}) drops with increasing hardness.

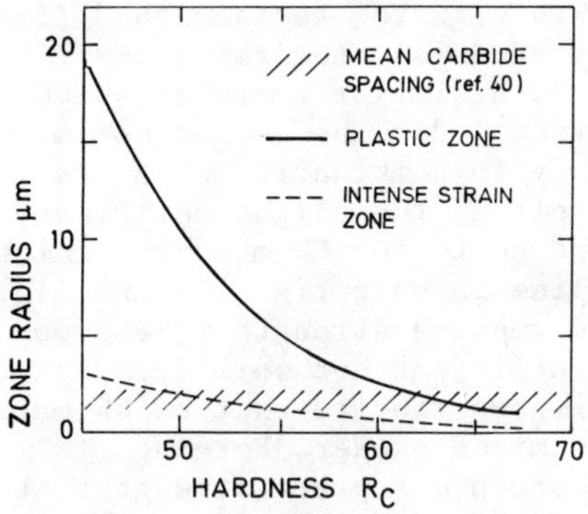

Fig. 9. Size of plastic zone compared to carbide spacings for varying hardness

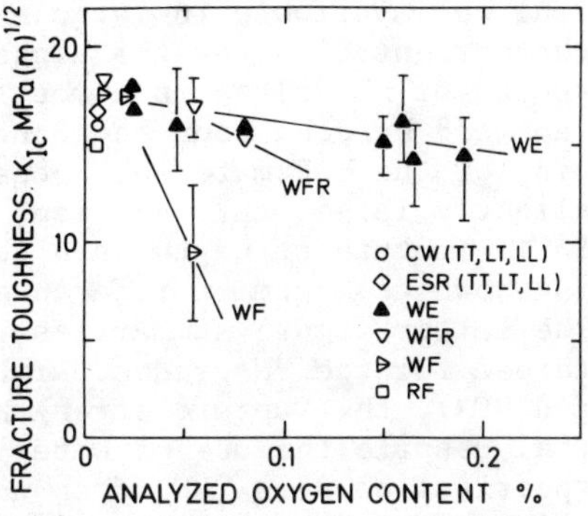

Fig. 10. Influence of oxygen content on fracture toughness of PM, ESR, and CW M2 at 65 R_c.

An important issue for the PM materials is the influence of oxygen content and hot work in fracture toughness. This is illustrated in fig. 10. Oxygen reduces the fracture toughness because the crack can partly run along imperfectly bound particle boundaries. However, the sensitivity to oxygen is strikingly different for the various processing routes. Water atomized powders give satisfactory fracture toughness (on a par with the average of CW grades) up to 1000 ppm of oxygen when strongly hot worked by extrusion (grade WE; typical extrusion ratios were of the order of 13:1). Densification by forging in gettered cans (10,11,18) with subsequent hot rolling (60 pct. reduction, grade WFR) puts a lower, but still comfortable limit on the oxygen content. A still lower limit holds if the powder is merely densified by forging (grade WF), but even then, water atomized powder with its irregular and fine particles can be used at higher oxygen contents than the spherical powder resulting from REP-atomization (grade RF). However, fig. 10 shows that all four processes yield fracture toughness levels fully equivalent to CW practice when the oxygen content is kept low enough. For water atomized powders, this can be achieved without difficulty by proper gettering, while for gas atomized and REP powders, low oxygen levels come naturally. (The high oxygen levels of some of the PM grades in fig. 10 were intentionally chosen for the purpose of this study.)

RUPTURE STRENGTH

Fig. 11 illustrates the strong directionality of rupture strength which is characteristic of high speed steels whose carbides are aggregated in stringers (32). Only the ESR material was available in large enough gage for tests with different orientation. At the highest hardness, the ratio of TT and LT or LL values is almost 2:1. A similar ratio has been reported by Johansson and Sandström (33). The modest hot working of the ESR material, possibly in combination with its slightly larger carbide size, explains the slight deficiency in TT rupture strength when compared to the CW material which is shown as a common reference line in both fig. 11a and 11b. The latter figure summarizes the rupture strength values obtained for the PM grades. With sufficient hot work (grades WE and WFR), the rupture strength is superior to that of CW material despite the use of water-atomized powder. Forging of spherical powders (RF) does not produce rupture strengths at all equivalent to CW material. This is attributed to occasional remaining pores and, perhaps, imperfect particle bonds.

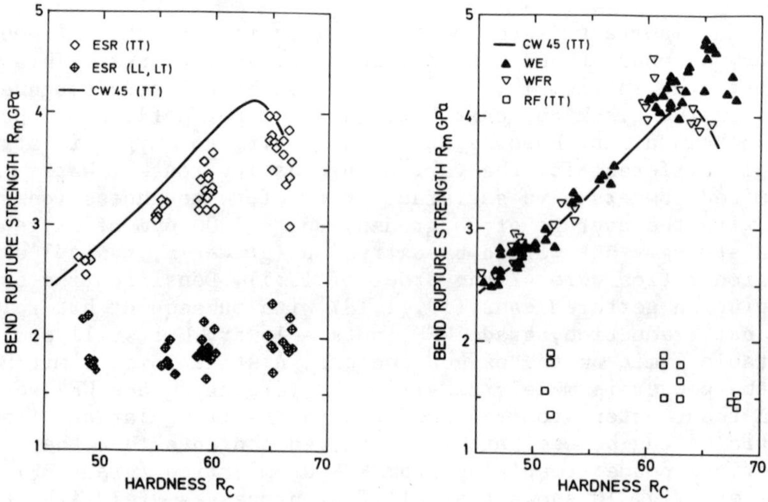

Fig. 11. Rupture strength vs. hardness for M2 steels made by (a) electroslag remelting, (b) PM processing (data for CM material from ref. 18 shown as curve)

The influence of oxygen content on the rupture strength of these materials is illustrated in fig. 12. As in the case of fracture toughness, oxygen tolerance is highest for extruded material, followed by forged-and rolled. The limiting oxygen contents where these powder materials still have rupture strength equivalent to conventionally produced material, are actually slightly above those for equivalent fracture toughness. Consolidation by mere forging would require extremely pure powders in the case of water atomization, and is obviously impractical for spherical powders.

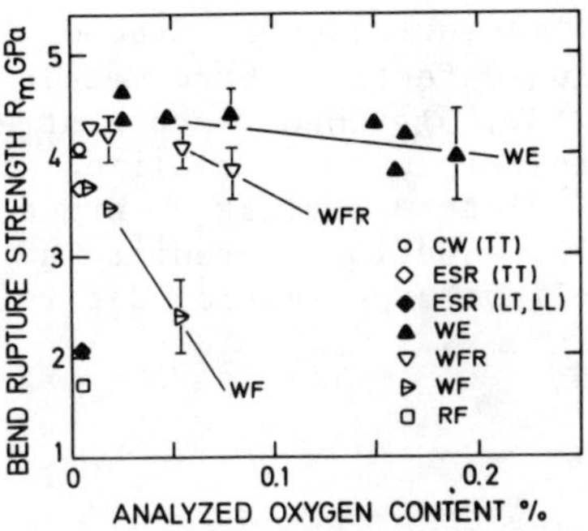

Fig. 12. Influence of oxygen content on rupture strength of PM, ESR, and CW M2 at hardness 65 R_c.

In a fracture mechanical view, bend rupture is determined by the influence of preexisting microcracks or flaws which can nucleate cracks when subjected to stress. Such flaws could be nonmetallic inclusions, large carbide particles or aggregates, pores or machining defects on the surface. They must be imagined to be distributed throughout the volume of the specimen, and to show a distribution of sizes and shapes as well. When the specimen is deformed, the stress acting on a given flaw will depend on its site with respect to the support and load points and depth below tensile surface. - Rupture will occur when the local stress at some flaw-associated crack has become strong enough to provoke unstable crack growth. The required stress is related to the crack size by

$$\sigma_c = \frac{K_{Ic}}{\sqrt{a/Q}} \qquad (1)$$

where a = crack size and Q = a factor depending on the flaw geometry and the ratio of nominal to yield stress. For an internal circular crack, Q = 0.70, and for a half-circular surface crack, Q = 0.58 (in both cases the crack plane is assumed perpendicular to stress, and a = crack radius).

In a population of locally and size- as well as shape distributed flaws, rupture will be caused by the one with the maximum product $\sigma \cdot \sqrt{a/Q}$. Thus *rupture stress is given a new meaning in that it reflects the distribution of flaws in the material.* With known fracture toughness, the flaw severity (a/Q) can be calculated from the effective stress at the site of fracture origin. If Q can be established, one can even derive the distribution of flaw sizes.

This fracture mechanical interpretation of bend rupture strength seems to have been applied first by Exner, Walter and Pabst (34) to sintered alumina. Its first application to high speed steels is due to Johnson (22).

Several authors have actually measured, by SEM fractography, the size of the fracture-initiating defects in hard metals (33-38) and in high speed steels (39,40). When rupture stress, corrected for the site of fracture origin and normalized with respect to fracture toughness, is plotted against observed flaw size in the manner of fig. 13, equation 1 predicts a linear plot with slope $\sqrt{Q}$. This allows experimental determination of the shape factor Q.

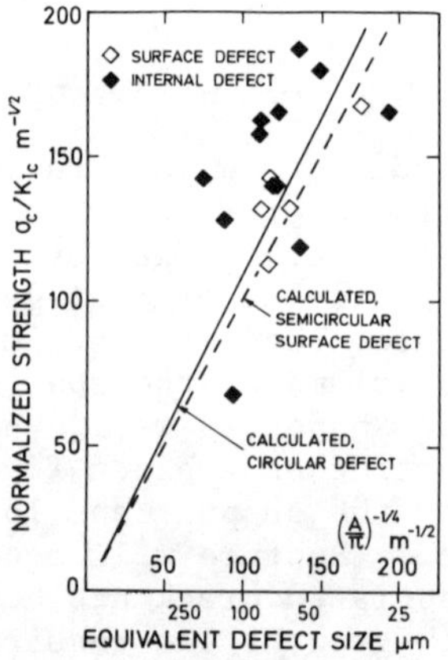

Fig. 13. Local flaw stress at rupture, normalized for fracture toughness, related to measured flaw size, plotted as (flaw area)$^{-1/4}$.

For hard metals, this procedure yielded two distinctly different Q-values, one of which was clearly related to sharp, crack-like flaws, while the other (lower) pertained to rounded pores and inclusions (38). For the present materials, the lines in fig. 13 were drawn with slopes corresponding to the theoretical shape factors for a sharp half-circular surface crack, and for an interior full-circular crack. Their fit with the data points indicates that rupture seems to start from sharp, crack-like flaws. Sharp crack models with theoretical shape factors have also been used successfully by Johansson and Sandström (28,33,39) in correlating HSS rupture strength and fracture toughness data. These observations encourage us to trust the 'sharp crack' model of rupture initiation in high speed steels. Extending the flaw size calculations with Q = 0.70 to all rupture stress values in our material - irrespective of whether the fatal flaw could be identified or not - yields a range from 9 to 160 μm for all grades and processing routes combined. This happens to be inclose agreement with the range 7 - 130 μm calculated (also with a 'sharp crack model') from observed rupture strengths in PM M3:2 by Johansson and Sandström (28).

Low rupture strengths (large crack nuclei) occur especially in PM materials made with insufficient hot working. With particle diameters of the order 200 - 500 μm, the calculated crack sizes may be associated with occasional patches of poor particle bonding.

Low rupture strengths were also observed in the ESR material for those fracture orientations which allow the crack plane to follow (LL) or include (LT) the stringer direction. The maximum directionality ratio in fig. 11 was 2:1. This would require crack nuclei in the ratio of 4:1. It is unlikely that the size of preexisting microcracks associated with stringers should so strongly depend on direction. - Nucleation from microcracks also appears unlikely in the light of the strong effect of matrix hardness on rupture strength (22) in a given steel: higher tempering can hardly create larger microcracks.

Thus one has to look for a likely process of *crack nucleation under stress*, which also must admit a strong influence of stringer direction. Matrix debonding from nonmetallic inclusions would account for the observed directionality, and when inclusions are numerous, this may be the true reason: Johansson and Sandström (28) found inclusions of fitting size at the fracture origin of their rupture specimens of a PM M2:3 steel. However, in our own experiments with an ESR grade of very low inclusion content, directionality was strong and yet, SEM-fractography ruled out inclusion-nucleated fracture in most rupture specimens.

Our hypothesis (fig. 14) is that most rupture-initiating cracks in clean materials (or in materials whose nonmetallic inclusions are finely dispersed, as in the grades made from water atomized powders) *nucleate* by debonding (or cracking) of large carbides, and that they have a *first stage of easy (subcritical) growth* following carbide/matrix boundaries along the stringers, and cracking occasional small bridges of matrix phase between the carbides in a stringer.

This stage of easy growth will stop at substantial interruptions of the stringer, at large carbides, or when the crack runs entirely out of the stringer. Since the stringers form a

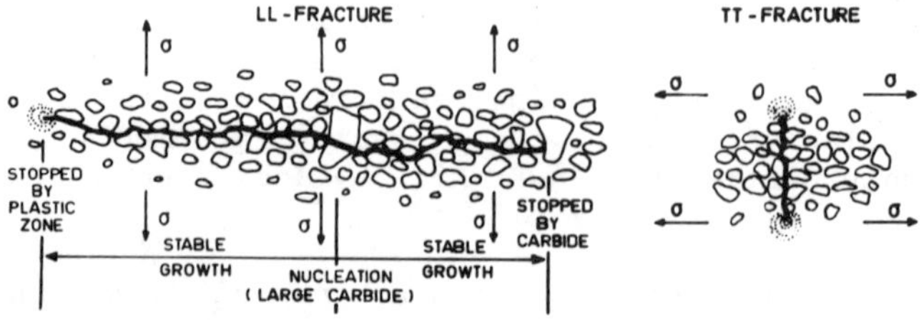

Fig. 14. Hypothetical model of sub-critical crack growth in stringers.

continuous network in space, easy growth could extend to areas belonging to more than one stringer, but in any case the easy growth stage would be much more limited in the TT orientation than in LT or LL.

Rupture strength is known to depend strongly on the size, shape, and distribution of the carbides. It is strongly reduced by coarse stringers in materials with little hot reduction (31,41). Cast materials normally have only one-fourth the rupture strength of wrought materials, but by a suitable carbide spheroidization anneal, this can be partly repaired (42). It seems to be especially important that the carbides should be well isolated from each other (42); sintered steels seem to be less sensitive to the presence of large carbides than cast-and-wrought material (14).

The dependence of rupture strength on chemical composition has been explained in terms of carbide volume fraction and (for Co grades) matrix solution strengthening (43).

Overheating produces angular carbides at grain corners and carbide plates in the grain boundaries, which have long been known to reduce rupture strength (44). Hellman (23,41) has demonstrated the special sensitivity of PM grades to overheating. The faster dissolution of their fine carbides during austenitization, and their generally somewhat increased carbon level, shift the regime for grain boundary carbide formation to lower temperatures. This is the reason why PM materials must be hardened at lower temperatures than CW.

DISCUSSION:
FRACTURE TOUGHNESS, RUPTURE STRENGTH AND TOOL PERFORMANCE

First, rupture strength is obviously related to the risk of breakage of tool *shafts* (drills, taps, long unsupported cutters) - but this is seldom critical.

More to the point is the widely held opinion, based perhaps as much on intuition as on empirical fact, that rupture strength is important for the resistance of cutting edges to *chipping or crumbling* - but coupled with the uneasy feeling that the correlation is fuzzy and seems to be disturbed by hitherto unidentified effects.

One such effect may be that tool wear is seldom due to one isolated mechanism. Almost 20 years ago, Wilmes (44) emphasized that chipping, through its effects on edge geometry and friction, might be the real cause of many failures ascribed to hot blunting - and thus, might be more important than generally realized. Wright and Trent (45) have documented a mechanism called "attrition wear" in which the tool materials is broken away in larger or smaller fragments adhering to the work or to the swarf. This occurs predominantly when built-up edges are formed. More recently, Berry and Al-Tornachi (26) have identified tool hardness regimes in a single-point milling test in which wear was partly (700 - 820 HV $\approx$ 59 - 64 R_c) or wholly (> 820 HV) due to chipping. In the latter range (which is that of normal tool hardness), chipping involved the breakaway of large fragments.

The fracture-mechanical interpretation of rupture suggests a second factor which might mask the relation between rupture strength as determined in a testing machine and tool performance even in cases of dominant chipping: in the rupture test, the stressed volume is about 400 times greater than in a tool tip. Thus rupture testing brings out the distribution of the really large flaws - which are far between - while chipping must be caused by much more closely-spaced crack nucleation sites. Whether these are of the same kind as the ones which cause rupture in large specimens remains to be established, but the mechanism of fig. 14 seems plausible for them. Further study of the distribution of flaw sizes along the lines of Exner (34) and Johansson (28) should prove rewarding if attention was focussed on the small tail of the flaw distribution. Evaluation by Weibull statistics would enable the volume distribution of flaws to be characterized, and the metallographic study of fracture-initiating flaws could be improved by the precision-sectioning technique of Kerr, Eylon and Hall (46).

Laboratory tests with small stressed volume are available in the form of notch bend or impact tests, but they have to be conducted in large enough numbers to provide reliable statistical information on the tail of the flaw distribution. Dlouhy (47) has devised an impact test and Almond (48) a static test in both of which a specimen with actual cutting edge geometry is subjected to a point load close to its edge until it crumbles. This seems a good way of testing realistic small volumes, and the necessary large-number-testing would be facilitated by readily available, cheap specimens.

In addition, it will be necessary to get a precise grip on the influence of temperature and loading rate. For cemented carbides, Suzuki (49) has recently shown that above 700°C fracture is due to a different kind of flaws than at lower temperatures. Irrespective of flaw type, the effect of temperature and loading rate on fracture toughness would have to be known for all fracture-mechanical evaluation of rupture or crumbling tests. To our knowledge, there is only one set of elevated fracture toughness measurements (made, however, at normal, low loading rates): Ingelström's (25) results on the PM material ASP 23 (M3:2), which extend up to 700°C, indicate a maximum of fracture toughness at about 350°C of ca. 1.5 times the room temperature value and a return to that level at around 600°C. A similar trend is suggested by Johansson and Sandström's analysis of hot rupture strength in terms of flaw size for the same steel (28). Fracture toughness being a matrix property, one could hope for similarity of temperature and load rate dependence in all high speed steels if the comparison was based on states of equal hot hardness, but further experiements are definitely needed.

The few measurements of hot rupture strength available (50-52) also indicate a maximum around 300°C, with an abrupt fall at 600°C. - Almond's test could easily be adapted to elevated temperature work.

A further point of difference between rupture and chipping could be the enlargement of sub-critical cracks by fatigue in the case of intermittent cutting. Data on fatigue crack propagation in HSS have been published by Berry and Al-Tornachi (26) for CW M2 and for PM M3:2, and by Rescalvo and Averbach (21) for ESR T1. Crack growth followed the Paris-Erdogan relation, $da/dN = const.(\Delta K)^m$, with $3 < m < 6$ for M2 and M3:2, and $m = 4$ for T1. Rescalvo and Averbach point out that the threshold for crack propagation is particularly low ($\Delta K_o < 5$ MPam$^{1/2}$), and that the stage of strongly accelerating growth, with rapidly increasing m, begins already at low stress intensity peaks, in consequence of the low fracture toughness. - Like fracture toughness, fatigue crack propagation seems to be governed by the state of the matrix, with little influence of carbide distribution (at least at normal hardness levels). Again, the decisive measurement would be fatigue crack propagation at tool tip temperatures, and such experiments are urgently needed.

Summarizing then, the points discussed here - stressed volume, distribution of the small and numerous as opposed to the largest flaws, fatigue crack growth, together with the temperature and loading rate dependence of fracture toughness and fatigue - would suffice to explain why the empirical correlation between tool performance and rupture strength as tested in the laboratory is fuzzy. They also indicate the remaining gaps in our knowledge, all of which could be filled by experimentation with established techniques. All in all, the situation seems to hold promise of a satisfactory understanding of the toughness of cutting tools in the foreseeable future, based on fracture-mechanical concepts which exist today.

ACKNOWLEDGEMENTS

Our own work in the field reviewed here spans a period of seven years. Partial support by Uddeholms AB and the Swedish Board of Technical Development in the early phases, and by the Austrian Ministry of Science and the Austrian Science Research Council (Project S-14) in the later phases, is gratefully acknowledged.

REFERENCES

1. O.W. Reen, Mod. Dev. Powder Metall. (Proc. 1965 Int. Powder Metall. Conf.) Vol. 2, 1966, p. 182.
2. E.J. Dulis and J.A. Neumeyer, "Materials for Metal Cutting", Iron and Steel Institute (London) Publ. 126, 1970, p. 112.
3. P. Hellman, H. Larker, H. Pfeffer and J.B. Strömblad, Mod. Dev. Powder Metall. (Proc. 1970 Int. Powder Metall. Conf.) Vol. 4, 1971, p. 573.
4. H.F. Fischmeister, Annual Reviews Materials Sci., Vol. 5, 1975, p. 151.
5. T.A. Neumeyer and A. Kasak, Metall. Trans. Vol. 3, 1972, p. 2281.
6. A.N. Stroh, Advances in Physics, Vol. 6, 1957, p. 418.
7. K.H. Zum Gahr, Z. Metallk., Vol. 67, 1976, p. 678, Vol. 68, 1977, p. 381.
8. P.U. Gummeson, Powder Metall. Vol. 15, 1972, p. 67.
9. J.J. Dunkley and R.J. Causton, Powder Metall. Int., Vol. 8, 1978, p. 115.
10. L.R. Olsson, V. Lampe and H.F. Fischmeister, Powder Metall. Vol. 17, 1974, p. 347.
11. L.R. Olsson, V. Lampe and H.F. Fischmeister, PM 78-SEMP 5, Preprints of 5th Europ. Powder Metall. Symp., Stockholm 1978, part I, p. 237.

12. A. Okayama, H. Ando, K. Someno and M. Tamura, Tetsu to Hagane, Vol. 59, 1973, p. S253 (H. Brutcher Transl. 9146).
13. F.L. Jagger and W.J.C. Price, Powder Metall., 3rd Europ. Powd. Met. Symp. Conf. Suppl. Vol. 1, 1971, p. 407.
14. A.J. Wronski, L.B. Hussein-al-Yasiri and F.L. Jagger, Powder Metall., Vol. 22, 1979, p. 109.
15. R.W. Bratt, G. Steven and B.H. Seaton, Soc. Mfg. Eng. (Dearborn) Tech. Pap. TE-73-203 (1973).
16. D.O. Harris and H.L. Dunegan, J. Materials, IMLSA, Vol. 3, 1968, p. 59.
17. K. Eriksson, Scand. J. Metall., Vol. 4, 1973, p. 182.
18. L.R. Olsson and H.F. Fischmeister, Powder Metall., Vol. 21, 1978, p. 13.
19. "Metals Handbook", 8th ed., Vol. 9, "Fractography and Atlas of Fractographs", Amer. Soc. for Metals, Metals Park, Ohio, 1974.
20. B. Karlsson and K. Wright, 13th Ann. Techn. Meeting Int. Metall. Soc., Brighton, August 1980.
21. J.A. Rescalvo and B.L. Averbach, Metall. Trans., Vol. 10A, 1979, p. 1265.
22. A.R. Johnson, Metall. Trans., Vol. 8A, 1977, p. 891.
23. P. Hellman and H. Wisell, Colloque Int. sur les Aciers à Coupe Rapide, St. Étienne, Nov. 1975, Bulletin Cercle d'Études des Métaux, Spec. No., 1975.
24. K. Eriksson, Scand. J. Metall., Vol. 2, 1973, p. 197.
25. N. Ingelström, Swedish Institute for Metals Research, Report IM-1072, Stockholm, April 1975.
26. G. Berry and M.I. Kadhim Al-Tornachi, Metals Technol., Vol. 4, 1977, p. 289.
27. R.A. Queeney and R.A. Heffelfinger, Metall. Trans., Vol. 8A, 1977, p. 309.
28. H. Johansson and R. Sandström, Mat. Sci. Engg., Vol. 34, 1978, p. 165.
29. A. Mason and P. Kenny, Metallurgia, Vol. 82, 1970, p. 205.
30. J.L. Chermant and F. Osterstock, J. Mater. Sci., Vol. 11, 1976, p. 1939.
31. P. Hellman, Werkstatt und Betrieb, Vol. 108, 1975, p. 5.
32. G.A. Roberts, J.C. Hamaker and A.R. Johnson, "Tool Steels", ASM (Metals Park, Ohio) 1962.
33. H. Johansson and R. Sandström, Mat. Sci. Engg., Vol. 36, 1978, p. 175.
34. H.E. Exner, A. Walter and R. Pabst, Mat. Sci. Engg., Vol. 16, 1974, p. 231.
35. H. Suzuki and T. Takase, Planseeber. Pulvermet., Vol. 23, 1975, p. 271.
36. H. Suzuki and T. Takase, Planseeber. Pulvermet., Vol. 24, 1976, p. 271.
37. E.A. Almond and B. Roebuck, Metal Sci., Vol. 11, 1978, p. 458.

38. H. Fischmeister, J. Diblik and W. Kompek, to be published.
39. H. Johansson, Metall. Trans., Vol. 9A, 1978, p. 96.
40. L.R. Olsson, Dissertation, Chalmers Univ. of Technology, 1978.
41. P. Hellman, Cutting Tool Engg., Vol. 28, No. 5-6, 1976, p. 3.
42. S. Wilmes and E. Weber, Industrieblatt, Vol. 64, No. 1, 1964, p. 8.
43. S. Wilmes, Archiv Eisenhüttenw. Vol. 35, 1964, p. 649.
44. S. Wilmes, Stahl u. Eisen, Vol. 81, 1961, p. 676.
45. P.K. Wright and E.M. Trent, Metals Technol., Vol. 1, 1974, p. 13.
46. W.R. Kerr, D. Eylon and J.A. Hall, Metall. Trans., Vol. 7A, 1976, p. 1477.
47. M. Dlouhy, Pokroky Praškove Metalurgie, Vol. 4, 1971, p. 9.
48. E.A. Almond, Private communication, to be published.
49. H. Suzuki, Y. Taniguchi, Planseeber. Pulvermet., Vol. 25, 1977, p. 23.
50. Yu. Geller, A.G. Gordeziani and L.S. Kromnev, Steel in the USSR, July 1972, p. 571.
51. R. Siegel and G. Fritzsch, Neue Hütte, Feb. 1976, p. 110.
52. R. Sandström, Swedish Institute for Metal Research, Report IM 1224, Stockholm 1977.

[Discussion is presented on the next page.]

DISCUSSION

Dr. V.A. Tipnis (Tipnis Assoc.)

Have you considered creep as one of the contributing modes for tool material/edge failure?

Professor Fischmeister

Not for toughness-related failure modes (chipping). Judging from experience in a related field (thermal fatigue resistance of hot work die steels) cyclic creep can be very important or indeed decisive for crack nucleation in cases where thermal cycling (with high rates of heating or cooling produces steep temperature gradients with concomitant plastic flow) constitutes the major load on the material. When the main load is mechanical (eg. cyclic impact as in slot bar turning), fatigue and fracture toughness seem to me to be the decisive properties of the tool material. The answer to your question could also be couched in terms of high temperature fatigue theory: At low frequencies and high temperatures, there is enough time for creep to make a noticeable contribution to the accumulation of damage. At high frequencies and intermediate temperatures (which are the conditions prevailing in most interrupted cutting operations), the contribution of creep in limiting life becomes small.

A.J. Bahmiller (Cleveland Twist Drill Co.)

Have you measured any effect of sulphur added to high-speed steel on the fracture toughness of this material?

Professor Fischmeister

We have not studied resulphurized materials, nor was there enough variation of sulphur levels in the steels we used to give you an answer based on fact. However, I can extrapolate from what we know: Sulphur, when present in the form of elongated inclusions, must be expected to have a strong effect on the rupture strength, especially in fracture orientations containing the direction of elongation (LL, LT in the nomenclature of our paper). Its effect on fracture toughness should be very small at hardness levels of interest for cutting operations (i.e., above 63 R_c).

EFFECT OF MATERIAL CHARACTERISTICS ON PERFORMANCE OF HIGH SPEED STEEL CUTTING TOOLS

Richard W. Bratt

Crucible Specialty Metals Division

ABSTRACT

High speed tool steels get their name from their ability to cut other materials at high speeds during which considerable heat is generated at the cutting edge. In order to be successful, high speed steels require:

(1) Capability of high room temperature hardness - Rc 63 to 68 for good wear resistance.

(2) Capability to maintain high hardness at elevated temperatures - called red hardness.

(3) Adequate toughness for the application.

Material quality characteristics which affect these requirements, may affect tool performance. The important quality characteristics discussed are chemistry, macro and microstructure, heat treatment, surface condition, and general workmanship.

Further, the more uniform the high speed material is, the tool may perform at a higher level and with more uniform results. Particle metallurgy high speed steels provide a more uniform raw material from which to make high quality tools having more uniform response to heat treatment, more toughness, easier grinding, and improved tool performance.

Particle metallurgy also lends itself to the production of high alloyed grades which are not available through conventional steelmaking practices. These grades can provide higher red hardness or wear resistance for difficult cutting applications, or can save critical alloys such as cobalt, by appropriate alloy substitution and chemistry balance.

WHAT ARE HIGH SPEED STEELS?

High speed steels are high-alloy steels designed to cut other materials efficiently at high speeds, despite the extreme heat generated at the tool's cutting edge...heat involving temperatures up to and beyond 1000°F. Actual temperatures will vary according to cutting conditions, coolants used, and other factors of production.

THREE BASIC REQUIREMENTS

To give good cutting tool performance, a high speed steel must have three basic requirements:

1. Capability for heat treatment to high room temperature hardness...Rc63 to Rc68, or even higher. High hardness is necessary for good wear resistance.

2. Capability for maintaining high hardness at elevated temperatures. This is called red hardness. Because they are high speed steels, they are subject to considerable heat where the cutting edge meets the work piece chip interface, so red hardness is necessary.

3. Impact toughness enough for interrupted cut tooling applications. High speed steels are also notably tougher than carbide materials.

Naturally, in addition to these three basic requirements, high speed steel must have a capability for being fabricated, hot worked, machined, ground, and whatever else is necessary to make the cutting tool.

Material characteristics which affect these requirements, will affect tool performance. One of the primary characteristics of any material is its chemistry.

ALLOYING ELEMENTS...WHAT THEY DO

The major alloying elements used for high speed steel are carbon, chromium, tungsten, molybdenum, vanadium and cobalt - all added to the base material iron.

Carbon, of course, is common to all steels. High speed steel has a high carbon content, ranging from .70% in grades such as T-1 on up to about 1.60% in grades such as T-15. Carbon is required for heat treatment that will achieve the necessary high room temperature hardness. It forms wear resistant carbides and helps promote the "secondary hardening" which occurs when high speed steel is tempered.

Chromium is present in all high speed steels, practically always within the narrow range of 3.50% to 4.50%. It provides hardenability: the ability to harden through large sections. It also forms wear resistant carbides and promotes a certain amount of scaling resistance.

Tungsten is used in most high speed steels. In the T (tungsten) types, it is present in large amounts...from 12% to about 19%. Tungsten promotes high red hardness. It forms wear resistant carbides and helps promote secondary hardening through carbide precipitation, which occurs during tempering. In the M (molybdenum) types of high speed steel, tungsten may or may not be present in significant quantities, ranging from 0.5% to about 7%.

Molybdenum is found in many high speed steels. In the T types there is very little molybdenum...from zero up to about 1.25% maximum. The M types contain from 4% to about 10%. As an alloying element, molybdenum has the same effect that tungsten has. In fact, they are actually interchangeable as elements in high speed steel, except that only about one half as much molybdenum is required to do the job tungsten does.

Vanadium is in all high speed steels in amounts from 1% to 5%. Its major characteristic is the formation of a highly wear resistant carbide. Vanadium carbide is the hardest carbide found in high speed steel, which is why high vanadium grades are difficult to grind, except in P/M high speed steel. These same carbides also help maintain a very fine grain size at the high heat treating temperatures used for high speed steels.

Cobalt in high speed steels ranges from 0% on up to about 12%. Its principal characteristic is the promotion of high red hardness by retarding the breakdown of martensite at elevated temperatures. It does not form carbides.

HIGH SPEED STEEL FAMILIES

There are, of course, many grades of high speed steel. In some respects they may seem unrelated, but actually most grades can be fit into a "family" which starts with a basic general purpose steel. By taking this basic grade and adding various alloying elements to improve certain grade characteristics, we can form a high speed steel family.

To illustrate family formation, let's use the M1 type as our basic general purpose grade. This is a moly base high speed steel, with good toughness as a major characteristic.

M1

Best Toughness

Increased: V → M7 → Wear Resistance

Increased: Co → M33 → Red Hardness

Increased: V and Co → M46 → Wear Resistance and Red Hardness

Figure 1. Building the M1 family of High Speed Steels

As shown in Figure 1, by increasing the content of vanadium - which forms wear resistant carbides - in M1, we get M7, a grade that offers improved wear resistance. Or we can add cobalt - which promotes high red hardness - and get M33, a grade with outstanding red hardness. Or you can add both vanadium and cobalt to M1 and wind up with a combination of increased wear resistance and increased red hardness. This is M46, a super high speed steel.

Compare the chemical composition of the four grades in the M1 Family in Table 1. Note that whenever you add vanadium you increase carbon content along with it. This is because vanadium forms carbides and you must have extra carbon to go along with the extra vanadium. You may or may not increase carbon significantly along with increased cobalt.

Table 1. M1 High Speed Steel Family Compositions

Type	C	W	Mo	Cr	V	Co	Feature
M1	.84	1.75	8.50	3.75	1.20	-	Toughness
M7	1.02	1.75	8.50	3.75	1.90	-	Wear Resistance
M33	.88	1.75	9.50	3.75	1.20	8.25	Red Hardness
M46	1.25	2.00	8.25	4.00	3.00	8.25	Wear Resistance and Red Hardness

It must be remembered that when vanadium and/or cobalt is added, toughness will decrease.

In terms of user benefits, where you are using M1 but want more wear resistance, the first step would be to use M7. Or, if toughness is no problem at all, you could use M46 and get better wear resistance along with high red hardness. On the other hand, if toughness is a problem and you are using one of the grades with increased alloy content, drop back to M1 for improved toughness.

Of course, recommendations for changing grades of steel are usually made only after exploring heat treatment and hardness variations which can also affect wear resistance and toughness.

In Figure 2, we have the opportunity to examine the relative toughness, wear resistance and red hardness of high speed steels; and also to compare them with three common tool steels. Note that the general purpose grades have the highest toughness of the high speed steels. As you add vanadium or cobalt, or both, toughness decreases.

It's evident that the general purpose grades have the lowest wear resistance, and that as vanadium content increases, wear resistance goes up. Cobalt alone does not provide wear resistance, but notice how increasing vanadium and cobalt in the super high speed steels boosts wear resistance to its highest level because of the vanadium carbides and the high attainable hardness.

The red hardness curve shows that general purpose grades have the lowest red hardness. Some of the wear resistant grades do show a little improvement in red hardness, but decided improvement comes with added cobalt in the cobalt grades, and again in the super high speed steels with added cobalt and vanadium content.

Another point to be made regarding chemistry, is that balancing of the chemical elements is important to insure proper response to annealing for good machinability, and response to heat treatment for fine grain size and the correct hardness. Specialty steelmakers take this into account in producing high quality high speed steel. There is also a limit to how much alloy can be added to steels and still have them workable, machinable and grindable. We will talk more about that later.

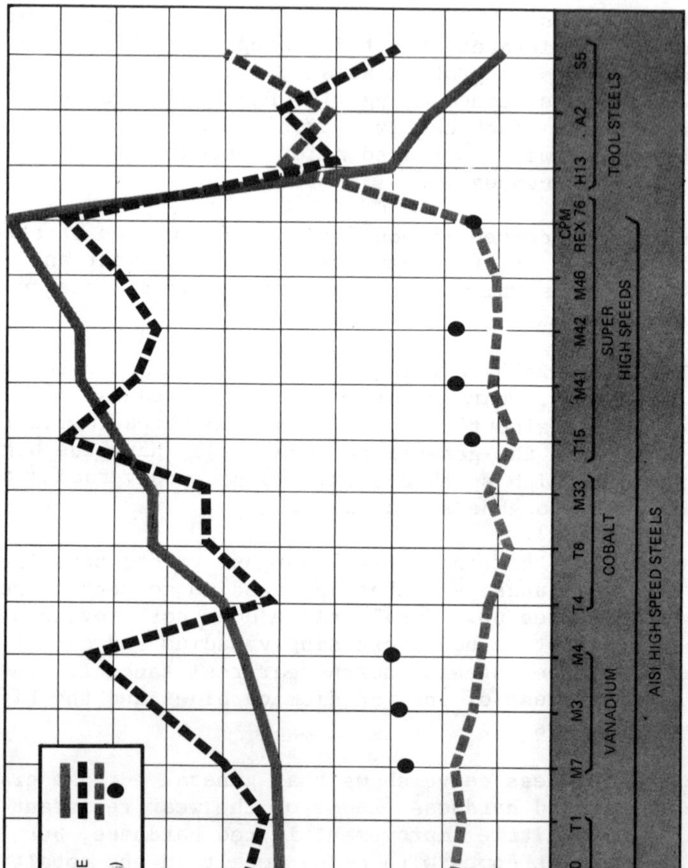

Figure 2. Relative Properties of High Speed Steels

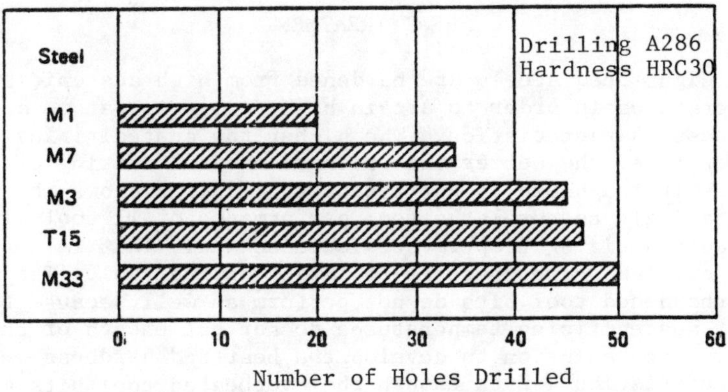

Figure 3. Relative Performance of High Speed Drill Steels

Figure 3 illustrates very nicely the effects of adding vanadium and cobalt to drill steels. The general purpose grade M1 (1.20 V) drilled 20 holes with increasing performance from M7 (2.0 V), M3 (3.0 V), T15 (5.0 V, 5.0 Co). The best performance was with M33 (1.20 V, 8.0 Co) which indicates that this application required good red hardness along with wear resistance which would be expected drilling A286 superalloy. Figure 4 similarly illustrates the better performance of T15 (5.0 V, 5.0 Co) premium high speed steel over M7 (1.20 V) end mills.

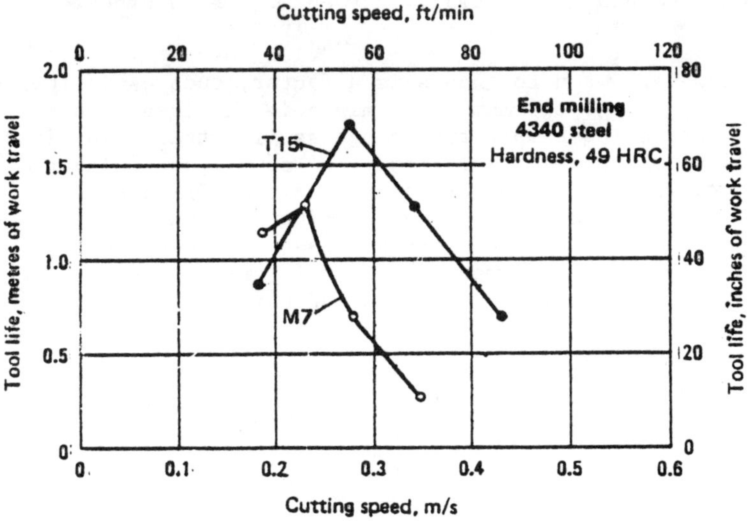

Figure 4. Relative Performance of Two High Speed End Mills

HEAT TREATMENT

High speed steels are hardened from high austenitizing temperatures in order to attain high room temperature and red hardness characteristics. The higher the austenitizing temperature, the better the red hardness and cutting properties - if toughness is not a factor. Table 2 shows the increasingly better lathe tool performance of T1 tool bits as the austenitizing temperature is increased. This is continuous cutting where there is no impact on the tool. The underhardened tool bits do not perform as well because the lower austenitizing temperatures do not put enough of the alloys into solution to develop the best red hardness characteristics. Even though the overheated tool bits are very brittle, they perform very well since toughness is not a factor in this test.

Table 2. Relative Performance of T1 Tool Bits With Increasing Austenitizing Temperature (Navy Lathe Test-Continuous Cutting)

Austenitizing Temperature (F) Gas Furnace								
2250	2275	2300	2325	2350	2375	2400	2425	2450
90%	62%	105%	109%	126%	146%	157%	190%	188%
Underhardened			Normally hardened			Overheated		

However, when toughness is a factor, such as in intermittent cutting, optimum performance is attained within a narrow hardening temperature range as illustrated in Figure 5 involving CPM Rex 76 tool bits. The normal hardening temperature range for Rex 76 cutting tools is 2150/2185F (salt bath). Underhardening reduces cutting performance since hardness and red hardness are decreased. Overhardening reduces cutting performance in this intermittent-cut test because the tools lack toughness and chipping occurs.

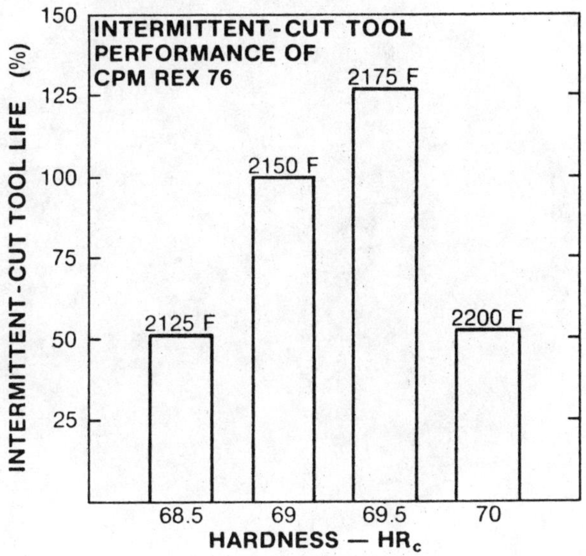

Figure 5. Relative Performance of CPM Rex 76 Tool Bits with Increasing Austenitizing Temperature in Intermittent Cutting

TOOL MATERIAL QUALITY CHARACTERISTICS

Both the high speed steelmakers and the tool manufacturers take special precautions to insure good quality material goes into cutting tools. However, even with these precautions, there are certain non-uniformities that are inherent in conventionally produced high speed high alloy steels to variable degrees. Two of the important quality characteristics are carbide segregation and pattern, and longitudinal carbide distribution.

Figure 6 shows some typical degrees of center segregation and ring pattern in high speed steel. These are taken from ASTM A561 - Recommended Practice for Macro-etch Testing of Tool Steel Bars which is part of ASTM A600 - Standard Specification for High Speed Tool Steel. The segregated

Figure 6. Typical Center Segregation and Ring Pattern in Conventional High Speed Steels

Rating

	Longitudinal Carbide Distribution, Mid-Radius		
Bar Size	Hook Conv.	Network Conv.	Streak Conv.
Up to 1/2"	2-1/2	2	2
1/2" to 1"	2-1/2	2	2
1" to 2"	3	2-1/2	3
2" to 3"	3-1/2	3	4
3" to 4"	4	3-1/2	4-1/2
4" to 6"	4-1/2	4	5
Over 6"	5	4-1/2	5-1/2

Magnification: X100

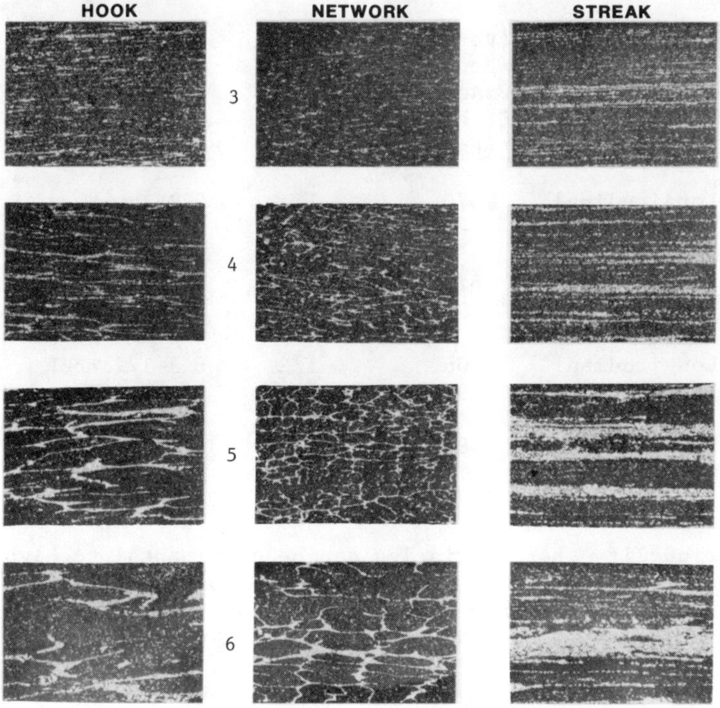

Figure 7. Longitudinal Carbide Distribution Ratings for Conventional High Speed Steels

areas are high carbon and alloy areas which are weaker and, when severe, may be prone to cracking in heat treating or tool service. Even with a good macroetch, the high speed microstructure may exhibit varying degrees of hook, network, or streak when longitudinal samples are examined at higher magnifications. A common rating system for these characteristics is shown in Figure 7. This is another form of carbide segregation and, when overly severe, can result in lower tool toughness. See Table 3. We will discuss more about this later.

Table 3. Effects of Carbide Distribution and Section Size on Mechanical Properties of Conventionally Produced M2S(HC) High Speed Steel Bars. Hardened and Tempered 2175F/4 Min., Oil Quench, 1025F 2 + 2 + 2 Hrs.

Bar Dia. (In.)	Specimen Orientation	Average Bend Fracture Strength (ksi)	Average Charpy C-Notch Value (ft-lb)	Longitudinal Carbide Distribution Rating (Fig. 7)
3/4	Longitudinal	670	11	2
	Transverse	-	-	
2-1/16	Longitudinal	571	11	3 streak
	Transverse	309	2	
4-9/16	Longitudinal	506	6-1/2	3-1/2 hook
	Transverse	290	2	
7-1/8	Longitudinal	487	6	4-1/2 hook
	Transverse	272	1-1/2	

Other quality characteristics can be detrimental to high speed steel tool performance as in any steel parts. The product should be free from injurious internal and surface defects such as pipe, bursts, excessive non-metallic inclusions, cracks, seams, laps, etc. Surface decarburization, if not removed from the bar stock, or if put on during heat treatment and not removed during finish grinding, can cause cracking during heat treatment or soft tool cutting edges which will wear out quickly. Likewise, carburization can also cause cracking during heat treatment or hard, brittle tool edges which may fail catastrophically. These do not need more discussion here, other than to say that good workmanship is required in the steel and tool making.

Conventionally produced high speed steels have been significantly reduced in section from the ingot to finished bar by hot working. This refines the ingot structure and improves the toughness of the bar stock. However, the toughness is not the same in all directions. High speed steel bar stock has directional characteristics or fiber similar to a log of wood. It is tougher in one direction than the other. In fact there are three directional characteristics that can have an effect on tool life. It may not always be possible to take full advantage of the optimum directional characteristics in tools to be manufactured, but there are definite advantages in trying to do so. This directionality effect has been generally recognized, but quantitative data on the subject is virtually non-existent. A few years ago, we did some interesting work to obtain some quantitative data.

Figure 8 illustrates schematically the three classes of relationships between the fiber axis of the material, the cutting edge, and the cutting force. Intermittent cut lathe tool tests were conducted with tools representing the three orientations or classes shown in Figure 8. Tools for this purpose were prepared from a 2" rd. bar of M2S(HC) from commercial stock. The results in Figure 9 show a clear distinction of performance among the orientation classes. The longitudinal Class 1 tools exhibited a relatively low wear rate and the longest tool life (.010" flank wear). The Class 2 and 3 tools wore at appreciably faster rates. The Class 3 tools not only wore rapidly, but also failed prematurely by nose breakdown rather than gradual wear.

We have also related these orientation classes and tool performance with bend fracture testing on samples taken from the same 2" rd. bar of M2S(HC). Table 4 shows the results of the bend fracture tests along with the intermittent tool performance data. The correlation of the data is very close.

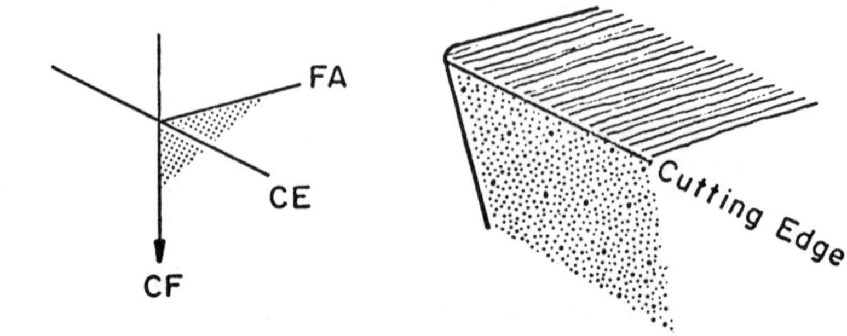

Class I. Longitudinal (CF, CE, FA mutually perpendicular)

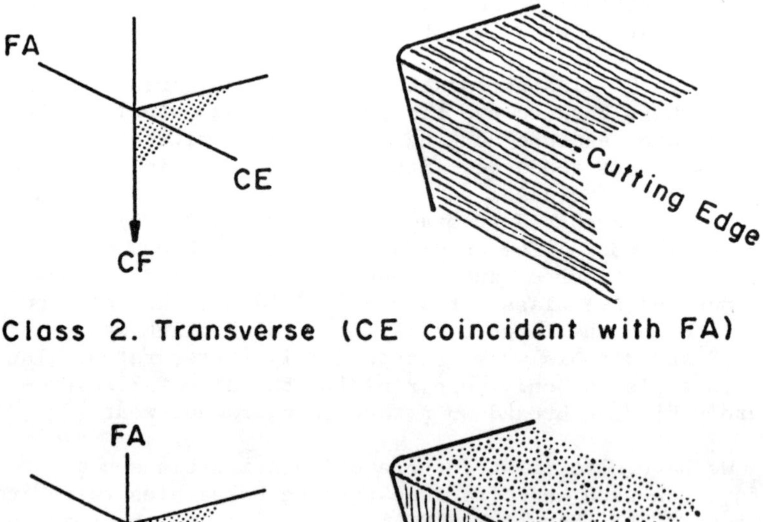

Class 2. Transverse (CE coincident with FA)

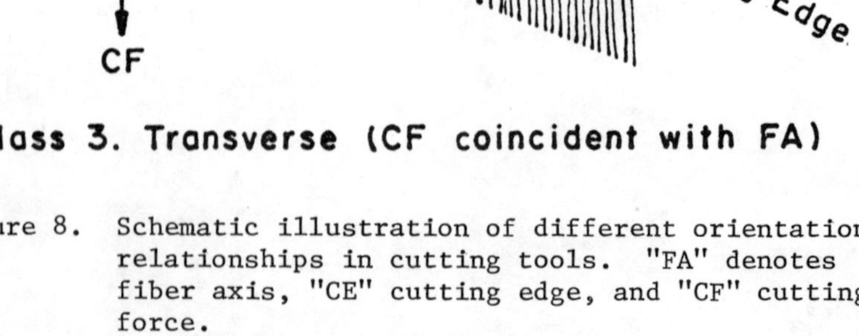

Class 3. Transverse (CF coincident with FA)

Figure 8. Schematic illustration of different orientation relationships in cutting tools. "FA" denotes fiber axis, "CE" cutting edge, and "CF" cutting force.

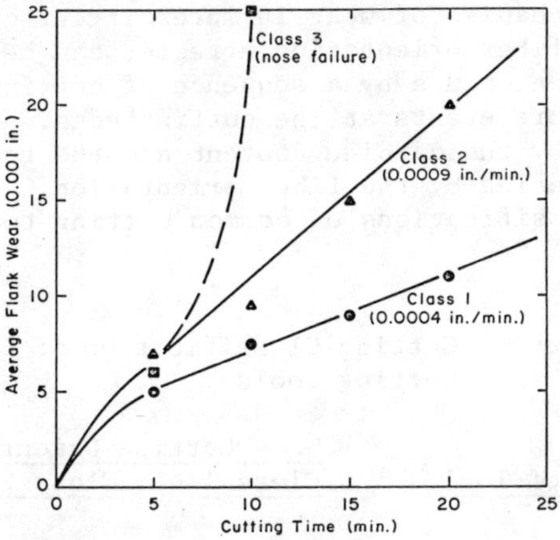

Figure 9. Effect of fiber orientation on the intermittent-cut tool performance of M2S (HC). The cutting conditions are given in Table 4.

Table 4. Effect of Fiber Axis Orientation on Bend Fracture Strength and Cutting Performance of a 2-In.-Dia. Bar of M2S (HC)

Specimen Fiber Axis Orientation	Bend Fracture Strength (ksi)	Intermittent-Cut Tool Performance* (Impacts to End Point)
Longitudinal (Class 1)	571	2610
Transverse (Class 2)	294	1380
Transverse (Class 3)	324	1150

*Cutting conditions -

Workpiece: 4-slotted bar of H13 @ Rc 33
Tools: 1/2-in. inserts, with geometry 3°, 6°, 10°, 10°, 10°, 10°, 0.030 in.
Feed: 0.010 in./rev.
Speed: 50 surface ft/min.
Depth-of- cut: 0.060 in.
Coolant: Sulfo-chlorinated mineral oil
End point: 0.010 in. average flank wear

The mechanism of wear in intermittent cutting, as related to fiber orientation effects, can be explained. Tool deterioration occurs by a sequence of continuous wear and micro-fracture events at the cutting edge. The wear rate is controlled by the carbide content and the tool hardness; the micro-fracturing by the fiber orientation. Table 5 shows the cutting classifications of common cutting tools.

Table 5. Cutting Classification of Common Cutting Tools

Type of Tool	Cutting Orientation		
	Class 1	Class 2	Class 3
Broach			x
Cut-off tool bit	x		
Drill	x		
End mill	x	x	
Form tool		x	
Gear shaper*			x
Hob	x	x	
Milling cutter	x	x	
Reamer		x	
Tap	x	x	
Tool bit		x	
Tool insert			x

*Often forged to minimize Class 3 cutting orientation

Grindability, or the ease of grinding hardened high speed tools is directly related to the material characteristics, and is worth a brief mention here. Since all high speed cutting tools are hard, high alloy materials, they require care in grinding to prevent damage to the tools. The primary factors affecting grindability are the vanadium content and the carbide particle size. Higher vanadium contents and large carbide particle sizes reduce grindability dramatically. Where conventional M2 will have a grindability ratio of 3.9 (ratio of metal removed to the aluminum oxide grinding wheel worn away), T15 with 5% vanadium has a ratio of 0.6, making it extremely difficult to grind. Poor grinding practices can result in actual cracking of tools, or softening the tool surface so that tool performance is poor. Figure 10 shows a case where abusive grinding practices cracked the T4 type cut-off tools.

PARTICLE METALLURGY PRODUCES IMPROVED MATERIAL CHARACTERISTICS

Particle Metallurgy has been commercially applied to high speed steels since 1970 starting with Crucible's Particle Metallurgy process (CPM). Other processes have also evolved and all are covered in the literature. CPM high speed steels are characterized by complete homogeneity and a very fine carbide particle size. Figures 11 and 12 illustrate these characteristics. From the previous discussions on material characteristics, it follows that this more uniform raw material will make better cutting tools - and that is just the case. Cutting tools made from CPM high speed steels have more uniform response to heat treat, are tougher, are easier to grind with no sacrifice in wear resistance, and give better tool performance. Four examples of these benefits are shown in Figure 13 and Tables 6-8. Field experience has shown that the CPM approach has been particularly beneficial when using, or upgrading to, the more highly alloyed high speed steels, because the improved tool performance can be obtained without the grinding or brittleness problems which usually are associated with these alloys conventionally produced.

The CPM process not only improves the quality characteristics of the conventional high speed steels, but also enables the producer to make more highly alloyed materials which would not be commercially feasible by conventional steelmaking practices because they would not be fabricable. Two such alloys are CPM Rex 76 and CPM Rex 25 shown in Table 9.

Magnification: X1 Unetched

Photograph of Rex AAA cut-off tools showing grinding burn and resultant failure in service by cracking and chipping.

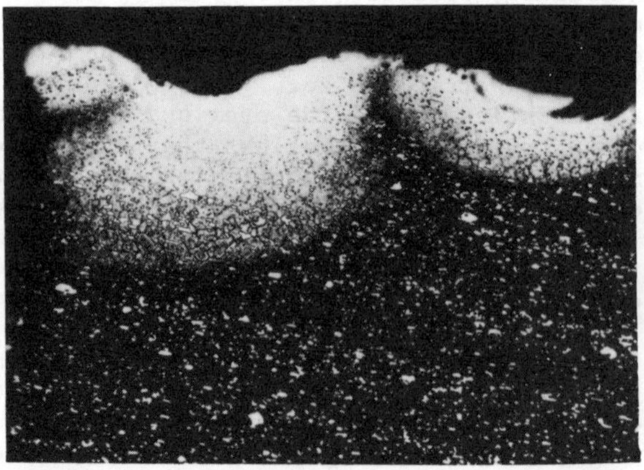

Magnification: X250 Etchant: 2% Nital

Photomicrograph of above burned cutting edge showing severe rehardening of surface layers. Hardness at cutting edge dropped to Rockwell C 59 compared to a normal hardness of Rockwell C65.

Figure 10. Effects of Abusive Grinding on High Speed Tools

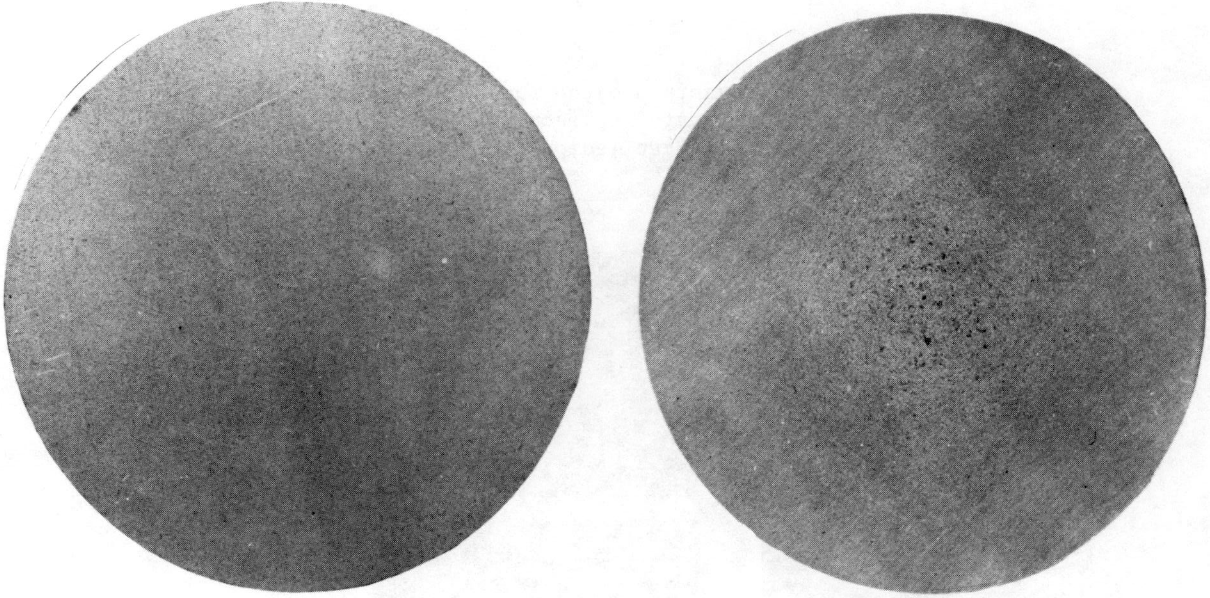

CPM Conventional

Figure 11. Comparison of the macroetched sections of M2S (1.0% C), both CPM and conventional bars, 5" diameter. The center segregation in the conventionally processed steel is not found in the homogeneously-etching CPM product.

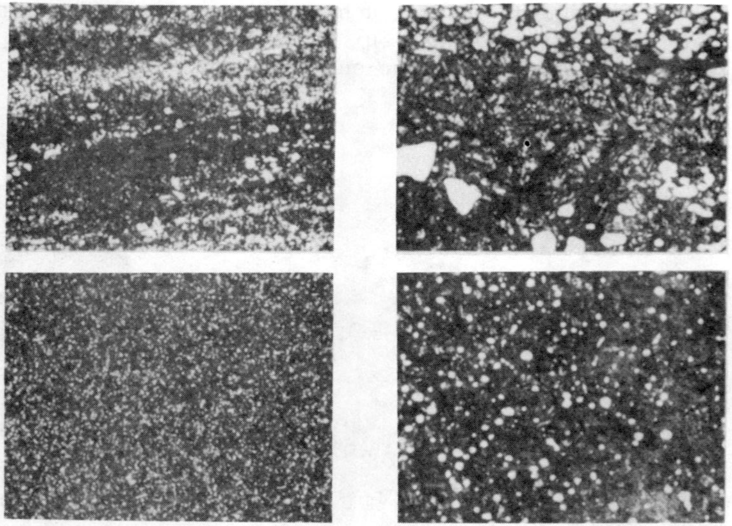

Figure 12. Comparison of the longitudinal carbide distribution and particle size in CPM and conventional 5" diameter M2S (1.0% C) bar. CPM is characterized by very uniform distribution of small carbides.

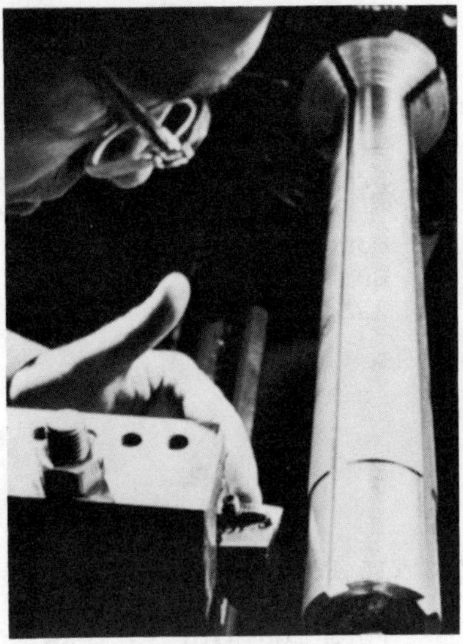

Rex M2S HC	Avg. No. of Impacts to Failure	Rating (% of Standard)
CPM	5700	260
Conventional	2200	100

Rex M2S HC	Spread of Tool Life Values (% of Average Tool Life)
CPM	± 16%
Conventional	± 36%

Figure 13. Slotted log test shows improved CPM M2 Tool Bit Performance

Table 6. CPM T15 Form Tools Perform Better and Grind Easier

Tool:	Form Tool
Material Machined:	T-347 Stainless Steel
Test:	Life and grindability vs. T15
Results:	Tool Life: 25% Improvement Grindability: "Twice as easy" Also a 10% improvement in surface finish

Table 7. CPM M4 Drills 25% More Holes

Tool:	Spade Drills
Material Machined:	4140, HT 28/32 Rc
Test:	Drilling: 4-1/8" dia. holes x 18" deep
Results:	Conv. M4: 4 holes CPM Rex M4: 5 holes 25% Improvement.

Table 8. CPM T15 Form Tools Give Best Performance

Tools:	Form tools of various sizes
Material Machined:	52100 bearing steel
Test:	Comparing tool life with previous M3S-2, M47 and CPM Rex M42
Results:	

Grade	Tool Life Between Grinds
M3S-2 & M47	4 hrs (100%)
CPM Rex M42	8 hrs (200%)
CPM Rex T15	11 hrs (275%)

Customer also reports better surface finish on parts machined with CPM Rex T15. Tool prints have been changed to specify CPM Rex T15.

Table 9. New Super High Speed Steels Can Be Produced Commercially

Standard AISI Grades	Nominal Composition, Percent						Total Alloy Content	Rc Hardness	Grind-ability	Toughness C Notch Ft.Lbs.
	C	Cr	V	W	Mo	Co				
M2	.85	4.2	2.0	6.4	5.0	-	16.40	65	3.9	17.5
M-41	1.10	4.3	2.0	6.8	3.8	5.0	22.85	67.5	1.3	8.5
CPM T-15	1.50	4.0	5.0	12.3	-	5.0	27.75	68	2.2	12
New Grades										
CPM Rex 76	1.50	3.8	3.2	10.0	5.3	9.0	32.80	69	3.8	10
CPM Rex 25	1.80	4.0	5.0	12.5	6.5	-	29.80	68	2.4	9.5

CPM Rex 76 was introduced commercially in 1976. With its high cobalt content and high heat treated hardness capability, this grade has better red hardness than any other commercial high speed steel. The 3% vanadium and high hardness give it excellent wear resistance. Because it is a particle metallurgy product, it has good grindability and adequate toughness. Rex 76 is designed for difficult machining applications such as heat treated materials, titanium, high temperature alloys, etc. One example of its tool performance is shown in Table 10.

Table 10. CPM Rex 76 Production Tool Bit Performance

Tool:	1/2" sq. Tool Bit	
Material Machined:	Turning 17-4 PH at Rc 40	
Test:	Compared to CPM Rex T15 and M46	
Results:	Tool Material	Tool Life
	M46 (Standard)	100%
	CPM Rex T15	156%
	CPM Rex 76	167%

High speed tool steels for severe duty cutting applications have traditionally contained cobalt to maintain cutting efficiency at elevated tool temperatures. The matrix strengthening effect of cobalt, together with its significant role in promoting elevated temperature hardness retention are so well documented in the literature that cobalt is considered an inherent and indispensable element in any super high speed tool steel composition. Incentives to abandon this traditional approach and pursue the development of cobalt-free super high speed steels were provided by major uncertainties in cobalt supply plus significant escalations in price. By utilizing the unique characteristics of the Particle Metallurgy process in combination with a modified approach to alloy design, a successful development of cobalt-free super high speed steels has been achieved. CPM Rex 25 has the metallurgical properties and performance capabilities of the cobalt-containing grade T15. It was introduced commercially in December 1979 and has proven its performance in both production and laboratory tests as shown in Tables 11 and 12.

Table 11. CPM Rex 25 Production Test

Tool: 3/4" square tool bit

Operation: Machining 1-3/8" hex. T-316 stainless

Test: Tool life versus CPM Rex T-15 and CPM Rex 76 using the following machining conditions:

Speed	228 RPM
Feed	.003" IPR
Depth of Cut	.274"
Coolant	Sulfur Chlorinated Oil

Results:

Grade	Avg. Tool Life	Avg. Pieces/Hr.
CPM Rex 25	7.83 hrs	19.0
CPM Rex T-15	8.19 hrs	20.3
CPM Rex 76	7.67 hrs	20.2

3 to 4 tests per tool

Grindability of all tools was reported as good.

Table 12. CPM Rex 25 Tool Manufacturers Lab Test

Tool: 1/2" dia. End Mills

Operation: Profile milling 4340 at 285 BHN

Test: Tool life compared to conventional T-15

Results: Inches drilled: 130"

Wear Land:	
CPM Rex 25	.008"
Conv. T-15	.011"

CPM Rex 25 exhibited 27% less wear

Additional work is about completed on an equivalent performance substitute for the 8% cobalt grades such as M42.

There will certainly be much more in the future from the particle metallurgy process toward improving the characteristics of high speed steels for cutting tools.

DISCUSSION

Dr. Marc Borom (General Electric Co. CRD)

The homogeneity which can be achieved with powder metallurgy is commendable. Since tests of conventionally cast high-speed steel tools show such an anisotropy in mechanical properties, is there any advantage to be gained by enhancing directional properties through either directional solidification or directional recrystallization, as employed in advanced turbine blade castings, and is anyone working in this area?

R.W. Bratt

We are not doing anything along that line and I don't know of anyone who is - at least in sizes that would make cutting tools.

INFLUENCE OF ALLOY CONTENT ON THE PERFORMANCE OF HIGH-SPEED STEEL TOOLS

W. E. Henderer

Vermont Tap & Die Co.
Vermont American Corp.
Lyndonville, Vermont 05851

ABSTRACT

A factor which strongly affects tool-life is the yield strength of the tool material. This finding is reached under cutting conditions where the tool-life of M1 HSS taps is determined by the cutting edge fracture. Theoretical estimates of the influence of the various microstructural features of heat treated HSS indicates that the precipitation of fine scale carbides is largely responsible for the yield strength and to a minimal extent on the primary carbide volume fraction.

One current approach to increasing tool-life is to increase the alloy content of HSS by utilization of the particle metallurgy process. These tool materials are characterized by a large volume fraction of well dispersed primary carbides. Tool-life data is presented for drills manufactured from conventional M7 and commerically available highly alloyed grades of P/M HSS. With some exceptions, all grades of HSS exhibited approximately equivalent tool-life in drilling both AISI 52100 alloy and 304 stainless steels.

In addition tool-life data of drills manufactured from low alloy grades of HSS is given which lends support to the hypothesis that residual carbide content has a less significant contribution to wear resistance than commonly believed.

INTRODUCTION

Alloy availability and cost is becoming increasinly recognized as a critical current and future problem. In an era where trade sanctions are increasingly used, national economies are becoming more vulnerable to the heavy dependence on foreign sources of supply of alloys essential for cutting tool materials. The situation which has arisen from the U.S. importation of cobalt is a well known example. In addition to cobalt, the United States currently has a net importation reliance of 50 percent or greater for three and the five alloying elements used in the manufacture of high speed steel (manganese 97 pct., chromium 91 pct., tungsten 56 pct.) (1). In addition to problems associated with the reliance on imports, alloys mined domestically are in short supply as well. In particular, associated with the current shortage of molybdenum has been a very substantial escalation in the cost of this alloy.

A reassessment of the role of alloy content as well as the potential for alloy substitution in high speed steel is demanded in light of uncertainties of alloy supply. As a prerequisite a better understanding must be developed of the wear mechanisms which determine the useful life of HSS cutting tools and subsequently an understanding of the effect of alloy addition on the microstructural characteristics which determine tool failure.

Wear of high speed steel tool can differentiated into two broad categories: continuous wear and wear by fracture or chipping of the cutting edge. The mechanism(s) responsible for continuous wear has been extensively debated but it is generally recognized to depend on the specific tool/work material pair as well as the general machining conditions (speed, feed, coolant, etc.) Mechanisms predominantly postulated to account for continuous wear include: diffusion, adhesion, and abrasion. Among these, the abrasive wear model is most consistent with experimental observation and plausible physical reasoning (2). Abrasive wear is in essence reverse cutting caused by the encounter of hard particles contained within the work material (usually oxide, carbides or nitride inclusions) with tool surface. On the other hand wear by fracture of the cutting edge generally occurs under conditions when the force exerted on the cutting edge is large (high work material hardness, heavy feed). Although high speed steel is commonly assumed to be entirely brittle, careful examination of the fracture surfaces of chipped cutting edges indicates that plastic deformation is associated with fracture. Therefore plastic deformation of the tool material is common to both continuous wear and wear by fracture although the deformation associated with continuous wear is highly localized.

HEAT TREATMENT OF HIGH SPEED STEEL

In the annealed condition, high speed steel is composed of a dispersion of large primary carbides in a ferrite matrix. HSS is subjected to two basic heat treatments: (1) an austenitizing-quenching treatment which provides a hardened martensitic matrix and (2) tempering for secondary hardness. During the austenitizing treatment, the primary carbides are dissolved either partially or totally and upon quenching a supersaturated solid solution results. During tempering at an intermediate temperature, the carbide forming elements precipitate on a very fine scale and this causes the strong secondary hardening reaction which makes HSS unique.

The influence of tempering temperature on the machanical properties of M1 HSS has been previously studied(3). At the same time parallel tests with M1 HSS taps has been conducted to determine not only the influence of tempering temperature on tool-life but in addition the pertinent correlation between tool performance and the various mechanical properties measured. The results of this investigation is shown in Fig. 1.

The response of hardness HV (1000 gram load) is typical of that normally obtained with M class HSS. Peak secondary hardness was found to occur when tempered at 510C.

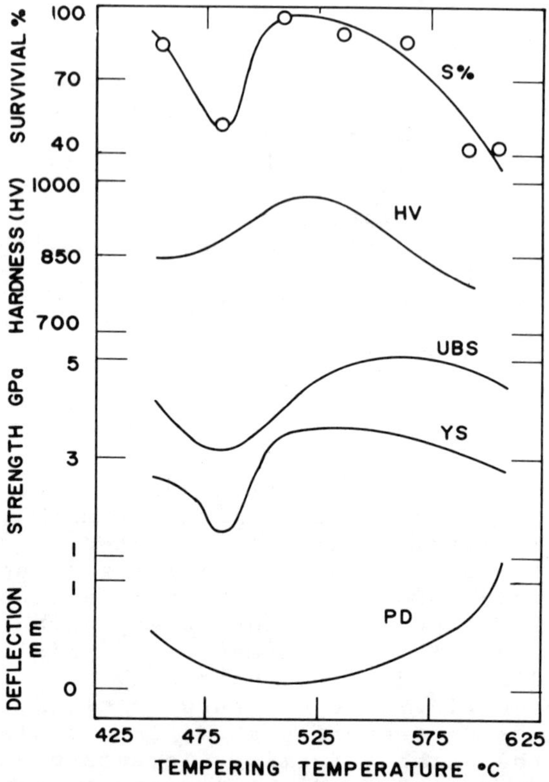

Fig. 1. Property and Tool-life Response to Tempering Temperature, M1 HSS.

The double point bend test as developed by Grobe and Roberts (4) was employed to determine the following mechanical properties: yield strength (YS), ultimate bend strength (UBS), and plastic deflection (PD). As evident in Fig. 1, the yield and ultimate bend strength respond to tempering temperature in a manner not normally apparent in the secondary hardness response. That is for temperatures less than peak secondary hardness there is a distinct minimum strength (475C). On the other hand, maximum yield strength corresponds to the temperature for peak hardness and if tempering is conducted at higher temperatures both decrease. Ultimate bend strength is the greatest at a temperature approximately 60 C greater than that where peak yield strength and secondary hardness occurs. Ductility measured here as plastic deflection (PD) responds as expected in relationship to hardness, that is minimum ductility occurs at maximum hardness and increases as hardness decreases.

Tool-life tests were conducted with taps manufactured from the same material and heat treated at the same temperatures and time as the bend test specimens. An alloy steel E52100 was used as the work material, hardened and tempered to 32 to 34 HR_C such that the mode of tool failure was due to the fracture or chipping of the cutting edges. The results of this testing is shown in Fig. 1 along with the mechanical properties. Since the machining conditions were severe, tool-life was extremely short. Rather than using the median life, the percentage of taps which survived three holes (S%) of operation was used as the criteria for comparing performance.

It is apparent in Fig. 1 that tempering temperature strongly affects the tool-life of taps under these conditions. Among the various mechanical properties, the strongest correlation to tool-life is the yield strength. In spite of the fact that tool failure was due to edge fracture, it can be inferred from the response of plastic ductility to tempering temperature that fracture toughness has no apparent relationship to tool-life under these conditions. On the other hand, tool-life depends primarily on the yield strength. Note in particular in Fig. 1 that both minimum and maximum values of yield strength and tool survival occur at the same tempering temperature and that the response otherwise is identical. Apparently the ability of the cutting edges to resist permanent deformation (shape change) controls the probability of early fracture and median tool-life.

Finding the strong correlation between yield strength and tool-performance under conditions where failure results from cutting edge fracture, an effort was made to determine the microstructural mechanisms responsible for strength (5). The yield strength of HSS is considered to be the result of two basic transformations: (1) martensitic strengthening due to carbon dependent solid solution strengthening concomitant with extraordinary small grain size (lattice distortion), and (2) the precipitation of small sized, closely spaced carbides. Based on dislocation mechanics, values for each contribution can be theoretically calculated if a value of 0.25 wt. pct. is used for the carbon content for martensitic strengthening, and if the spacing between precipitated carbides is assumed to be 322 Å (6):

Martensitic Strengthening	1.04 GPa	(151 ksi)
Precipitation Strengthening	2.70 GPa	(392 ksi)
Total	3.74 GPa	(543 ksi)

Experimentally the maximum yield strength of M1 HSS has been found to be 3.61 GPa (524 ksi).

INFLUENCE OF ALLOY CONTENT

The exceptionally high yield strength of M1 HSS is shown to be due primarily to the precipitation of fine scale secondary carbides 2.70 GPa (72 percent) and to a lesser extent the strength of the martensitic martix 1.04 GPa (28 percent). After heat treatment, most of the HSS compositions in use today have a significant amount of undissolved, residual carbides (10-20 percent by volume). It is conventionally assumed that these residual carbides contribute significantly to the wear resistance of HSS. However in terms of the yield strength, it can be shown that since the distance between residual carbides is so large (20 microns) relative to the spacing of precipitated carbide (0.032 microns) the contribution to strength is of little consequence (0.1 percent). In support of this Rescalvo and Averbach (7) recently showed the yield strength and hardness of M50 and T1 HSS were remarkably similar in spite of the vast difference in undissolved carbide content (2.6 and 15.2 vol pct. respectively). In this case, the composition of the matrices were similar and therefore no difference in these properties could be expected.

Through use of particle metallurgy processing, a current approach to improving tool-life is to dramatically increase the alloy content of HSS over that which may be obtained with conventional cast and wrought processing. Characteristic of such materials is a large volume fraction of well distributed residual carbides. Tool-life tests of drills manufactured from highly alloyed P/M high speed steels have been made to compare their life with drills manufactured from the lower alloyed commonly used grades of HSS (8). The results were obtained with 1/4 in. diameter drills in 32-34 HR_c AISI E52100 and AISI 304 stainless steel (Table 1).

Table 1. Drill Tool-life of P/M and Conventional HSS.

Grade	Alloy Content*	Median Tool-Life E52100	304
Conv. M7	15.7 pct.	176 holes	96 holes
Conv. M42	23.9	150	42
P/M M3-2	18.7	260	58
P/M M3-2+8.5% Co	26.8	256	39
P/M Rex 76[1]	31.4	156	92
P/M M4	18.0	155	80

* wt. pct. carbide forming alloys (W, Mo. V, Cr, Co)
E52100 steel, 32-34 HR_c, 12.7 mm, 61 m/min, 0.11 mm/rev
304 stainless, 180 BHN, 12.7 mm, 79 m/mm, 0.13 mm/rev

Although there are some exceptions, there is no conclusive improvement in tool-life of drills manufactured from the various grades of P/M HSS. For example drills manufactured from P/M grade Rex 76[1] with a total alloy content of 31.4 pct. performed no better than drills manufactured from conventional M7 with an alloy content of 15.7 pct.

In addition to providing data on the effect of high alloy content, the influence of cobalt on tool-life is shown. Among the metallurgical benefits, it is claimed that cobalt increases peak secondary hardness and temper resistance by increasing

---

[1]No AISI designation, Crucible proprietary alloy

the nucleation rate of carbide precipitation while at the same time decreasing the rates of growth and agglomeration. It should be pointed out that the most widely used cobalt bearing grade of HSS M42 not only contains a nominal 8.0 pct. cobalt but in addition a higher than normal carbon content, 1.07 pct.

Although the conditions used for the drill tests in Table 1 are typical for which cobalt HSS tools are often specified, it is obvious that the cobalt bearing grades M42 (8.0 pct Co), P/M M3-2 + 8.5 pct. Co, and P/M Rex[1]76 (9.0 pct. Co) perform no better than the grades which do not contain cobalt. In particular P/M M3-2 with no Co out performs P/M M3-2 with 8.5 pct. Co. in 304 stainless steel. The current approach of one tool steel manufacturer to the cobalt shortage is to develope cobalt free P/M HS steels by dramatically increasing the alloy content with respect to the tungsten equivalency (9). In light of the above data, this approach is questionable.

To further resolve the contribution of residual carbide content on the wear resistance of HSS drills, low alloy high speed steels M52 and Fagersta D950[2] were tested against conventional, more heavily alloyed grades of HSS M1 and M7 (6). Tool-life tests were conducted at two cutting speeds in 32-34 HR_c AISI 4340 with 1/4 inch diameter drills. The results of these tests are shown in Table 2.

Table 2. Tool-Life of Drills Manufactured From Low and Conventional Alloy Contents

Grade	Alloy Content*	Vanadium wt. pct.	Median Tool-Life 17 m/min	28 m/min
M52	11.0 pct.	1.70 pct.	459 holes	30 holes
M7	15.3	1.82	376	30
D950[2]	11.2	1.17	270	27
M1	14.9	1.22	272	21

* wt. pct. carbide forming alloys (W, Mo, V, Cr)

[2] No AISI designation, Fagersta proprietary alloy

It is readily apparent from the results of these tests that tool-life does not correlate with the content of carbide forming alloy. In fact M52 outperforms M7 by 22 percent at the lowest cutting speed in spite of the fact that M7 has a substantially greater content of residual carbides. The only apparent correlation which exists between composition and drill performance is vanadium content. M7 and M52 HSS with 1.70 and 1.82 pct. V respectively distinctly outperforms both M1 and D950[2] with 1.17 and 1.20 pct. V respectively. This may be related to the finer dispersion of precipitated VC carbides in contrast to the coarser size and spacing of precipitated MoC and/or WC carbides.

Although there is little published information which has systematically shown the influence of residual carbides on tool-life, there is some information which shows that primary carbides may be in fact detrimental to tool performance. For example Doyle has shown that during machining 1046 carbon steel at moderate speeds (30 m/min) with M2 HSS tools that wear occurs by shear of the tool surface and that observed abrasion wear tracks are the result of carbide particles being carried along in the sub-surface plastic flow zone in regions of crater wear of flank wear (10). In addition Gahr has concluded that if carbides have about the same hardness as the matrix, as in tool steels, they are in fact detrimental to wear resistance because they act as internal notches(11).

SUMMARY

These tool-life results illustrate the necessity of re-evaluating the traditionally held assumptions regarding the role of alloy composition to tool performance. The view presented here is that tool wear may be controlled principally by the yield strength of the tool material regardless whether wear is continuous or results from cutting edge fracture. The alloy content of the most commonly used high speed steels (M1, M2, M7) is 15-17 pct. carbide former. Of this, approximately 8-10 pct. remains as undissolved residual carbides after normal heat treatment and does not contribute to the yield strength. Although traditionally assumed to contribute significantly to the wear resistance of high speed steel, the tool-life data given here both for highly alloyed P/M and commonly used M1 and M7 HS steels contradict this belief. The high yield strength of HSS results primarily from the precipitation of secondary carbides. That it is sufficient to alloy only to the point which upon austenitizing results in a saturation of the matrix of elements which will precipitate as secondary carbides upon tempering has an economic importance in an era of alloy scarcities.

ACKNOWLEDGMENTS

The author expresses his appreciation to the Vermont American Corporation and the Vermont Tap and Die Company for their support. Mrs. Rhoda Weed is thanked for typing the paper.

REFERENCES

(1) A. G. Gray, "Materials Substitution: Challenge of the 80's," Metal Progress, 115 (1980) p. 33.

(2) S. Ramalingam, von Turkovich, B. F., "Process Optimization from Materials Viewpoint", Annals of the CIRP, 27 (1978).

(3) W. E. Henderer, B. F. von Turkovich, "The Influence of Heat and Surface Treatments on the Performance of M1 HSS Taps", Influence of Metallurgy on Holemaking Operations, ASM (1978) p.17.

(4) A. H. Grobe, G. A. Roberts, "The Bend Test for Hardened High Speed Steel", Trans. ASM, 40 (1948), P. 435.

(5) W. E. Henderer, "Strengthening Mechanisms in High-Speed Steel as Related to Tool-life," Trans. ASME, J. Engineering for Industry, 101 (1979), p. 217.

(6) W. E. Henderer, B. R. von Turkovich, "The Signifance of the Mechanical Properties of High-Speed Steel to Tool Fracture and Wear", Symposium on New Developments in the Processing and Properties of High Speed Tool Steel; 109th AIME Annual Meeting, February 24-25, 1980.

(7) J. A. Rescalvo, B. L. Averbach, "Fractured and Fatique in M-50 and 18-4-1 High Speed Steels", Metallurgical Trans. A, 10A (1979), p. 165.

(8) W. E. Henderer, B. F. von Turkovich, "Tool-life of P/M Drills and Tap", Proc. Sixth NAMRC, 1978, p.283.

(9) W. Stasko, F. R. Dax, "Cobalt-Free High Speed Steels", symposium on New Developments in the Processing and Properties of High Speed Tool Steel, 109th AIME Annual Meeting, Feb. 24-28, 1980.

(10) E. D. Doyle, "Effect of Different Heat Treatments on the Wear of High Speed Cutting Tools", Wear 27 (1974) p. 295.

(11) K.-H., F. Gahr, "Relation Between Abrasive Wear Rate and The Microstructure of Metals", Wear of Materials 1979, ASME (1979) p. 266.

DISCUSSION

V.A. Tipnis (Tipnis Associates)

I'd like to make the following comment. Electron Microscopic observations on worn H.S.S. tools indicate that the residual carbides can gouge the tool material matrix and hence contribute to wear rather than resist it. This is especially true of VC and M_xC_y type carbides in H.S.S.

R.W. Bratt (Crucible)

The first tool life test was designed to produce edge chipping. Were the tests on the lower alloy materials designed for the same purpose?

Dr. Henderer

In the tap tests we hardened the work material to 32 R_C, and under these circumstances the cutting edge will fracture rather than wear. The tool life was extremely short. The tests on the lower alloy materials were drilling applications where edge strength is substantially greater and the mode of failure occurred by wear rather than edge fracture.

FABRICATION AND EVALUATION OF POWDER METALLURGY HIGH-SPEED TOOL STEELS

Kishor M. Kulkarni and Mark T. Podob
Consolidated Metallurgical Industries

The FULDENS^{tm} process for cold pressing and sintering of high-speed steel cutting tool preforms is described. Current experimental and production applications are presented. Laboratory and shop tool life test results are summarized for turning, interrupted single tooth milling, end milling and gear cutting, as well as heat treatment processing parameters for optimum tool mechanical properties.

The FULDENS^{tm} process permits close-tolerance, complex shapes of high-speed tool steel to be made repeatedly with mechanical properties and performance characteristics equal or superior to parts made by machining conventional tool steel mill products, but at a significant cost savings. Both cold isostatic and conventional mechanical compaction techniques are used to manufacture preform shapes from irregularly shaped powders that for the first time require no additional working after sintering to achieve 100 percent density.

The resultant material possesses a microstructure which is representative of a more chemically homogeneous and isotropic material than produced by conventional techniques. The grain size is finer and carbides are small and uniformly distributed, resulting in improved grindability. The homogeneity produces a part which has a uniform and predictable heat treat response.

DISCUSSION

V.A. Tipnis (Tipnis Associates)

What is the lead time on this process?

Dr. Kulkarni

It depends very much on the part complexity. Isostatic parts may be on the order of 2-3 months. Mechanical parts would be more like 4-6 months. However, with the current economic situation tooling is obtained in little over a week, so the lead times are low right now.

SESSION III

COATED AND TREATED TOOL MATERIALS

THE INFLUENCE OF COATING THICKNESS AND COMPOSITION UPON METAL-CUTTING PERFORMANCE

Thomas E. Hale
Donald E. Graham
General Electric Company
Carboloy Systems Dept. - Detroit, Michigan 48232

INTRODUCTION

Coated cemented carbide cutting tools are now more than 10 years old, and their acceptance in the metal-cutting industry is still increasing at a steady pace. In the United States, about 35% of the cemented carbide tools currently used for metal cutting are coated.

During this first decade of coatings, there have been significant improvements made in the quality and performance of coated cutting tools. Some of these stemmed from improvements in the quality of the coating itself, namely, in better porosity and thickness control. Improvements have also been made in the control of the metallurgy of the coating-substrate interface, resulting in improved coating adhesion and suppression of strength degrading interface reactions. Additional improvements have been obtained by use of the cemented carbide substrate compositions designed to optimize the two most important substrate properties, namely, breakage resistance and thermal deformation resistance.

In addition to the TiC coating which was the original coating introduced in 1969-1970, a few other coatings are now offered; namely, TiN, HfN, and Al_2O_3. The use of multiple-layer coatings, such as TiN over TiC and Al_2O_3 over TiC, has also become popular.

The popularity of coated cutting tools stems from the superior combination of wear resistance and breakage resistance they have relative to uncoated tools. Figure 1 shows a plot of the wear resistance machining steel vs. bend strength for a series of uncoated and TiC-coated cemented carbide compositions. The 5-micron TiC coating used for this comparison caused a bend strength reduction averaging about 25%, while it increased the wear resistance by a factor ranging from 2.5 to 40. The resultant combination of wear resistance and bend strength is superior for some, but not all, TiC-coated compositions.

Industrial experience with coated cutting tools suggests that, in metal-cutting applications, the weakening effect of the coating is less than that indicated by static bend tests. The explanation for this difference is probably connected with the fact that surface cracks quickly develop on both coated and uncoated inserts during the cutting process, whereas in static tests the uncoated surface has a greater resistance to crack initiation than the coated surface. The breakage resistance of the cutting edge is then largely controlled by the propagation

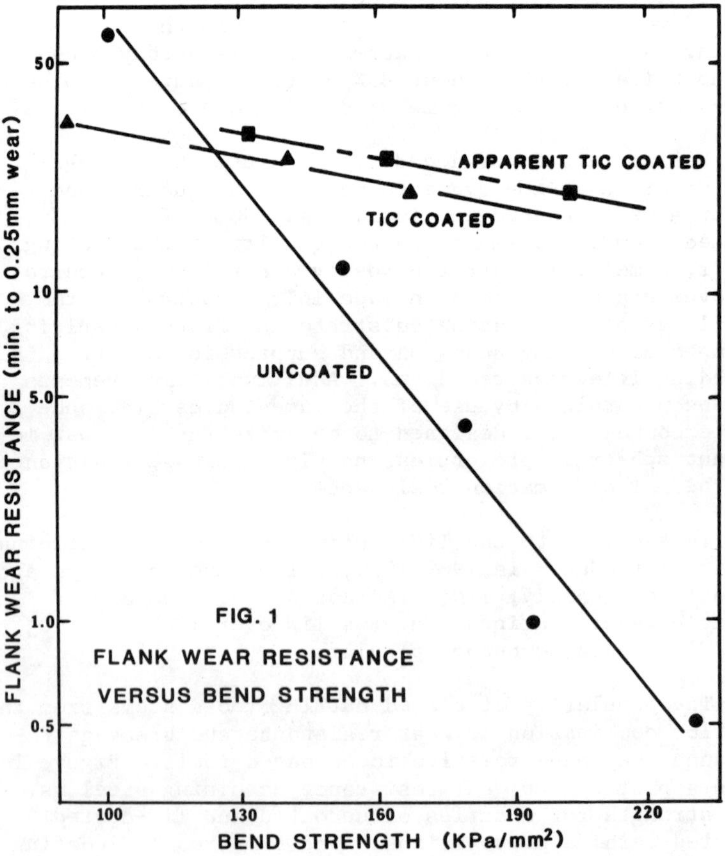

FIG. 1

FLANK WEAR RESISTANCE VERSUS BEND STRENGTH

of these cracks in the substrate at depths below the coating. If the coating is very thin and carefully deposited so as to avoid excessive interaction with the substrate, it has a minimal effect on dynamic crack propagation. When the apparent metal-cutting strength is estimated, the wear resistance-breakage resistance curve for TiC-coated compositions is shifted, as shown in Figure 1, to an even more favorable position relative to uncoated tools.

Perhaps the most important reason for the acceptance of coated tools is the significant increase in metal-cutting speed that can be obtained with no loss of life. Figure 2 shows tool life vs. cutting speed plots for a steel cutting carbide composition in the uncoated, TiC-coated and Al_2O_3-coated states. At a constant tool life of 10 minutes, the TiC coating permits a cutting speed increase of 50%, while the Al_2O_3 coating can run more than 90% faster. Such speed increases can significantly improve the productivity of the machining process and are the most cost-effective way to utilize coatings.

Despite the fact that coatings have been in use for 10 years, there still is no fully satisfactory explanation for how coatings with thicknesses nearly two orders of magnitude less than the dimensions of the wear scar can provide the substantial increases in wear resistance that are obtained in industrial machining applications. Some

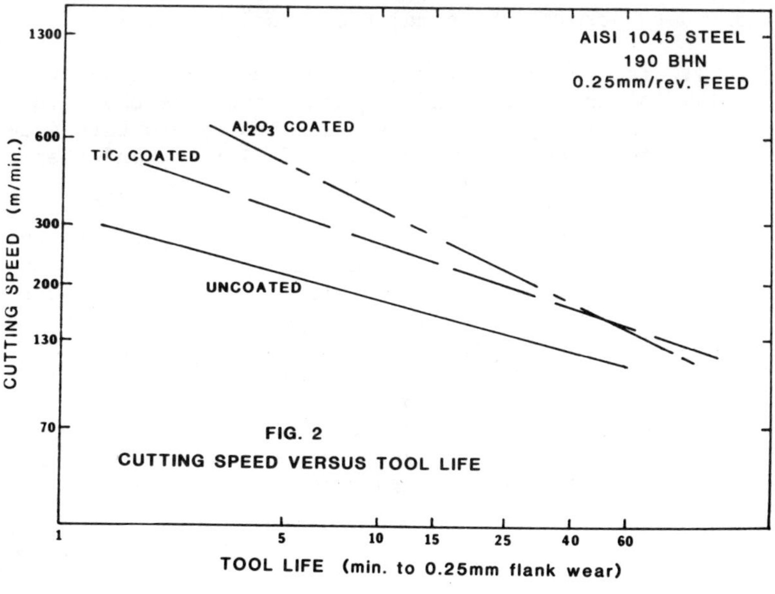

FIG. 2
CUTTING SPEED VERSUS TOOL LIFE

of the theories postulated are centered around the formation of lubricating films, the chemical stability of the coatings, and the confinement of the wear process to a shallow surface zone due to the manner in which surface asperities are heated by frictional contact with the workpiece. While most of the explanations offered seem plausible at first glance, some become less satisfactory as we learn more about the way coatings influence the wear process. The purpose of this paper is to provide detailed information about the way coatings improve wear resistance, with regard to coating thickness and compositional effects. Significant information is obtained by observing the change in wear rate as the coating is worn through. Since the wear behavior is apparently different for the crater wear vs. the flank wear, each of these wear zones will be considered separately.

CRATER WEAR

Figure 3 shows crater wear vs. time curves for a series of Al_2O_3-coated tools having different coating thicknesses when used to machine AISI 4340 steel at 183 m/min. (600 ft./min.). These data were used to construct Figure 4 which shows the influence of coating thickness on crater wear rate after the coating has been penetrated, and Figure 6 which shows the effect of coating thickness on crater wear resistance. Examination of these data permits three major conclusions:

1. The crater wear rate increases significantly after the coating is penetrated. As shown in Figure 3, relatively low rates were obtained up to crater depths of about 5 microns, and then the rates increased.

2. Even after the coating has worn through and the wear rate has increased due to contact between the chip and the substrate, the rate of crater depth development is strongly influenced by the coating remaining at the edge of the crater. Figure 4 shows that the wear rate during this period (measured between 25 and 50 microns' crater depth) decreases with increasing coat thickness, approaching a constant value as the thickness approaches 9-10 microns. At the

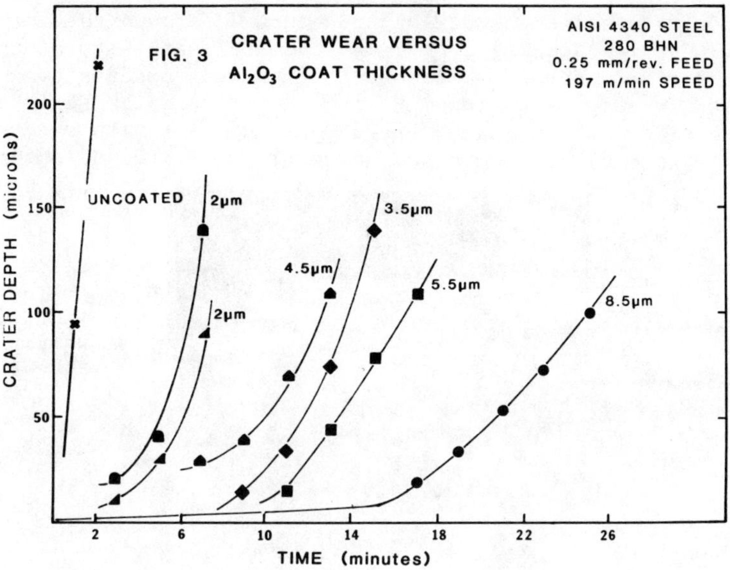
FIG. 3
CRATER WEAR VERSUS
Al_2O_3 COAT THICKNESS
AISI 4340 STEEL
280 BHN
0.25 mm/rev. FEED
197 m/min SPEED
CRATER DEPTH (microns)
200
150
100
50
UNCOATED
2μm
3.5μm
4.5μm
5.5μm
2μm
8.5μm
2
6
10
14
18
22
26
TIME (minutes)

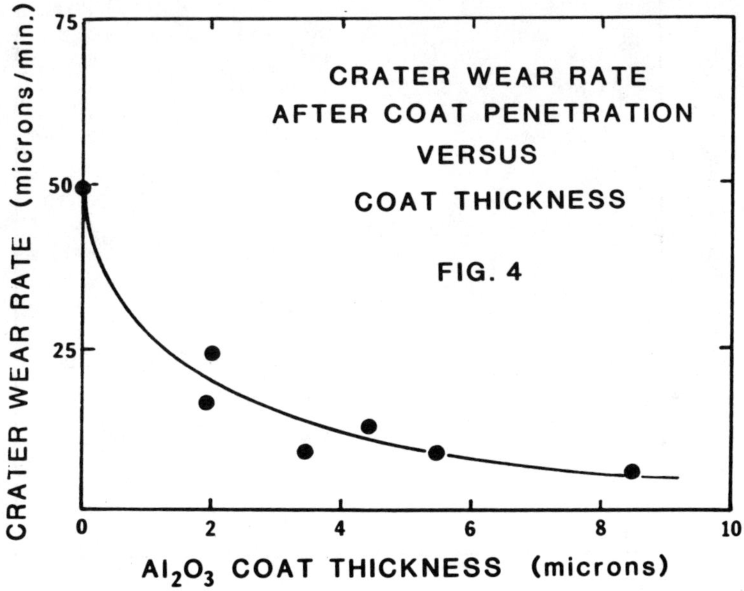
CRATER WEAR RATE
AFTER COAT PENETRATION
VERSUS
COAT THICKNESS
FIG. 4
CRATER WEAR RATE (microns/min.)
75
50
25
0
2
4
6
8
10
Al_2O_3 COAT THICKNESS (microns)

FIGURE 5

CRATER WEAR EDGE ON A TiC COATED INSERT

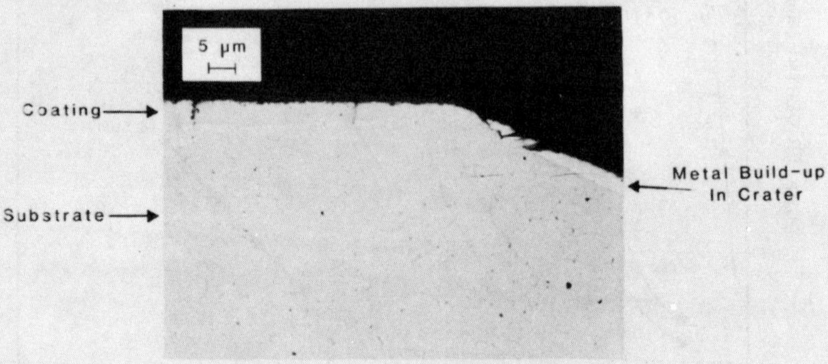

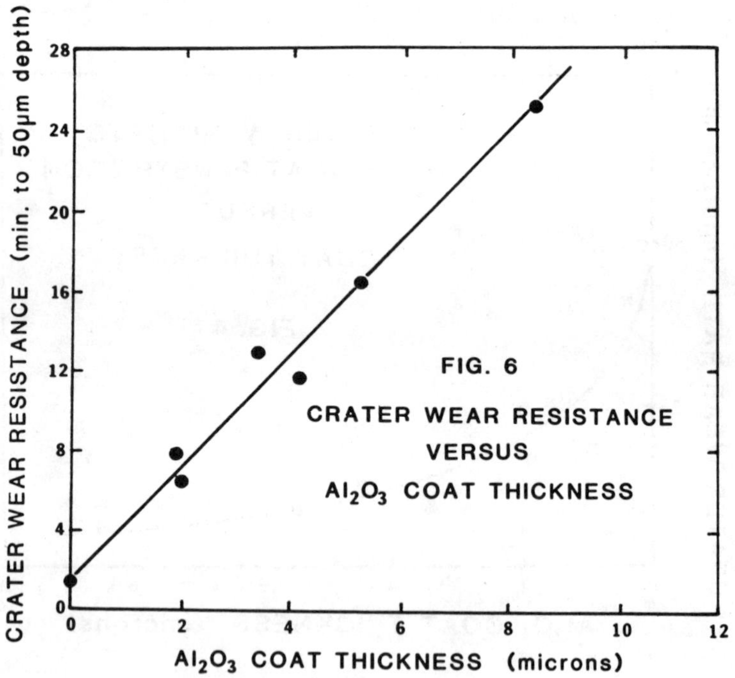

FIG. 6

CRATER WEAR RESISTANCE VERSUS Al_2O_3 COAT THICKNESS

9-micron thickness level, the crater wear rate during this phase of wear was about one-tenth of the uncoated rate. Figure 5 shows a micrograph of the edge of the crater for a TiC-coated insert. The manner in which the coating tapers into the wear pocket is indicative of the smooth wear occurring as opposed to a fracturing, pull-out wear mode. The coating thus exerts a considerable influence on crater wear rate, even after it is worn through.

3. Crater wear resistance is directly proportional to coating thickness. This linear relationship is illustrated in Figure 6 for Al_2O_3 coatings. Examination of TiC and TiN coating performance data shows a similar behavior, differing only in the slope of the lines, as will be shown later. This linear relationship holds true regardless of the crater depth criterion selected to represent crater wear resistance.

The influence of coating composition upon crater wear resistance is illustrated in Figure 7 where crater wear resistance is plotted against coating thickness for Al_2O_3, TiC, and TiN coatings. The inherent crater wear resistance of each coating composition is the slope of its curve. For the machining conditions employed, the inherent crater wear resistance of TiC was about equal to that of TiN. The Al_2O_3 coating had more than twice the inherent crater wear resistance of the TiC and TiN coatings.

A comparison of hardness values in Table I, at 1000° C (the approximate temperature in the crater zone), shows that Al_2O_3 may have slightly higher hardness than TiC (and probably TiN). When chemical stabilities are compared, using the free energy of formation values shown in Table I, it is apparent that Al_2O_3 has a significant stability advantage over TiC or TiN. When coupled with the fact that crater wear is very strongly influenced by cutting edge temperature, it would appear that the coating provides crater wear resistance mainly by acting as a barrier to chemical breakdown and diffusion reactions.

The crater wear behavior of a coated insert thus has two major stages. During the first stage, which lasts until the chip first penetrates the coating, the very

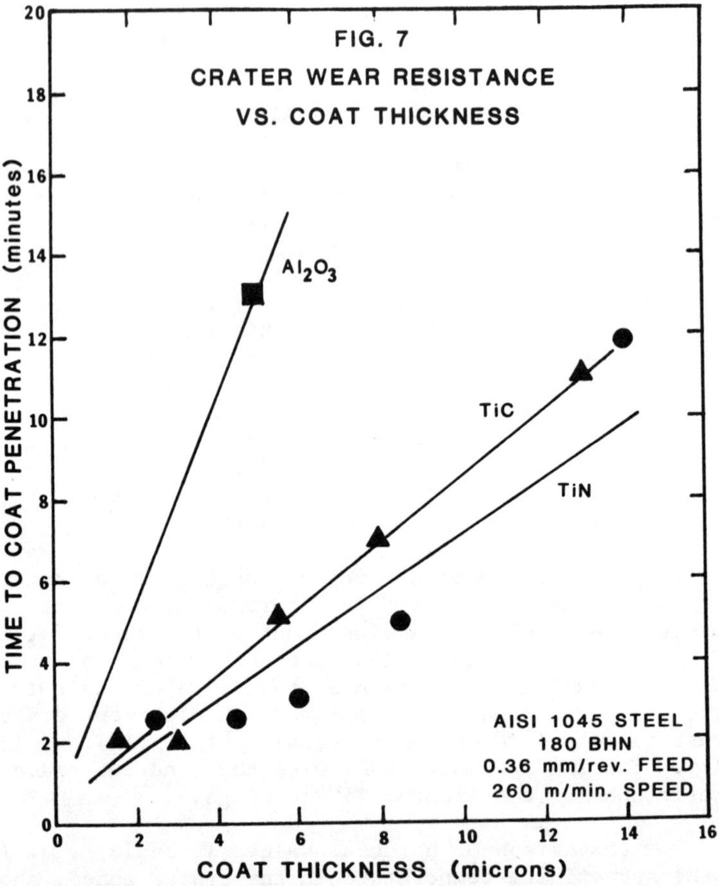

Table 1. Hardness and Chemical Stability of WC, TiC, and Al_2O_3

Compound	Hardness Kgm/mm Rm. Temp.	1000° C	Free Energy of Formation @ 1000° C Kcal/gm Atom O, C, or N
WC	1900	160	- 8.6
TiC	2500	200	- 40.7
Al_2O_3	2000	300*	-101.
TiN	2000	-	- 52.

*Extrapolated from 800° C.

high chemical stability of the coating substantially retards crater development. The duration of this stage of wear is directly proportional to coating thickness and is about twice as long per unit thickness for Al_2O_3 coatings as for TiC or TiN coatings.

Once the crater wear penetrates the coating, the crater wear rate increases rapidly due to increasing contact between the chip and the substrate material, with its inherently higher wear rate. During this second stage the coating at the edge of the crater exerts a strong retarding effect on crater wear rate, an effect that increases with coating thickness up to a thickness of about 10 microns.

FLANK WEAR

A typical flank wear vs. time curve for a coated tool is shown in Figure 8. This data was generated using a 5-micron Al_2O_3 coating to machine AISI 4340 steel. The arrow on the curve indicates the approximate point where the coating was first worn through on the flank side. It is apparent that, unlike the crater wear situation, there was no flank wear rate increase when the coating was pen-

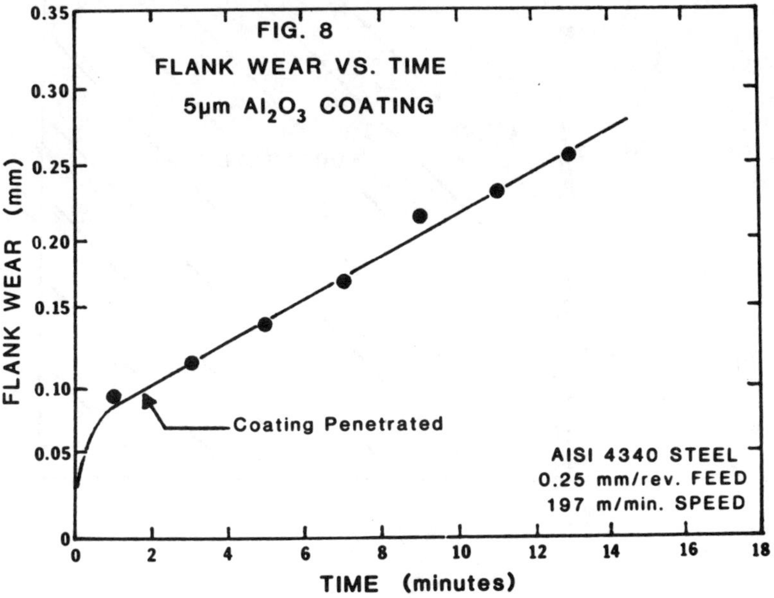

etrated. Flank wear curves for TiC coatings show the same behavior. The effect of coatings on flank wear behavior thus appears to be quite different than for crater wear.

Figure 9 shows a scale drawing of a cutting edge having a 50-micron-radius hone and a 5-micron coating at the point where the wear is .25 mm (.010"). From this perspective, it is apparent that most of the worn edge is uncoated, but yet the wear rate is the same as when the edge was fully covered by the coating. This suggests that region A, where coating remains at the top edge of the wear scar, or region B, where the bottom of the scar is coated, or a combination of these two regions, dominates the flank wear process.

FIG. 9

CROSS-SECTION OF FLANK WEAR REGION

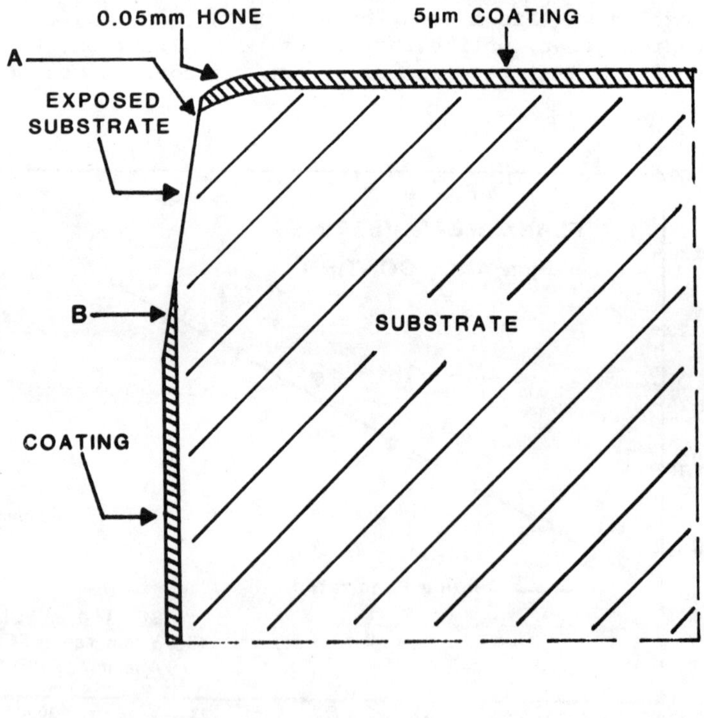

In order to determine the relative importance of these two regions, cutting tests were run using TiC-coated inserts from which the coatings had been partially removed, as shown in Figure 10. Figures 10A and 10B represent the fully coated and fully uncoated conditions, each having a 50-micron-radius honed edge. Figure 10C shows an insert with the coating removed only from the honed region of the cutting edge. In Figure 10D, the coating has been removed from the honed edge and the entire rake face, and in Figure 10E it was removed from the honed edge and the entire flank face. These configurations were obtained by grinding off the coating from the desired face and then rehoning the edge. These inserts were then used to machine AISI 1045 steel at 300 SFPM, and the flank wear was recorded as a function of cutting time. Figure 11 shows the resultant wear curves.

Curves A and B show the expected high rate for the fully uncoated edge and the low rate for the fully coated insert.

PARTIALLY REMOVED COATING CONFIGURATIONS

FIG. 10

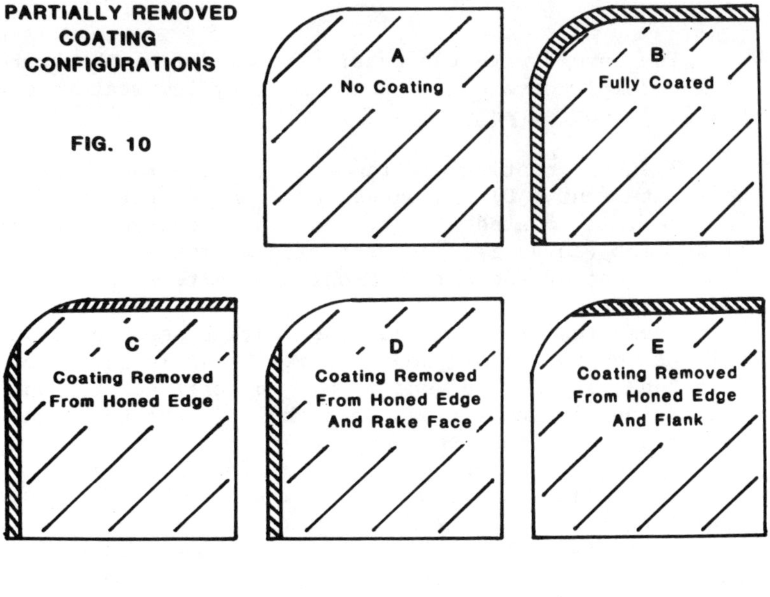

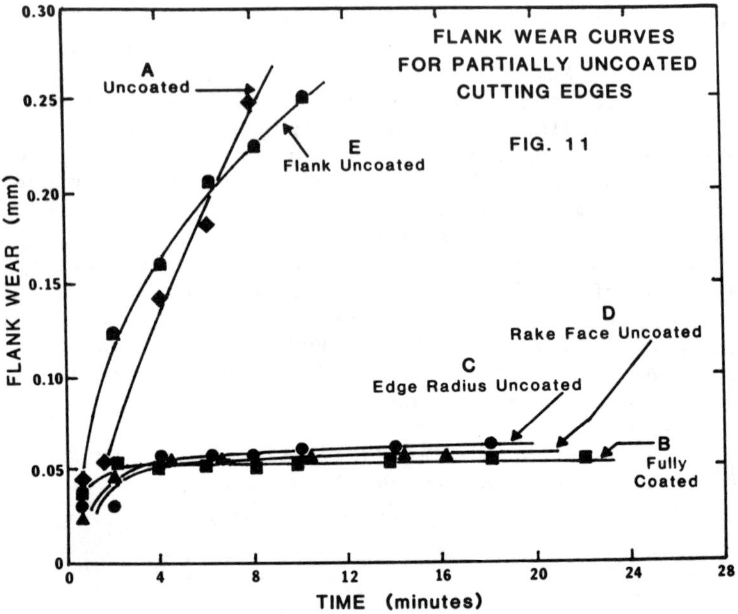

The insert with the coating removed from the honed region only, curve C, had the same very low wear rate as the fully coated edge.

When the coating was removed from the rake face, the wear rate (curve D) was unchanged from the rate observed for the fully coated insert, and, when removed from the flank face (curve E), the wear rate was essentially the same as that of the fully uncoated insert.

These results show that the coating need be present only at the bottom of the flank wear scar in order to obtain the fully coated wear rate. The critical region for generation of normal flank wear thus appears to be the bottom of the flank wear scar.

The influence of coating thickness upon flank wear resistance is illustrated in Figure 12 for Al_2O_3 and TiC coatings. In both cases the flank wear resistance first increases with increasing coating thickness and then

FLANK WEAR RESISTANCE VERSUS COATING THICKNESS

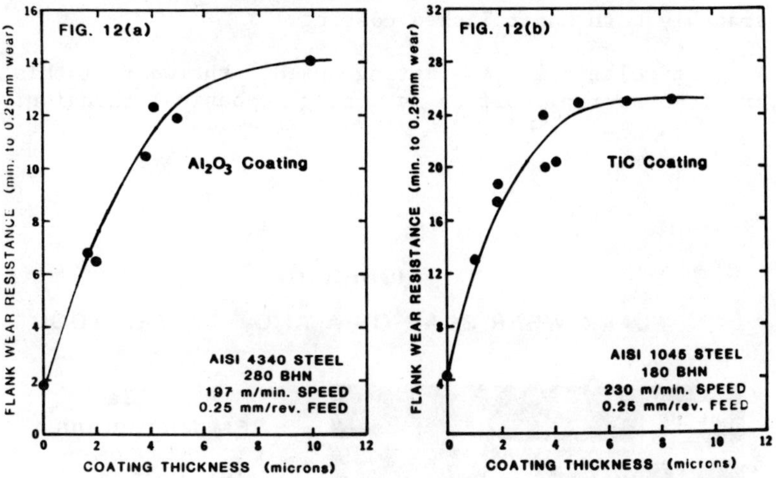

levels off at thicknesses greater than about 4-6 microns. This is again quite different from the crater wear situation where crater wear resistance shows no sign of leveling off up to thicknesses of at least 10-12 microns.

From the above-described observations, the following mechanism for the manner in which coatings reduce flank wear can be postulated. The critical contact zone between the flank side of the cutting edge and the workpiece is the bottom of the wear scar. When this region is coated with a compound having high abrasion resistance and high chemical stability and a thickness of at least about 5 microns, the critical contact between workpiece and cutting edge is borne completely by the coating which acts as an efficient bearing surface to support the feed force and greatly retards the enlargement of the wear scar. If the coating thickness is less than about 5 microns, partial contact with the substrate occurs in this critical wear region and a higher rate of wear results, perhaps proportional to the relative amount of substrate contacted.

This contact can be seen in Figure 13 which shows the flank wear scar for an Al_2O_3-coated tool used to machine AISI 4340 steel. Workpiece metal buildup was present along most of the wear scar, but the critical contact zone at the bottom was free of buildup and consisted of smoothly worn and polished coating.

At relatively low cutting speeds, the wear of this critical bearing surface is mostly mechanical abrasion;

FIGURE 13

FLANK WEAR SCAR ON A Al_2O_3 COATED TOOL

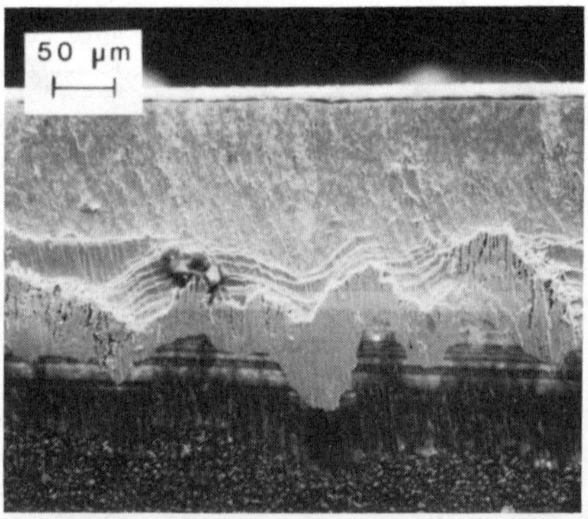

FIG. 13a
SEM Micrograph

Al_2O_3 Coating

Fe Build-up

Critical Wear Surface
Al_2O_3 Coated

Unworn Coating

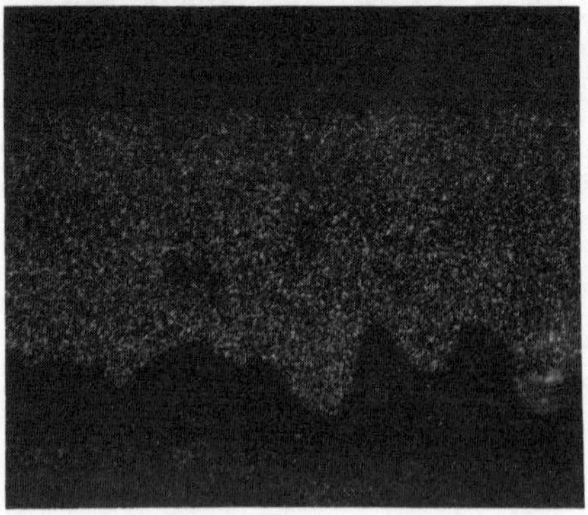

FIG. 13b
Fe X-Ray Map

High Fe Concentration

Critical Wear Region

thus, coatings with high hardness and abrasion resistance are most effective. At higher cutting speeds -- thus higher temperatures -- wear becomes increasingly influenced by chemical reactions between the workpiece and the contact zone, and thus coating compounds with high chemical stability become more effective; hence the crossover between Al_2O_3 and TiC coatings observed in Figure 2.

CONCLUSIONS

Crater wear resistance is directly proportional to coating thickness and strongly dependent upon coating composition. The inherent crater wear resistance of Al_2O_3 coatings is more than twice that of TiC or TiN coatings. The crater wear rate increases significantly when the crater depth exceeds the coating thickness, but, even when the coating is penetrated, the coating at the edge of the crater retards crater growth at a rate dependent upon coating thickness. Coatings appear to improve crater wear resistance by providing a chemical/diffusion reaction barrier at the chip contact zone.

Flank wear resistance is insensitive to coating thickness beyond a minimum value of around 5-9 microns, and the flank wear rate does not increase when the coating is penetrated. The critical region for flank wear appears to be a narrow zone at the bottom of the flank wear scar. The coating need cover only this zone to provide the improved flank wear resistance observed. Coatings thus improve flank wear resistance by providing a contact bearing surface between the cutting edge and the workpiece which wears slowly by a combination of mechanical abrasion and chemical reaction degradation. As cutting speeds increase, the chemical wear increasingly dominates. Thus, more abrasion-resistant coatings, such as TiC, work better at lower speeds, while more chemically stable coatings, such as Al_2O_3, resist wear better at high speeds.

REFERENCES

(1) E. M. Trent, "Wear of Metal Cutting Tools", Treatise on Materials Science & Technology, Vol. 13, 1979, pp. 443-488.

(2) E. Lenz, D. Pnueli, and L. Rozeanu, "The Effect of a Thin Coating of Cutting Tools", Wear, 53 (1979), pp. 337-344.

(3) T. Yamaguchi and J. V. Biggers, "High Temperature Reaction Between WC Tools and Ferrous Metals", Powder Metallurgy Int., Vol. 7, #2, 1975, pp. 75-81.

(4) O. Kubaschewski and E. LL. Evans, "Metallurgical Thermochemistry", 1958.

(5) N. P. Suh, "Coated Carbides - Past, Present and Future", The Carbide Journal, Jan.-Feb. 1977, pp. 3-9.

(6) J. I. Elgomayel, J. F. Radavich, and M. Tseng, "The Study of the Wear Mechanism of Titanium Carbide Coated Tools", Int. Journal Mach. Tool Des. Res., Vol. 19, pp. 205-219.

(7) S. Ramalingam, "Stoichiometry of TiC and Its Significance to the Performance of Hard Metal Compacts", Materials Science & Engineering, 29, (1977), pp. 123-130.

DISCUSSION

Dr. V.K. Sarin (GTE Labs)

Are the aluminum oxide coatings on the substrates pure Al_2O_3?

T.E. Hale

Yes they are pure aluminum oxide with an extremely narrow transition zone between.

R.D. Mills (Tungsten Carbide Mfg.)

Have you done any work with double and triple coating materials such as TiC, TiN, and Al_2O_3? If so, what effect do they have on the wear curves?

T.E. Hale

Yes we have done some work. You see the same general relationships. You get the average value of the three coatings if you are looking at crater resistance. The flank wear favors which one is best?

THE CUTTING PERFORMANCE OF PHYSICAL VAPOR DEPOSITED CEMENTED CARBIDES AND HIGH-SPEED STEEL

YOSHIHIKO DOI *
AKIRA DOI *
MITSUNORI KOBAYASHI *
YOSHIKATSU MORI *

1. INTRODUCTION

The ion plating process which is one of the PVD processes was employed for the present study of the coating of $Ti(C_{.7}\ N_{.3})$ compound-hereinafter TiCN-on to the cemented carbide and high speed steel. We investigated the metallographical characteristics of the coated film, adhesive strength on to the substrate, wear resistance and toughness in the cutting tests and the applicability for the steel milling operation is discussed. Ion plating equipment as developed for the present study is shown in Fig. 1.

2. EXPERIMENTAL PROCEDURE

The ion plating process consists of ion bombardment and plating. The ion bombardment was undergone for about 20 minutes in the argon atmosphere of 4 Pa. in order to clean the substrate of the cemented carbide or high speed steel at their surfaces. After the bombardment the vacuum chamber was evacuated to a pressure of $6x10^{-3}$ Pa. and then reaction gas, such as nitrogen or acetylene, was introduced to a pressure of 4×10^{-2} Pa. and the plating process was started. The plating conditions are then shown in Table 1.

Table 1 Ion plating conditions

SUBSTRATE MATERIAL	IONIZING VOLTAGE (V)	ACCELERATING VOLTAGE (V)	REACTION GAS PRESSURE ($\times 10^{-2}$ Pa)	COATING THICKNESS (μm)
CEMENTED CARBIDE	40	700	4	3 OR 5
HIGH-SPEED STEEL	40	250	4	3 OR 5

* Sumitomo Electric Industries Ltd.

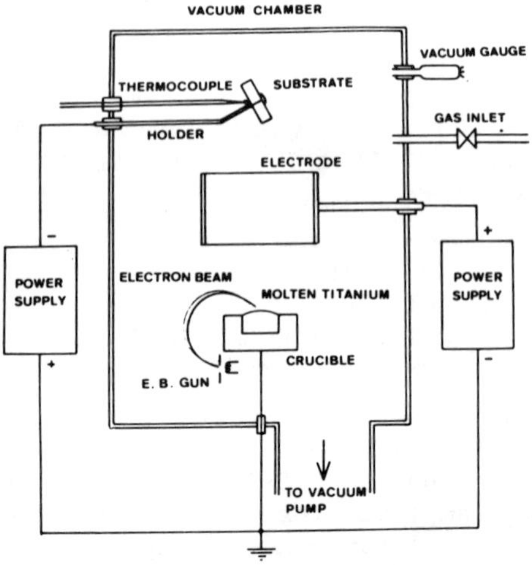

Fig. 1 Schematic of an Ion Plating System.

3. COATING ON CEMENTED CARBIDE

3.1. The Properties of Coated Cemented Carbides

3.1.1. The Metallographic Structure

A coated cemented carbide by CVD method has gained a solid reputation in the tool market as the most popular and useful cutting tool material. This CVD process, however, often gains a brittle decarburized layer (eta phase, for example) between the substrate and the coating layer. The quality of CVD cemented carbide such as toughness, depends upon the thickness of the decarburized layer, which nowadays is seldom observed by the optical microscope because of the progress in CVD technique. PVD, on the other hand, never yields a decarburized layer in the substrate thanks to a lower deposition temperature, which usually is thought to be below 500°C.

Fig. 2 shows the typical metallographic structure of the CVD and the PVD cemented carbide grade. The fracture surface of the PVD does not show grain boundaries in S.E.M. photos and looks like glassy structure. The X ray diffraction analysis found the large half-value diffraction width at the specific diffraction angle (220) TiCN, which is not usually observed in CVD case.

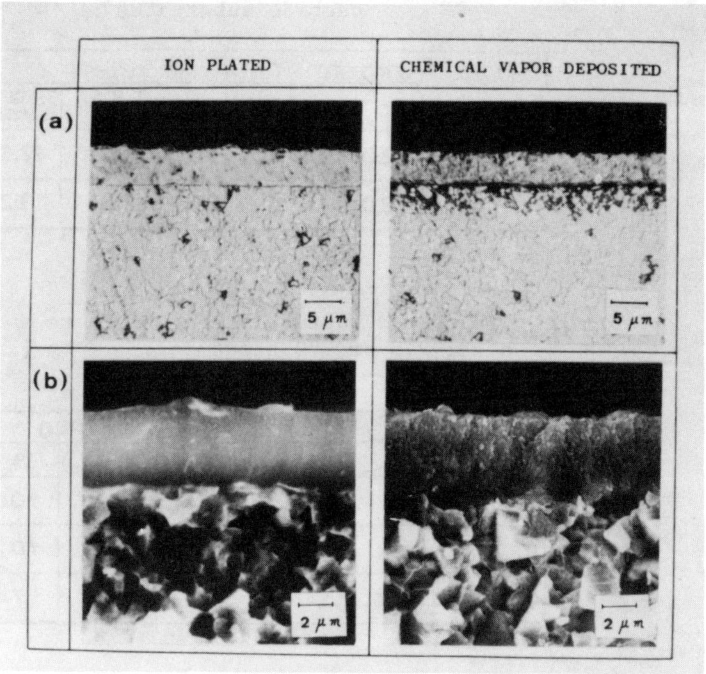

Figure 2. A Cross-Section of Polished(a) and Fractured (b) Surfaces of the TiC Layer of Carbides Coated by the Ion Plating Process and the CVD Process.

3.1.2. The strength

Table 2 shows the physical properties and chemical composition of substrates employed.

Table 3 shows the comparison of T.R.S. (Transverse Rupture Strength) values obtained for various samples. T.R.S. was measured for the 4μm - TiCN coated cemented carbide of orthorhombic shape (^{W}8mm x ^{T}4mm X ^{L}30mm) with the span length of 20 mm. The CVD specimens having little or no eta phase showed T.R.S. values lower than that of the substrate by 20 to 40% (2), however, the PVD specimens showed T.R.S. values almost equal to that of the substrate.

Table II Characteristics of carbide substrates

Substrate (ISO grade)	Composition (wt%)			Properties		
	WC	(Ti·Ta·Nb)C	Co	HRA	T. R. S. (Kg/mm²)	SPECIFIC GRAVITY
P30	72.0	19.5	8.5	90.6	216	12.61
P40	78.0	11.5	10.5	89.5	232	13.26

Table III Transverse rupture strength before and after PVD or CVD treatment.

T.R.S. \ ISO GRADE	P30		P40	
	(Kg/mm²)	%	(Kg/mm²)	%
Substrate	216	100	232	100
PVD (5μTiC) coated	222	103	237	102
CVD (5μTiC) coated	145	67	172	74

(Each data is the mean value from ten specimens.)

The cause of the reduction in T.R.S. by CVD is assumed to be due to the following reasons:

a) The diffusion of carbon atoms into the coating layer (TiC, TiN or TiCN) from the substate weakens the very surface of the substrate in such a manner that the substrate is degraded in its strength. The brittle surface layer may contain so-called eta phase to some degree even though eta phase can hardly be observed by an optical microscopy or by X ray diffractometer.

b) The as-said carbon-diffusion causes the contraction of surface layer in volume, resulting in the residual tensile stress created in the surface layer of the substrate. This stress lowers the apparent bending strength of the substrate (cemented carbide).

The as-said phenomena may not be observed in PVD because the diffusion of carbon atoms is not so much accelerated owing to the relatively low treating temperatures.

c) The residual tensile stress tends to be created in the coating layer (i.e. TiC or TiN) as the furnace is cooled down to the room temperature, which is due to the difference in the thermal expansion coefficient between the coating layer and the substrate. The CVD process is assumed to cause much high residual tensile stress than the PVD process because the coating temperature for CVD is about 1000°C and that for PVD is lower than 500°C.

d) The CVD films are, in most cases, chemically bonded to the substrate of cemented carbide, the bonding being so strong that cracks can propagate into the substrate immediately after their birth at the surface of the coating layer. The PVD film is thougt to be physically bonded to the substrate, cracks being blocked at the interface between the coating layer and the substrate. The strong bonding may result in the degradation of the substrate as will be discussed in next section.

3.1.3. The adhesive strength

The diamond indentation method (Rockwell A type indentator) was used to examine the adhesive strength of the coated film on to the substrate. The observation by the optical microscope indicates that the CVD specimen is superior to the PVD specimen as far as the adhesive strength is concerned, but the cracks initiated in the coating layer seem to propagate much further in the case of CVD.

3.2. The Cutting Performance

3.2.1. Toughness

We prepared TiCN coated inserts having the coating layer of 4μm on to the substrate of P30 and P40 grade (cemented carbide) for the interrupted cutting test.

The testing method and the result are shown in the Fig. 3 The chamfer (-25° x 0.15 mm) was put on the cutting edges of every insert. The present result indicates that the toughness of the insert is slightly improved by the PVD treatment, though the toughness of the insert being extremely lowered by the CVD treatment. PVD treated insert shows better interrupted cutting toughness than the untreated one, which may be attributed to the well-known fact that the coating layer significantly suppresses the adhesion between the insert and the work material.

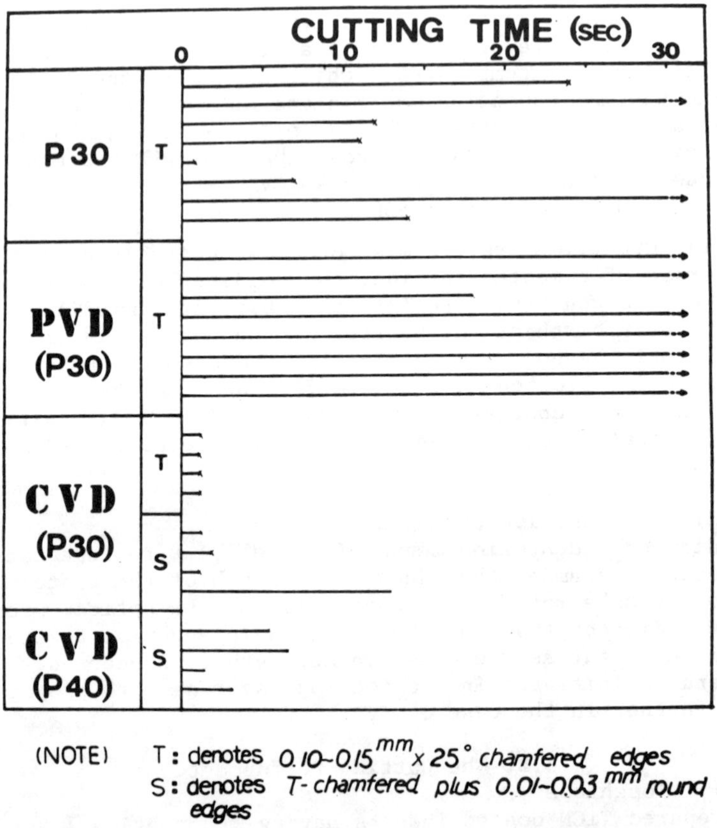

Fig 3. Cutting time until failure in the interrupted turning test : work material, AISI 4135 ; tool, 0,5,11,6, 15,15,0.8 mm ; cutting speed, 80m/min ; depth, 2 mm : feed rate, 0.2 mm/rev.

3.2.2. The steel milling applications

The CVD coated cemented carbide inserts, though very popular in turning use, are very rarely used for the milling use. The application of CVD inserts for the milling use is still limited to cast iron milling with relatively high cutting speed or steel milling with negative rake angle. CVD inserts of positive rake shape can hardly be used for the standard steel milling applications because of chipping. Present work on PVD proved that the coating inserts are able to be used even for the steel milling with positive rake angle as far as

the ion plating process is used for the coating process. The cutting test results obtained from steel milling application are shown in Fig. 4, 5. and 6. In these figures, VB means the average line of flank wear width and VB' means the local maximum flank wear width. From Fig. 5 one finds that the CVD grades tend to show the abnormal and local chipping wear. On the other hand the PVD grade with P30 substrate shows normal wear with significantly small wear rate when compared with P30. The difference in wear rate increases with cutting speed as shown in Fig. 4 or 5. When one chooses low carbon steel (say AISI 1006) as work material, the wear rate is drastically lowered by PVD treatment even at relatively slow cutting speed (Fig. 6).
The machining conditions for various tool materials in steel milling or cast iron milling use will be reported.

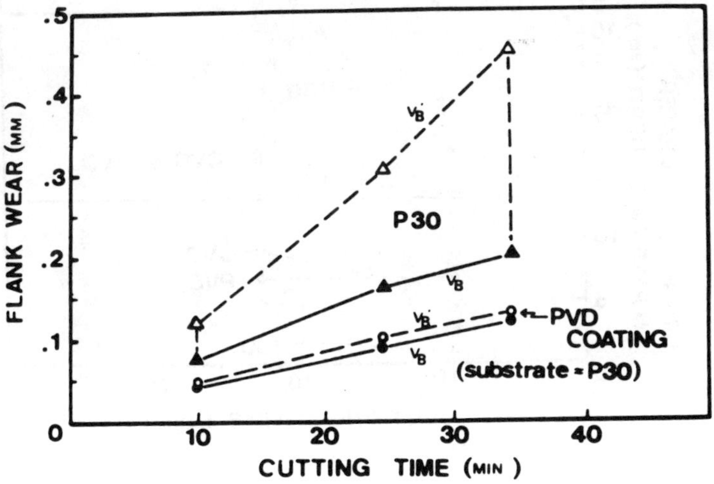

Fig. 4 Wear rate of the PVD treated and untreated insert in steel milling : work, AISI 4135 (Hs 35 to 40) ; cutter, ϕ6 inches, axial rake angle = +8° , radial rake angle =0° , lead angle = +15° ; insert, SPG422 ; cutting speed, 100m/min. ; depth, 3 mm ; feed rate, 0.29mm/t.rev.

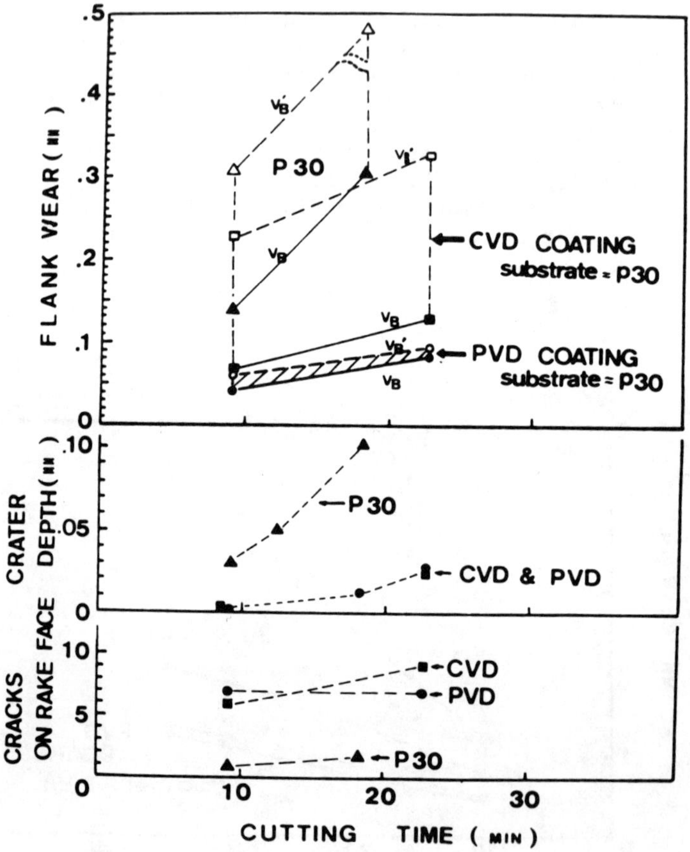

Fig.5 Wear rate and thermal crack progress of PVD treated, CVD treated and untreated inserts in steel milling : work material, AISI 4135 (Hs 35 to 40) ; cutter, ϕ6 inches, axial rake angle = +8° , radial rake angle = +0°, lead angle = +15° ; insert, SPG422 ; cutting speed, 180 m/min ; depth, 3mm ; feed rate, 0.18 mm/t.rev.

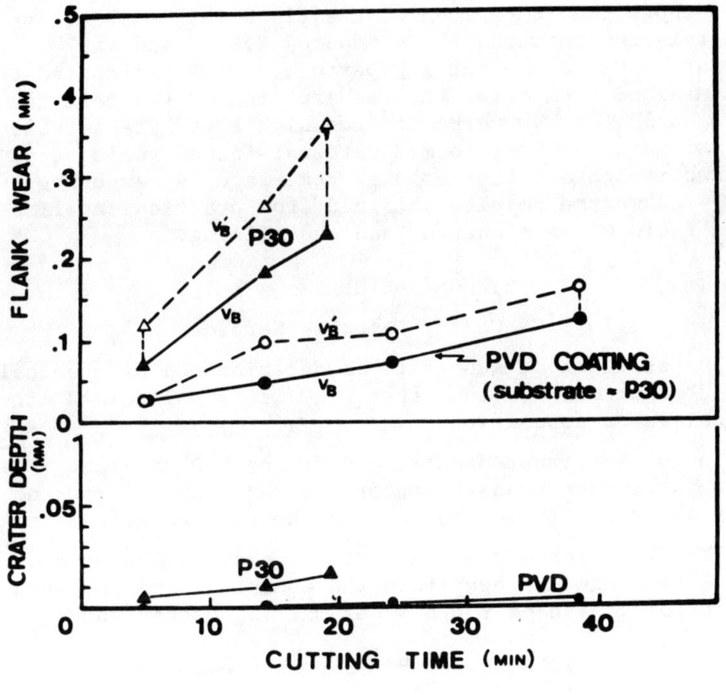

Fig. 6 Wear rate of the PVD treated and untreated inserts in steel milling : work, AISI 1006 ; cutter, ϕ6 inches, axial rake angle = +8° , radial rake angle = +0°, lead angle = +15° ; insert, SPG422 ; cutting speed, 124 m/min ; depth, 2 mm ; feed rate, 0.23 mm/t. rev.

4. COATING ON HIGH SPEED STEEL

The PVD process makes it possible to deposit TiCN film on high speed steel at the temperatures enough to avoid the annealing of the substrate.
Due to the relatively low treating temperature, the precise tools (i.e. hobs or forming tools) can be coated without any loss in the dimensional precision.
Fig. 7 shows the comparison of the flank wear and cutting speed ralation between the PVD coated H.S.S. and its substrate. Since the cutting performance of the coated tool is maintained even after the resharpening of the rake face, we can apply the PVD method to the high speed steel tools, such as, gear cutters, formed cutters, formed tools, taps and else, which are the tools of rake-face resharpening type.
Table 4 shows the results obtained from practical application in the field of gear cutting and end milling.

5. CONCLUSION

5.1 TiCN-Coated Cemented Carbide

(1) The strength of substrate is deteriorated by chemical vapor deposition. On the other hand it is maintained after physical vapor deposition.

(2) In the interrupted cutting test the PVD treated insert is tougher than the substrate, but the CVD treated insert shows poorer toughness than the substrate.

(3) The PVD treated cemented carbide shows a longer tool life, 2 to 5 times longer than the cemented carbide, when it is used for finish or rough steel milling operations.

5.2 TiCN-Coated High-Speed Steel

(1) The PVD treatment keeps the precision of tool shape and hardness of the substrate thanks to the low treating temperature.

(2) The PVD treatment gives 2 to 6 times longer tool file, and allows higher cutting speed than the non-treated tool.

(3) The PVD treated tool holds its tool life after a resharpening of the rake face, therefore, the treatment is well applied to the gear cutters, formed cutters, formed tools and taps.

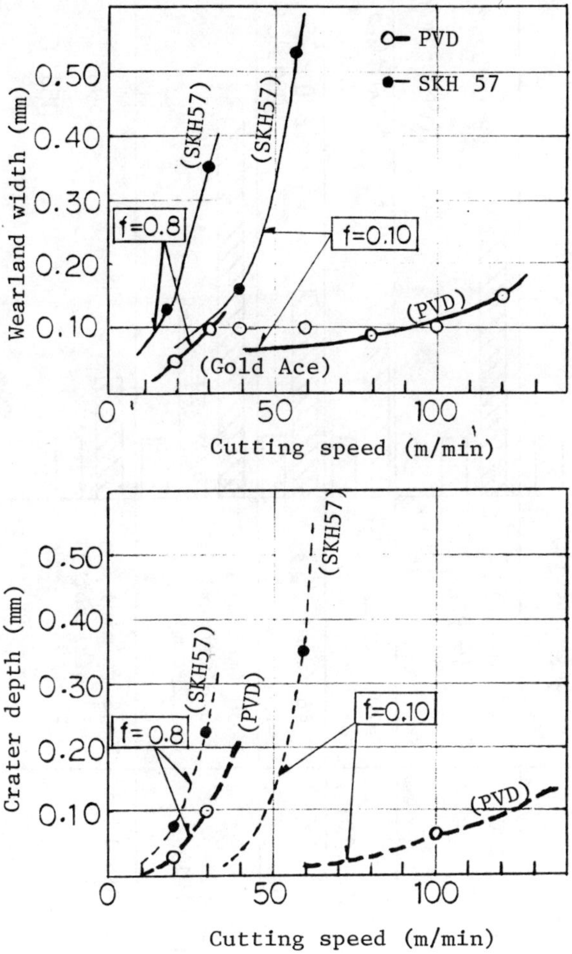

SKH 57
corresponds
to AISI M44

Work ; SCM3(Hs30) (AISI 4135)
Tool ; FP11R-44, SPG422 Wet
Cutting con.; d =2.0(mm), T =15(min) f =0.1 and 0.8(mm/rev)

Fig. 7 The influence of cutting speed on the wear rate of a high-speed steel with and without PVD treatment.

Table 4. Results of field tests (End mills and Gear cutters)

NO.	Tool	Work	Cutting condition	Relative tool life (against H.S.S.) 3 6 (time)
1.	End mill ϕ22 (4N)	Heat resisting steel (HB225)	V=14 (m/min) f=0.05(mm/tooth) d=6 (mm) Wet	16 Passes (1 Pass)
2.	End mill ϕ25 (4N)	Steel mold (SKD61,HRC43) (AISI H13)	V=18 (m/min) f=0.04(mm/tooth) Wet	
3.	Milling cutter ϕ250	(SUS304, HB200) (AISI 304)	V=25 (m/min) f=60(m/min) d=5 (mm) Wet	V_B 0.06 mm (V_B 0.35 mm)
4.	Hob ϕ32,PA20°	Spindle (S45C)(AISI 1045)	V=106 (m/min) Wet	200 P.C.S (70 P.C.S)
5.	Hob m=2,PA20°	(SCM 3) (AISI 4135)	V=46 (m/min) Wet	680 P.C.S (180 P.C.S)
6.	Hob ϕ75	Carrier (SC,HB260) (cast steel)	V=30 (m/min) f=0.8(mm/rev) d=3.2 (mm) Wet	
7.	Spiral bevel Gear Cutter	(SCM21,HB160) (AISI 4130)	V=68 (m/min) f=32(sec/tooth) Wet	360 P.C.S (150 P.C.S)

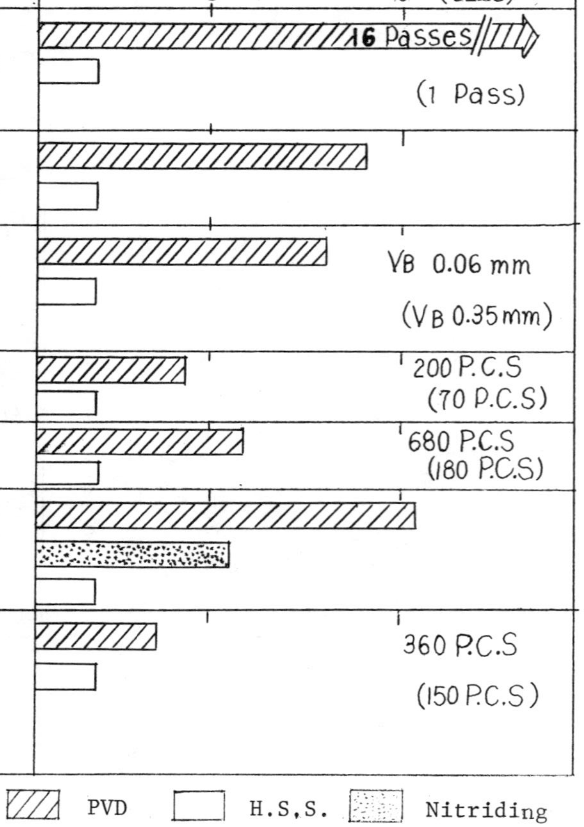

References

(1) M. KOBAYASHI, Y. DOI, " TiN and TiC coating on Cemented Carbides by Ion plating ". Thin Solid Films, 54 (1978) 67-74

(2) KLAS G. STJERNBERG, "Fracture toughness by the Three Point Bending Test " , 5th European Symposium on Powder Metallurgy (1978), 365-369.

(3) YOSHIO HIROMATSU, YOSHIHIKO DOI, MITSUNORI KOBAYASHI, MOBUHIRO TAKAHASHI, " Cutting Performance of the Gold Ace ", SUMITOMO ELECTRIC REVIEW NO. 18Dex. 1978 66-73.

[Discussion is presented on the next page.]

DISCUSSION

T.J. Skingle (Acme-Cleveland Corp.)

What is the price premium for coated H.S.S. tools over uncoated tools? Are you currently in production on coated H.S.S. tools, and, if so, for what length of time?

Yoshihiko Doi

We do not have a price comparison. Our company has been in production with this process since 1977.

Professor Bunshah (UCLA)

I am very pleased to see that you've taken the CVD process and put it into practice and note that the configuration you chose for experimental apparatus is extremely similar to the process called Activated Reactive Evaporation that we developed in 1971. We have reported data starting in 1975 on coated H.S.S. tools and we have found in continuing study that the TiC.'s, TiN's, HfN's coatings yield improvements in tool life very similar to what Mr. Doi shows. Recently we have been experimenting with twist drills and the results are very encouraging. Using 1/4" twist drills and AISI 4150 resulphurized steel (300 BHN), the uncoated drills gave an average of 19 holes. The familiar black oxide coated drills give an average of 25 holes and the TiN coated drills give an average of 150 holes. I wanted to ask Mr. Doi what is their experience on coated drills.

Yoshihiko Doi

We have much experience with coated drills. Usually the life improves 2 to 6 times. The problem is regrinding. After regrinding the effect can't be maintained. We have adapted this process to a throw-away type drill.

Professor Bunshah (UCLA)

The Russians reported that drills coated with MoN and CrO_2 yield an improvement of 1.5 to 2 times which could be maintained after resharpening.

PROPERTIES AND CUTTING PERFORMANCE OF NITRIDING CEMENTED CARBIDE AND TiC CERMET TOOLS

Sokichi Takatsu
Kunio Shibuki
Hirotaka Kiso
Toshiba Tungaloy Co., Ltd., Kawasaki, Japan

1. INTRODUCTION

Hard alloys such as cemented tungsten carbides and TiC cermets are extremely important cutting tool materials and demand for them is increasing year by year. These alloys have a higher hardness and wear resistance than high speed steels and higher strength and chipping resistance than ceramics. They have a higher cutting speed ability than high speed steel cutting tools. Moreover, they can be used in interrupted and heavy cutting conditions where ceramic tools would break.

However, from the viewpoint of productivity and machining economy, there is great demand for longer tool life and a higher metal removal rate. With this in mind, many improvements in tool materials and tools have been made. Carbide coated by a chemical vapor deposition method is a typical example. At present, a large amount of carbide inserts coated with TiC, TiN, Al_2O_3 etc. are being used. These coated inserts give several times longer tool life than uncoated inserts for the same cutting condition and they can be used at higher cutting speeds and have higher metal removal rates. However, there are some limitations for the coated inserts such as:

1) Not recommended for fine finishing.
2) Dimension is changed by coating.
3) Brittle decarburized layer and cracks formed near the coating layer reduce chipping resistance.
4) Peeling-off of the coating layer.
5) Expensive investment cost and high controlling technology.
6) Effective coating on TiC cermets is difficult.

Therefore, it is very worth-while to try to develop a new

surface treating method which will cover the limitations of coating. Nitriding is one of the hopeful methods, because it is believed to form a hard layer inside of the tool surface without changing its dimensions. Wear resistance of the tool will be improved by the formation of nitrides. Nitriding of steels by gases and salts has been widly used for a long time and new methods such as ion nitriding and plasma nitriding have been developed recently. Hardness and wear resistance of the nitrified steels are increased by solution hardning and the formation of nitrides.

On the other hand, nitriding of WC cemented carbides and TiC cermets is not as common as steels. Only a little work has been reported on them. Nitriding of TiC cermets with nitrogen gas[1] and nitrogen ion by glow discharge[2] has been studied. In these cases, TiN is formed at temperatures above 1173K and a considerable increase in the hardness and the wear resistance is observed when the cermets are treated at temperatures from 1173K to 1373K. However, the effect of the composition on the properties, the change of the microstructure etc. are not clear. At the same time, nitriding WC cemented carbide has been commercialized in Japan.[3] It is said that there are no detectable changes in the alloys, but that tool life is increased by several times. The details such as the products of nitriding, reasons for the long tool life etc. are as yet unknown.

In the present project, nitriding of WC cemented carbides and TiC cermet with ammonia gas and nitrogen gas was carried out. Effects of the nitriding gas, the nitriding temperature and composition of the base alloy on the microstructure, properties and cutting performance were investigated.

2. EXPERIMENTAL PROCEDURE

Three kinds of sintered hard alloys were used in this experiment. (1) W1: TiC free WC-TaC-Co alloy which contains 98mol % of WC in the carbides. (2) WT2: WC-TiC-TaC-Co alloy containing 45mol% WC and 44mol% TiC in the carbides. (3) T3: WC free TiC-TaC-Mo_2C-Ni alloy containing 88mol% TiC in the carbides. Chemical composition, density and hardness of these alloys are given in Table 1. The sintered specimens were ground with diamond wheels into the shapes of an insert tip for a cutting tool (SNG 432) and a 4x8x25mm rectangular prism.

The specimens were rinsed with methyl alchohol and acetone in an ultrasonic cleaner prior to the nitriding. A horizontal type furnace with a 70mm inner diameter of the inconel furnace tube was used. It was heated by a Kanthal heater, and the tem-

Table 1 Chemical Compositon and Properties of Specimens.

alloy	composition , wt%						density Mg/m^3	hardness RA
	WC	TiC	TaC	Mo_2C	Co	Ni		
W1	92	-	2	-	6	-	14.7	91.8
WT2	60	18	14	-	8	-	10.6	92.0
T3	-	58	10	16	-	16	6.5	92.0

Table 2 Work materials for the Cutting Tests.

S48C	carbon steel for machine structual use, 0.45-0.51%C
S55C	carbon steel for machine structual use, 0.52-0.58%C
SNCM8	nickel-chromium-molybdenum steel, 0.36-0.43%C, 1.60-2.00%Ni, 0.06-1.00%Cr, 0.15-0.30%Mo
FC25	gray iron castings, tensile strength 196MPa
FCD40	spheroidal graphite iron castings, t.s. 392MPa
FCD55	spheroidal graphite iron castings, t.s. 539MPa

perature was controlled by the on-off system within ±5K. After evacuation of air with a rotary pump, the specimens were heated in a protective gas. The gas was H_2 below 723K and NH_3 or $1N_2$-$3H_2$ mixed gas above 723K. The dew point of H_2 and N_2 was 183K (-90°C) and 193K (-80°C), respectively. The purity of NH_3 was 99.99% with 10ppm H_2O and 5ppm O_2. The heating rate was 8K/min. The nitriding was carried out at temperatures from 773K to 1473 K. Then the specimens were coold in the furnace to room temperature.

X-ray diffraction and knoop hardness (0.98N load) of the nitrified specimens were measured at the surface and in each section, removing the surface layer step by step with diamond paste. Microscopic observation and EPMA analysis of the nitriding layer were carried out on the enlarged section at the nose radius of the insert tip, since the layer was very thin. Change of the strength of the specimens by the nitriding was investigated by transverse rupture strength with the 3-point method and crack length formed at an indentation of a Vickers diamond pyramid (98N load). Cutting tests were carried out on various steels and cast irons. The work materials are listed in Table 2. Flank wear, crater wear, deformation of the cutting edge and chipping property in turning and face milling were measured. Details of the cutting condition are given in the figures on the results. These properties and cutting performance were compared to those of the untreated specimens and the effects of the nitriding were discussed.

Table 3 X-ray Diffraction Lines on the Surface of Nitriding Alloys

(relative intensity)

d	phase	W1 A	W1 B	W1 C	W1 D	W1 E	WT2 A	WT2 B	WT2 C	WT2 D	WT2 E	T3 A	T3 B	T3 C	T3 D	T3 E
2.90	W.N g				10	18				10	3					
2.84	WC	35	35	29	13	8	16	13	5	3	1					
2.75	L.C t									5	8					
2.73	L.C e			19	1				18	3						
2.52	WC	100	98	87	33	21										
2.51	(W,Ti)C						100	97	73	48	19					
2.49	TiC											100	95	93	65	70 (TiC+TiN)
2.46	TiN								8	16	6			5	23	
2.37	W.N b,g				95	100				44	8					
2.26	L.C t				4	4				5	10					
2.23	L.C e,c		3	24	4	1			26	3	3					
2.16	(W,Ti)C						56	56	50	35	13					
2.16	TiC											90	88	85	60	70 (TiC+TiN)
2.12	TiN								2	10	16			8	20	
2.13	L.C e,t,c				5	4			5	11						
2.09	L.C c		4	3	9			52	6	8						
2.06	W.N b,g		5	44	26	38				16	3					
2.05	Co	3					4	4								
2.03	Ni											5	15	23	25	30
1.95	L.C t					3				3	6					
1.92	L.C e,c			13	3				15	3	3					
1.88	WC	88	88	76	33	26	34	32	11	10	6					
1.83	W.N g				3	4				3	2					

(nitriding) A:RT-1073K, B:1173K, C:1273K, D:1373K, E:1473K
b:beta tungsten nitride, g:gamma tungsten nitride,
e:eta-W_3Co_3C, t:theta-W_4Co_2C, c:chi-$W_9Co_3C_4$

3. EXPERIMENTAL RESULTS

3.1 STRUCTURE AND PROPERTIES

The X-ray diffraction lines on the surface of the alloys nitrified with NH_3 gas are listed in Table 3. All alloys do not show remarkable changes up to 1073K. Above 1173K, the intensity of the carbides such as WC, TiC etc. decreases with an increase of the nitriding temperature, though they are found even at 1473 K. In the case of W1, diffraction lines of W-Co lower carbides (W_3Co_3C, W_4Co_2C etc.) are found at 1173K. These lines become stronger at 1273K, but decrease above 1373K. Strong lines of tungsten nitrides such as beta nitride and gamma nitride[4] are observed above 1373K. In the case of WT2, the tungsten nitrides are found at 1373K, but the intensity decreases at 1473K. The lower carbides and TiN appear at 1273K. The intensity of the TiN increases as the temperature rises. The intensity of some lower carbides decreases at higher temperatures and some of the other lower carbides increase in intensity. In the case of T3, TiN is formed at 1273K. The lines of TiN are more remarkable at 1373K and they join the lines of TiC at 1473K. The lines of the Co in W1 and WT2 disappear according to the formation of the lower carbides, while the intensity of the Ni lines of T3 increases with the rising nitriding temperature.

Fig.1 shows changes in the relative intensities of the diffraction lines ($I_{component}/I_{total}$) and knoop hardness of the alloys according to the distance from the surface. The nitriding temperature was 1373K. The surface of W1 is rich in tungsten nitrides. Tungsten nitrides and lower carbides exist up to a depth of about 4µm and 5µm, respectively. Metallic Co does not exist near the surface, but it is observed on the inside layer where the lower carbides disappear. The hardness of the layer near the surface is about Hk 200 lower than that on the inside. The hardness increases with a decrease of the tungsten nitrides and the value of the inner part, about 4µm and deeper, is equal to that of the untreateu. The tungsten nitrides are also a major phase at the surface of nitriding WT2. But the amount and the depth of the tungsten nitrides and the lower carbides are less than W1. The behavior of the Co is the same as that of W1. TiN is found up to a depth of 2 to 3µm, and the hardness of this layer is about Hk 200 higher than on the inside and of the untreated WT2. No lower carbides are observed in T3. The intensity of the Ni lines at the surface and in the layer 6 to 8µm from the surface is stronger than on the inside. On the other hand, it is weaker than on the inside layer 1 to 5µm from the surface. TiN is observed up to a depth of about 5µm. The TiN layer corresponds to the low Ni layer. Distribution of the Ni and Ti near the nitriding layer is confirmed with an EPMA. Hardness of the TiN layer is about Hk 400 higher than on the inside and in the untreated T3, while hardness of the Ni-rich layer is about Hk 100 lower than on the inside.

Photo.1 shows the optical microstructures of the nitriding layers treated at 1373K. The depth of the nitriding layer formed by N_2 gas is about half of that formed by NH_3. A tungsten nitride layer is found on the surface of W1 and WT2 and its inside is rich in the lower carbides which are etched darkly in the photographs. A low Ni nitriding layer is formed near the surface of T3. The Ni is concentrated mostly at the surface and on the inner layer close to the nitriding layer. The Ni forms droplets on the surface of the nitrified T3 as shown in Photo.2. The drop-

Table 4 Transverse Rupture Strength and Total Crack Length of the Specimens.

alloy	rupture strength, MPa		crack length(98N), µm	
	untreated	treated	untreated	treated
W1	1830	1630	180	195
WT2	1460	1210	236	202
T3	1430	1190	213	190

lets become larger as the nitriding temperature rises, while in W1 and WT2, no droplets of Co are formed by the nitriding.

Table 4 shows the transverse rupture strength and the total crack length of the specimens. The transverse rupture strength is lowered by the nitriding at 1373K, while the total crack length is not changed.

3.2 CUTTING PERFORMANCE

Fig.2 shows flank wear of the alloys nitrified at various temperatures in the turning of steel and cast iron. No effect of the nitriding is found on W1 in the cutting of FC25 gray cast iron at 150m/min. Flank wear is almost equal, irrespective of the nitriding temperature. Flank wear of WT2 nitrified from 773 K to 1073K is also the same as that of the untreated in the cutting of S48C carbon steel at 150m/min. The effect of the nitriding is observed at temperatures above 1173K. Flank wear decreases with the rise in the nitriding temperature and the value of WT2 nitrified at 1473K is about 2/3 of the untreated. When S48C steel is machined at 180m/min with T3, the change in flank wear shows nearly the same tendency as in WT2, and the wear of T3 treated at 1473K is 1/3 of the untreated. In cutting FCD40 ductile cast iron at 100m/min with T3, the effect of the nitriding is the most remarkable. That is, the flank wear of T3 is reduced half of the untreated by nitriding at 973K. It becomes smaller at higher nitriding temperature and reaches 1/9 of the untreated by nitriding from 1173K to 1473K.

Fig.3 shows a cutting length with the nitriding specimens in face milling of steel or cast iron. The value given is the average of 3 tests. Chipping resistance of these alloys does not change up to the nitriding temperature of 1173K, but a reduction is observed at and above 1273K. The reduction in chipping resistance of W1 and WT2 is remarkable and the cutting lengths are 60% and 58% of the untreated with the nitriding at

Table 5 Flank Wear and Crater Wear of T3 in Dry Turning.

v=200m/min, f=0.40mm/rev, d=1.5mm, t=20min,
S48C steel(HB 190), SNG432, nitriding: 1373K

specimen	flank wear, mm		crater wear
	mean	maximum	mm
untreated	0.21	0.29	0.08
treated	0.15	0.19	0.05

1273K and 40% and 22% after the nitriding at 1473K,respectively. At the same time, the reduction of the chipping resistance of T3 is smaller than W1 and WT2. The cutting length hardly changes up to 1273K, and is 92% and 71% of the untreated after the nitriding at 1373K and 1473K, respectively.

Fig.4 shows flank wear of T3 nitrified with NH_3 gas and N_2 $-H_2$ mixed gas. In the dry turning of steel at 150m/min, both treated specimens show smaller flank wear than the untreated as shown in Fig.4(a). The effect of nitriding with NH_3 is more remarkable than with N_2. The flank wear of the T3 nitrified with N_2 is 86% of the untreated after 20min cutting, while with NH_3 it is 57% of the untreated. Fig.4(b) shows flank wear of T3 in wet turning of the same S48C steel at a low speed of 30m/min. In this case, the nitriding shows a more remarkable effect on the flank wear than on high speed cutting. The flank wear of the T3 treated with NH_3 is only 7% of the untreated after 10min cutting. However, the effect of N_2-treating is considerably smaller than with NH_3, where the flank wear is 34% of the untreated. In dry turning of FCD55 ductile cast iron, the untreated T3 shows a flank wear larger than 0.3mm after only 3min cutting as given in Fig.4(c). With nitriding, wear resistance of T3 is improved remarkably and flank wear is less than 0.3mm after 15min cutting. The flank wear of the T3 treated with NH_3 is about half of the T3 treated with N_2. In all cases shown in Fig. 4, NH_3 shows a better effect from the nitriding than N_2.

Maximum and mean flank wear and crater wear of T3 in steel cutting at 200m/min are given in Table 5. Both maximum and mean flank wear of the nitrified T3 are about 70% of the untreated after 20min cutting. The effect of the nitriding on crater wear is somewhat greater with the treated T3 being about 60% of the untreated.

Fig.5 shows deformations of the cutting nose of T3 in the SNCM8 Ni-Cr-Mo steel at a variety of cutting speeds and feed rates. At a feed rate of 0.39mm/rev, no deformations are observed during cutting for 3min at 150m/min. At higher cutting speeds, the deformations increase with the cutting speed as shown in Fig.5(a). But the deformation of the nitriding T3 is smaller and the value is 60% of the untreated at 200m/min. The flank wear and the crater wear of the nitriding T3 are smaller, too. Fig. 5(b) shows the relationship between cutting speed and feed rate where 0.03mm of the deformation occurs after 3min cutting. As shown in the figure, the conditions at which the nitriding T3 shows equal deformation to the untreated are at a 0.1mm/rev higher feed rate at the same cutting speed and at a 25 to 30 m/min higher cutting speed at the same feed rate.

4. DISCUSSION

Carbide and nitride are soluble in each other, forming carbonitride solid solutions with various compositions. In the process of nitriding, nitride is formed through substitution of the carbon by nitrogen. Therefore, the lower carbides and nitrides formed in the nitriding are not a pure carbide or nitride but are considered to be solid solutions with a variety of C/N ratios. Some diffraction lines of the products of the nitriding, especially those of the lower carbides, are not necessarily the same as those given in the ASTM card. This disparity in the lattice parameter is possibly mainly due to the difference of the C/N ratio in the carbonitride solid solutions.

The formation of lower carbides depends on the composition of the base alloy. The amount of the lower carbides decreases with the decrease of WC content and increase of TiC content. The lower carbides are formed by decarburization of WC with H_2 and impure oxygen (O_2, H_2O) in the treating gas.[5] TiC has better oxidation resistance than WC. Moreover, TiC_x has a wider carbon range than WC_x. The experimental results show that the preceding reaction is decarburization and the nitrides are formed subsequently. The amount of TiN formed by the nitriding depends on the TiC content of the base alloy, the nitriding temperature and the activity of the nitriding gas. The amount of TiN formed by N_2 is less than that with NH_3 gas because the activity of N_2 is lower.

In T3, Ni is thrown up from the nitriding layer to both the outside and inside of the layer. This process has not been clarified. However, it is supposed that it is related to the lower wettability of Ni to TiN than TiC and WC, since the oozed Ni forms droplets on the surface. On the other hand, Co in W1 and WT2 forms the lower carbides and does not leave the nitriding layer.

The hardness of tungsten nitrides is not as yet known. However, it seems to be considerably lower than WC from the low hardness of the surface layer of nitrified W1. In general, the hardness of cemented carbides increases when the lower carbides are formed. But in W1, the reduction in hardness by the tungsten nitrides exceed the increase in the lower carbides. The high surface hardness of the nitrified WT2 is due to the formation of the lower carbides and TiN. Changes in the hardness in the nitrified layer correspond to the amount of products of the nitriding. Hardness of TiN (Hv 2000) is lower than TiC (Hv 3000). However, the hardness of TiC cermets is increased by the addition of TiN.[6] The hardness of the nitriding layer of the T3 is increased by the formation of TiN and the decrease of Ni.

The decrease in the transverse rupture strength is caused by the formation of a hard layer and the lower carbides on the surface. The crack length on the nitrified layer is equal to or somewhat shorter than that of the untreated. The supposition is that compressive stress is generated on the surface because of a smaller coefficient of thermal expansion of the nitriding layer compared with that on the inside.

The nitriding has no effect on the flank wear of W1, though the lower carbides increase the wear resistance in general. This may be due to the low hardness and consequently the low wear resistance of the nitriding layer. The flank wear of WT2 and T3 is small at a high nitriding temperature and high TiC content. That is, the wear resistance is nearly proportional to the quantity of TiN formed by the nitriding. Formation of the lower carbides and decrease of Ni in the nitriding layer are also effective on wear resistance.

It is well known that the chipping resistance of cemented carbides is decreased by the formation of lower carbides. The chipping resistance of W1 and WT2 is mainly affected by the formation of lower carbides. The high surface hardness of WT2 may be another reason for the low chipping resistance. The reduction of the chipping resistance of T3 is not so large, in spite of the increased hardness and reduced Ni content of the nitriding layer. The reasons for this are thought to be that there are no lower carbides and the high Ni tough layer supports the nitriding layer.

Generally, TiC cermets show greater wear in cast iron cutting and in low speed cutting of steel, since TiC is weak in simple abrasion. The effects of the nitriding on the wear resistance in these cuttings considered to be due to the formation of Ti(C,N). TiN has a good oxidation resistance and a low reactivity with iron. Therefore, crater and flank wear resistance at a high cutting speed and a high feed rate where a large quantity of heat is generated are improved by the formation of TiN. The low Ni layer is also effective on crater wear. The hard nitriding layer protects the cutting edge from plastic deformation at high metal removal rates.

5. CONCLUSIONS

Hard nitriding layers are formed in cemented carbides and cermets containing TiC by the nitriding and their cutting performance can be improved remarkably. The effects are nearly proportional to the quantity of titanium nitride and carbonitrides formed by the nitriding. The TiC content of the base alloy, the nitriding gas and the nitriding temperature are the main factors influencing the amount produced. Brittle decarburized phases are apt to be formed when the base alloy contains WC. However, these will be eliminated by controlling the base alloy and the nitriding conditions. The nitriding of cemented carbides and TiC cermets is considered to be a promising new method to produce superior cutting tool materials.

REFERENCES

(1) H. Ito, M. Tanaka and K. Taguchi, "Nitriding on the TiC Base Cermet Cutting Tool," Nippon Tungsten Review, Vol. 6, Sep., 1973, p. 29.

(2) H. Tanaka and M. Watanabe, "Nitriding Method for TiC Cermet," Japan Patent (open), No. 51-63327, 1976.

(3) "I.NIX Treating on Cemented Carbides," Oyo Kikai Kogaku, Feb., 1978, p. 32.

(4) R. Kiessling and Y. H. Liu, "Thermal Stability of The Chromium, Iron and Tungsten Borides In Streaming Ammonia And The Existance of a New Tungsten Nitride," J. Metals, Vol. 3, No. 8, 1951, p. 639.

(5) S. Takatsu and Y. Arasaki, "Changes of Carbon and Oxygen in H_2 Atmospheres," J. Japan Soc. Powder and Powder Metallurgy," Vol. 19, No. 5, 1972, p.207.

(6) H. Suzuki, K. Hayashi, T. Yamamoto and W. J. Lee, "Effects of Additional Titanium Nitride on the Strength of TiC-Mo_2C-Ni Alloy," J. Japan Soc. Powder and Powder Metallurgy," Vol. 23, No. 7 1976, p. 224.

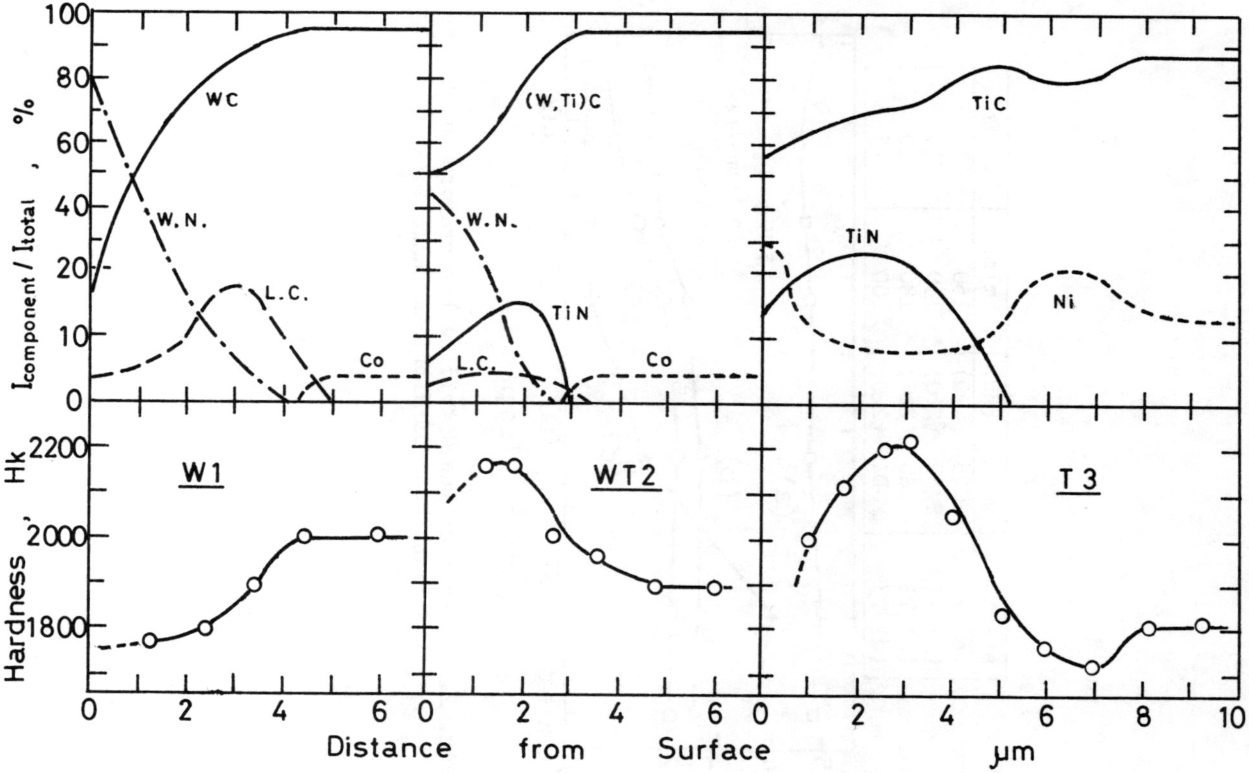

Fig. 1. Changes in the Intensity of X-Ray Diffraction Lines and Hardness with the Distance from the Surface.

specimen	work (HB)	v, m/min	t, min
(a) W1	FC25 (190)	150	10
(b) WT2	S48C (260)	150	10
(c) T3	S48C (260)	180	10
(d) T3	FCD40 (220)	100	5

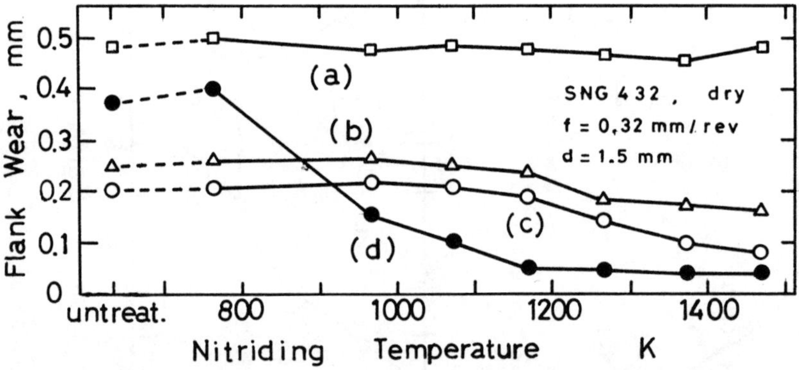

Fig. 2. Effects of the Nitriding Temperature on the Flank Wear of Nitriding Alloys.

specimen	work (HB)	w, mm	v, m/min
W1	FCD55 (240)	50	150
WT2	S48C (280)	80	100
T3	S55C (240)	80	100

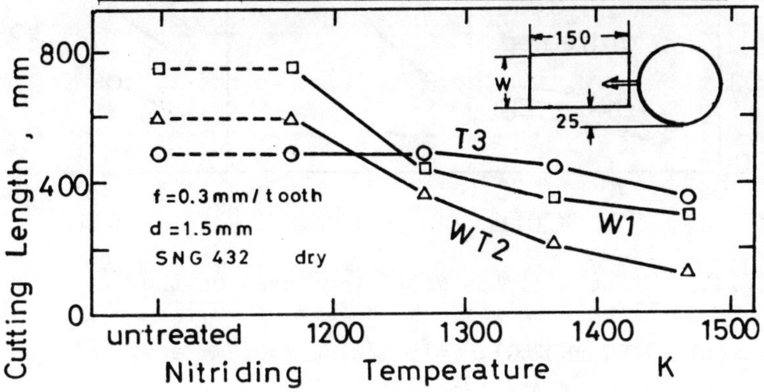

Fig. 3. Effects of the Nitriding Temperature on the Chipping Resistance of Nitriding Alloys.

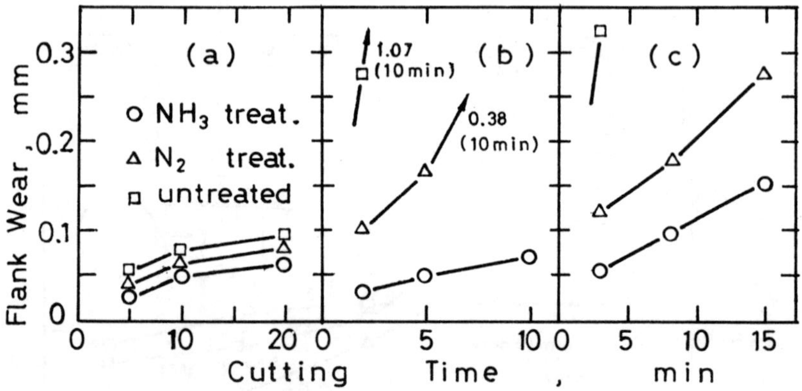

Fig. 4. Effects of the Nitriding Gases on the Flank Wear of T3 Nitrified at 1273K.

(a) S48C,(HB260), v=150m/min, f=0.3mm/rev, d=1.5mm, SNG432, dry turning.
(b) S48C(HB260), v=30m/min, f=0.1mm/rev, d=1.5mm, SNG432, wet turning (emulsion).
(c) FCD55(HB220), v=100m/min, f=0.3mm/rev, d=1.5mm, SNG432, dry turning.

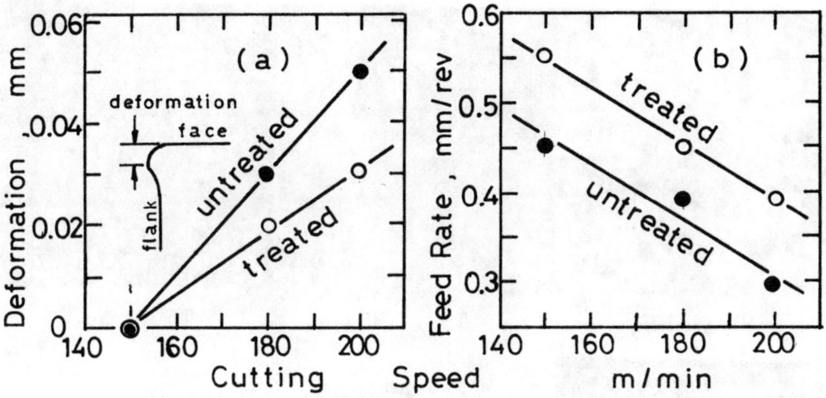

Fig. 5. Deformation of Cutting Nose in Dry Turning of Steel at Various Cutting Speeds and Feed Rates.

SNCM8(HB 200), d=1.5mm, t=3min, SNG432.

(a) Effect of Cutting Speed. f=0.39mm/rev.

(b) Relation between Cutting Speed and Feed Rate Where 0.03mm of the Deformation is Found.

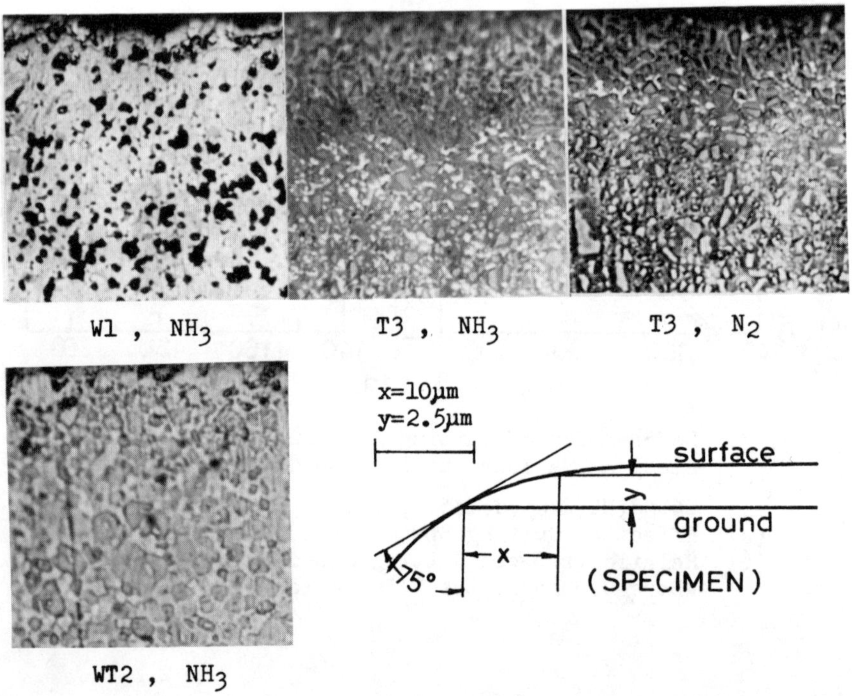

Photo. 1. Optical Microstructure of the Nitriding Layer Formed at 1373K.

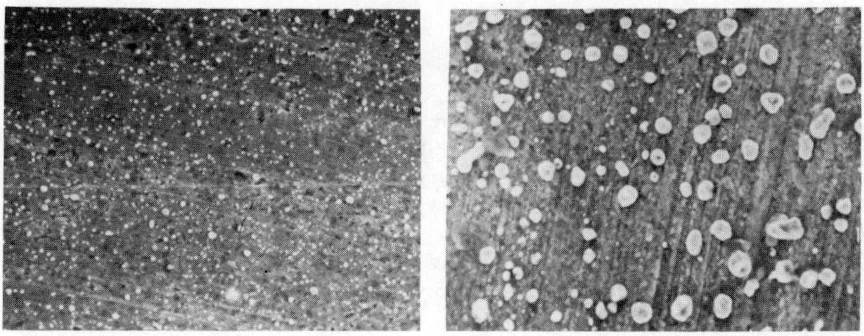

1273K 1473K

10μm

Photo. 2. Ni Droplets Formed on the Nitrified Surface of T3.

DISCUSSION

T.E. Hale (General Electric Co.-Carboloy)

What was the nitriding gas pressure?

Dr. Takatsu

About 1 atmosphere.

G.F. Wilson (General Electric SMBD)

When machining nodular cast iron was there any tendency for built-up-edge?

Dr. Takatsu

No b.u.e. was observed for the cutting conditions of this experiment. TiN will reduce adhesion of workpiece to the cutting edge, but Ni dropplets on the surface of the nitrided TiC cermet might increase adhesion. This should be studied in the future.

Dr. V.K. Sarin (GTE Labs)

What is the depth of the nitride coating?

Dr. Takatsu

In this case the nitriding temperature was 1373K and the time was 3 hours. The resulting thickness was 3 to 4 microns, and is very uniform.

Dhiraj Shah (Valeron Corp.)

Have you determined the ratio of C/N solid solution layer formed by nitriding? If so, what ratio is best?

Dr. Takatsu

The ratio has not yet been determined.

SESSION IV: PANEL DISCUSSION

QUALITY CONTROL AND APPLICATION OF CUTTING TOOL MATERIALS

PANEL DISCUSSION

QUALITY CONTROL AND APPLICATION OF CUTTING TOOL MATERIALS

Moderator - Dr. V. Tipnis:

This panel will discuss the problems and opportunities in the area of quality control and application of cutting tools. As you know, the cutting tool is the basic item in Metal Removal Technology. When a new cutting tool material is introduced additional requirements are placed on machine tool rigidity and horsepower, on workpiece materials and on part design.

As we enter the 1980's we have a number of opportunities to increase productivity thru the application of new cutting tool materials. There are two problems we face: How are we to choose the proper tool for a particular application? How do we maintain the quality of that tool so as to obtain uniform perfromance from lot to lot, batch to batch.

Our panel will address these two problems and then you will have a chance to ask questions of them. Our first speaker is Mr. David McCray, director of staff Engineering, Kennametal Inc.

Dr. McCray (Kennametal Inc.)

The application of and quality control in carbides had traditionally been subjective since many of the standards have been developed through an evolutionary process where the material performance specs and expectations were gradually refined through experience. By this procedure, we have been able to effectively correlate such factors as binder contents and sintering specifications to speeds, interruptions, transverse rupture strength, etc. These and other characteristics make up what we call the quality control and application system for carbide tooling. In the last few years, however, we have seen some major changes that have placed severe demands on this system. I would like to spend more time discussing these changes and how Kennametal and our counterparts are reacting to them.

Raw Materials

The four primary raw materials used in significant amounts in tungsten carbide tooling are cobalt, tantalum, tungsten, and titanium. Two of these, cobalt and tantalum, have received continuing attention regarding their long term availability. Cobalt is, of course, the binder used in most tungsten carbides, and tantalum is a key ingredient in controlling the thermal properties of the cutting material. In general, the tooling industry has reacted to these potential shortages by reducing the tantalum content in grades where ever possible.

Metalworking Industry Expectations

At the same time as we've been contending with these materials challenges, there have been some significant changes in the metalworking industry. Some of these changes are very profound and long term in nature.

Probably the most significant single factor affecting manufacturing technology today is the increased emphasis on productivity (thus automation) in virtually all phases of manufacturing in the U.S., and, for that matter, the world.

Up to five years ago, we all recognized the fact that machine tools in use in American industry were badly in need of updating and were quite often antiquated. This is no longer the case. Virtually every industry has been updating or replacing obsolete equipment with modern machinery, which is capable of challenging the metalworking materials of today.

Because of the high cost per hour of a metal removal operation, there has been a general trend toward higher metal removal rates and efficiencies through increased speeds and feeds. Increasing speeds and feeds is a direction the tooling industry has been pushing for quite a few years, and we now find that we are more frequently being challenged by customers running closer to the 'ragged edge' of tooling capabilities than ever before.

In any high productivity (particularly automated manufacturing) system, the best laid out metal removal plan and system can be devastated by uncontrolled or dangerous chips. Until recently, the traditional way to improve chip control was to crowd the chip by brute force, causing higher power consumption and reduced metal removal efficiency. This approach no longer is acceptable because of the new importance of higher metal removal efficiencies.

Kennametal and our counterparts have observed an acceleration in the use of hard-to-machine materials in metalworking industries. The concern for light weight, for example, in the automotive industry, has caused the emergence of high silicon aluminum as a material to contend with. High temperature alloys, which Mr. Gorsler and many of you at this conference are familiar with, certainly challenge the tool materials of today.

Predictability of performance is really a new dimension which has been added to tooling expectations of our customers. Not only must a tool perform well in a particular application, it must also be predictable. This is particularly important in a hands-off machining situation where an operator is not or cannot be available to continuously observe the operation and be ready to shut down and change the tool when it becomes dull or fails in some manner. This is, I believe, the most challenging of all the items we will be dealing with over the next few years. Predictability or variability as you know, is also a quality assurance or a statistical property.

Tooling Industry Changes and Reactions

I think you will all agree that not only has the metalworking industry changed, the tooling industry has changed along with it. I would like to discuss some of these changes and reactions and how they have impacted our quality assurance and application efforts.

Generally we and our competitors have reacted to the tantalum situation by adjusting our emphasis toward those materials that utilize less tantalum. This has not been without some difficulty, since tantalum is a significant additive in controlling the thermal properties of tungsten carbide cutting tools.

We have introduced new coatings, (aluminum oxide, titanium carbide, titanium nitride, etc.) and ceramics of various forms. These have certainly supported the move toward higher cutting speeds, but generally each of them shows a trend toward increased brittleness. Brittleness, of course, brings out chipping and fracture as the prominent tool failure mechanism, particularly in heavier feeds.

In answer to the demand for an improved chip control, there has been a profileration of new chip control geometries. All of these bumps and dimples are intended to efficiently and predictability control chips. I suspect that many of these geometries will gradually fade out in favor of those that offer more reliable chip control at reasonable machining efficiencies.

Predictability

I would like to talk about why predictability is such a challenge to us and how we plan to answer this situation.

In the past few years, many new coatings and materials have been introduced to answer the challenges I've already mentioned. As you know, the intoduction of new products and processes typically increases, temporarily, the variability in the process or product characteristics. We and our competitors are answering this situation by devoting more attention to optimizing and refining the manufacturing and quality assurance processes than we ever have before, in order to accelerate the learning curve as much as possible.

The raw material situation I discussed before, plus the emergence of new and exciting cutting materials, has shown generally a trend toward more brittle materials which are more susceptable to chipping and breakage than conventional carbides. Since breakage or chipping or fracture is basically a statistical phenomenon, and becomes more dependent upon the statistics of the material and the application as brittleness increases, a general effect of this trend has been to add variance to the material application characteristics. We are answering this challenge by an increased emphasis on the analysis and characterization of the failure of these more brittle materials through a variety of highly sophisticated, analytical, and experimental techniques.

At one time, Taylor's equation (which is based primarily on a wear mechanism) was considered the most sophisticated and accurate means of characterizing a cutting tool's performance. We are finding that this equation no longer adequately characterizes performance of these new material. We're reacting to this by investigating and applying new analytical techniques to characterizing total tooling system performance.

Although, we in the tooling industry have our work cut out for us with these challenges I've mentioned, we're excited about our place in industry's drive for increased productivity and welcome them as opportunities. Thank you.

Frank Gorsler (General Electric Co. - Aircraft Engine Group)

I'd like to take about 5 minutes to discuss Q.C. and application concerns for us as aircraft engine manufacturers. As for Q.C., when machining parts worth $10,000 and up, we are not going to take a significant chance of breaking an insert. We throw it away long before its worn out, and hence see little variability. We are not going to continue to do that as tool materials become more expensive and computer-aided planning systems acquire the capability of more accurate life prediction. Then we will be looking for a greater degree of predictability.

To illustrate our current needs, consider this graph of turning speeds of various engine alloys. As newer materials are developed, the trend is to zero productivity. We're fighting this trend thru new tool materials. Not too many years ago we were involved in a major program to apply CBN to turning of nickle base alloys. This has the capability of a 5-10 fold increase in metal removal rate. We still use some of it but its cost is considerable.

This next slide summarizes the major workpiece and tool materials in use today. It gives the cost per pound of removal of four alloys with several applicable tool materials. For example Rene'95 may cost us over $3/pound to remove with carbide. However, because it machines so slow, one man can run two NC machines and labor cost can be shared. The introduction of CBN got us down to around $1/pound but because of the high speed, required 2 operators for 2 machines, cutting into the saving. Hot pressed ceramics, which yield almost the same productivity, but have a much lower insert cost, have cut this in half. Ceramics, however, are not the answer to everything. They do not cut titanium effectively and their still relatively low strength does limit application to round inserts. It is expected that future ceramics will have the strength to be used in other geometries. In the iron-base alloy area (Marage 250, A286, M152 etc.) we don't see a lot of cost advantage to coated carbides. Tests show that slight improvements in productivity are eaten up by higher insert cost. The ceramics may offer some potential if we could run our machines that fast. Some day we will have that capability with totally enclosed automatic machines.

I hope the above summary will provide a brief sketch of our status in applying advanced cutting tool materials and formulate a basis for discussion latter in the session.

Sven Ekemar (Sandvik AB, Stockholm Sweden)

The development of cemented carbide grades is based on the needs of the users. The first part is to interprete the demands the tool material has to satisfy. Then the researcher in his work will make use of all the possibilities and resources available as to raw materials, process materials, established processes, and production equipment. Performance tests will sort out the final grade. The transfer from laboratory scale to production is the final part. In this connection the quality control has to be fixed.

Its purpose is to secure the materials properties to be in accordance with the application demands. Thus the quality control utilises a number of control parameters.

The control methods do not reveal the true state of a cemented carbide. This is overcome by monitoring the quality by means of mastering raw materials and process materials, of using standardized equipment, and standardized processes and of applying a quality control system.

The bluntness of the control methods can be illustrated. Coercivity and magnetic saturation measurements are most sensitive to minor iron content variations within cemented carbides. Such variations have a negligible influence on the performance properties.

Hardness is measured with the "Rockwell A" method in the USA and with the Vickers method in Europe. There is no simple relationship between Rockwell hardness values and Vickers hardness values.

A microstructure is an artificial two-dimensional picture of a three dimensional reality. The artificialness is the consequence of the metallographic preparation procedures and of the limitations of the microscopes. The interpretation of metallographic observations contains subjective elements.

Quality control data on cemented carbide products give a consistency information. If they are used standing by themselves for the assessment of cutting performance properties, the risks of misinterpretations are high.

Quality control on cemented carbides concern room temperature properties. Metal cutting means heat generation. The force, temperature and friction conditions are determined by an interplay between the operation, the cutting data, the edge geometry, the workpiece, the workpiece material, and the tool material. The demand spectra embrace wide ranges of variations. Frequently occurring contact temperatures are about 1000°C (1800°F). The heat flow into the insert causes temperature gradients. The contact pressures are high. When intermittent cutting processes are used thermal and mechanical fatique often will cause edge failures. Occasionally a cutting edge has to meet load shocks, which may be absorbed through plastic deformation. However, a deformed edge may fracture as a combination of reduced strength at elevated temperatures and of enhanced stresses owing to damaged edge geometry.

The wear mechanisms consist of adhesion, abrasion, diffusion, chemical reaction as well as microfracture processes. The contributions of these processes vary from operation to operation as well as within the operations.

When machining a simple carbon steel a common cutting speed is about 150 m/min (450 sfm). The contact time of a passing over the rake face is approximately 1/1000 second. When the chip leaves its temperature may be 1000°C (1800°F). Thus the active part of the chip material in contact consists of a ductile, low carbon, and low strength austenite containing a dispersion of cementite particles. This austenite dissolves easily the constituents cobalt and tungsten carbide.

This led us to develop coated cemented carbide inserts having a sophisticated geometrical and materials design. Optimum performance is achieved by balancing the coating and substrate to each other. The properties of the coating have their emphasis on wear resistance. The substrate is designed to enhance the wear resistance of the coating and to have balanced bulk properties - toughness behaviour, hot hardness etc. A coated cemented carbide has lower transverse rupture strength than an uncoated cemented carbide. However, they show in metal cutting almost the same toughness behaviour. Machining with coated carbides lowers the contact forces as well as the contact temperatures.

In conclusion:

1. Quality control of cemented carbides when used as a built-in part in a quality monitoring system gives a reliable documentation of the state and the consistency of the materials produced.

2. The multitude of deterioration mechanisms determining the lives of cemented carbide edges vary and interact to infinity.

Harold Hansen (The Boeing Co.)

I'm pleased at the emphasis placed on H.S.S. at this conference. It seems most focus on the higher-removal rate materials such as coated carbides etc. To give you some prospective of our business, we are now purchasing H.S.S. end mills at a rate of $6,000,000 per year which is 6 times the dollar volume of carbide tools.

My discussion will concern the evaluation of H.S.S. end mills. It will include justification for end mill testing, methods in use and Boeing's procedures. Our purchasing people tell me that the cheapest end mills run about 70% of the most expensive. This would imply that there is a variation in quality. The using shops noted a significant difference in performance. Since purchasing would normally buy on price it became the responsibility of the shops to provide backup data to justify the tool with higher initial cost. So we must determine the total cost of using the end mills.

There are 2 basic methods for evaluating end mills, utilizing them on long-run production jobs, and subjecting them to simulated tests, a standardized procedure where all manufacturers tools are subjected to the same cutting conditions. Boeing uses this latter method due to its low volume production. The parts are complex, but with a production of (for example) 5 airplanes per month, we don't make many of them. Another factor is that most of our machining is done on NC profile mills where operators tend to over ride the tape, which is justified in certain cases due to low part volume. This, then, is another reason for a standardized test.

From this chart you can see that about 20% of our end mills are used on aluminum. Low alloy steels use the greatest number although we also mill stainless steels and Titanium. Obviously more testing will be done on those materials where tool usage is greatest. We do little life testing on Aluminum due to the long lives and volume of material that would be required. We run tests with two sizes of end mills. One, a 1" dia, 2" flute length, four flute cutter is used for 10-12 separate tests. We then run the second size, 2" dia., 4" flute length, 6 flutes. It is possible that some manufactures make their larger end mills out of different material. We evaluate performance by measuring wearland. In addition we set a chipping factor which is sometimes the life limitation. We also run a breakage test which consists in measuring the load required with a rotating tool. Some end mills were found to break at very low loads.

This chart shows the scatter in tools from various manufacturers, under identical cutting conditions. If the best tool is rated 100%, some tools are in the lower 40's. We've run three series of tests, beginning in 1962 and are in the process of completing a fourth. In all three the scatter ranged from about 60 to 100%.

This chart shows the consistency of performance at 3 manufacturers over the 3 time periods of the tests. The first manufacturer had consistantly good performance in all 3 periods, for both H.S.S. and cobalt tools. The second also was consistant but the cobalt tools were at a much lower rate of performance. The H.S.S. tools from that manufacturer would qualify although the cobalt would not. The third example is of a much lower performing supplier and he would have been rejected all 3 times. Each year we conduct the tests we are more sure of the validity of the results.

We have considered writing a specification to assure performance, but our data suggest that it would not be practical. On this chart we evaluated both cutting tool material and radial rake. With all else constant M-42 varied from 95% down to 30%. M-7 ranged from 100% down to 46%. Again, ordering M-7 would not assure performance. M-2 ranged from 64% down to 30%. Radial rake also varied with manufacturer. When comparing 12° radial rake, performance ranged from 90% - 37%. At 20° it ranged from 89-46%. Therefore, we could not relate performance to radial rake.

The bottom line is cost saving which is made up of procurement costs, resharpening and shop life. In 1976 the estimated end mill procurement cost savings resulting from these evaluations was 32%. Applying that rate to the current buy of $6 million is a sizable saving.

To summarize, we would like to buy performance. We don't care what the tool is made of, or what some of its configurations are, as long as it will cut. We feel we cannot obtain this performance thru a specification, and must have some type of evaluation procedure.

Arthur Bahmiller (Cleveland Twist Drill Co.)

My company is a producer of rotary H.S.S. cutting tools including drills, end mills, reamers, taps etc., with emphasis on the first two of these.

One major variable which can effect the perfromance of cutting tools is imput raw material quality. We sample lots of H.S.S. coming into our plant and rate it according to various characteristics. This data is accumulated annually and supplied to purchasing as a guide.

As far as material grades, the major non-cobalt is M-7. Certain taps and drills are made of M-1 where more toughness is required. Most cobalt tools are M-33 although M-46 is used for some single point cutting tools. One of the important performance-related characteristics of a H.S.S. is it's ability to respond to heat treatment. We compare materials in a standard heat treatment for R_c hardness and fine grain size. For example, M-7 heat treated at 2200°F usually yields a hardness of 66+ and grain size of 12-16. One of the problems we've seen is the scale-up from small to large ingots, which has lowered this response. The bigger ingots tend to have larger carbides which result in lower hardenability. This has been improved by modifying the original grades, raising carbon. We've also developed ways of "balacing" our grades. Increasing Carbon and Vanadium in the high Moly H.S.S. presents a problem in obtaining normal carbide structure. M-7 yields M_2C carbide formation if you're not in good balance. This may be balanced with Tungsten, Silicon and Nitrogen. In the case of M-7, we've increased carbon, lowered Vanadium, and increased the levels of Silicon and Nitrogen.

Another area that is a struggle is the grinding of H.S.S. tools. The metallurgist has a responsibility of helping to obtain burn-free tools. We now do a lot more grinding as this yields a more accurate tool and is in some cases a cost reduction over milling. This also increases the possbility of grinding damage and processes must be audited and carefully controlled to prevent it.

Professor Tetsutaro Hoshi (Toyohashi University of Technology)

First I'd like to discuss a new carbide-tipped twist drill* marketed in Japan and a report from industry on its application. The use of HSS drills often becomes a production bottle neck. The common problems in drilling steels with carbide are edge fracture and poor hole accuracy. The first slide shows the design of a new carbide-tipped drill. It is provided with a special edge shape and negative land along its edge. Those two items are key in preventing edge breakdown. The two carbide brazed tips are of a special grade, the only one available on the market suitable for this application. The manufacturer of the carbide claims the high toughness is obtained by using higher Titanium content. It is a stock grade, but produced by a special process. It is said that this grade is unrivaled in its resistance to low-temperature fracture.

The drill geometry is designed for hole accuracy, and the flute design promotes smooth ejection of the chip. As a result, feeds of 5 to 10 times that of H.S. are possible with superior accuracy.

One of the earlier applications of this new drill has been reported to this conference by Mr. Diamon (Mr. Diamon's paper is reprinted in these proceedings, but was summarized in Panel by Professor Hoshi, who goes on to describe several production applications. The reader is referred to Mr. Diamon's paper for details-Editor).

Dr. J. Hazra (General Electric - Carboloy)

The carbide industry, in general, has made gigantic developments, starting with the brazed tool. The molded chipbreaker concept, coated tool materials and now generations of new coated tool materials.

Requirements have become very strict on the cutting tool performance from the stand point of the type of material to use, the type of chip control geometry to use, what type of tool systems to use and what type of predictability or tracking system one can generate.

* The drill to which Professor Hoshi is referring is described in detail in the paper by Dr. Horiuchi - Editor.

There a number of important parameters in the material approach to performance. The first is coatings capable of withstanding the high speeds dictated by productivity goals and made possible by newer machine tools with greater rigidity and H.P. The next consideration is the substrate to support the coatings. Thirdly, the insert geometry plays a key role. This includes the edge prepration and chip groove geometry required to meet the challenges of the work material and machine tool. Considering these three parameters, the challenge becomes one of selecting systems for specific applications. The Carboloy approach has been to use TiC and Al_2O_3 coatings with specific insert geometry combinations to meet challenges in the Automotive industry associated with the new Aluminum/Silicon alloys and in the Aircraft industry with it's difficult-to-machine Titanium and nickel-base alloys.

We have identified eight geometries to satisfy the various machining requirements. These are compatible with the two coatings and the substrates used. Because the systems become very complicated, quality control becomes critical. Performance, therefore, depends not only on the material/geometry but on the total engineered system, including Q.C.

Howard Kline (Caterpillar Tractor Co.)

Caterpillar Tractor Company is a major producer of heavy earth moving equipment, based in Illinois, with plants throughout the World. We are also in the diesel engine manufacturing business.

The latest cutting tool material applied at this company is aluminum-oxide coated carbide. It is used for turning, boring and facing cast iron. This slide shows a rough and finish part. Casting break out around holes is not acceptable. We're roughing it at about 900 ft/min. with depths of 1/4 to 3/8 inches. Finish cuts are made at 1200 ft/min. The water seen is not used to cool the tool, but only to wash the chips away. As a comparison of tool life at 1200 ft/min., the next slide shows that a better TiC coated carbide lasts a little over one minute. The "first generation" of Aluminum-Oxide coated carbides ran about 4 minutes while the second generation, about 11 minutes. This progress in tool materials was accomplished in the last 2-3 years.

Higher cutting speeds make part holding in rotating chucks more critical. In one case (slide shown) we lost 25%

of the chucking force at the required R.P.M. In others it can run as high as 70%. If speeds are going to increase, someone must build a better chuck.

As speeds increase chip disposal becomes more of a problem, particularly on large parts that must be mounted on vertical machines. This can be helped with chip deflectors and external guarding.

We have a few ceramic tool applications. Operator education and more extensive machinability data are required before they can be fully utilized, however. The next slide shows a good ceramic application. The tooling is stubby and internal chucking means speed is no problem. Roughing is done at 2000 ft/min. and finishing in excess of 3000. We used a pentagon insert for increased edge strength, and the wear is hardly discernable.

Our needs are not only for dependable tools but for better castings. Some of the main problems in transfer lines today are due to castings with sand and other inclusions which make tool life unpredictable. Caterpillar is interested in Adaptive Control due to the large amount of stock removed and its dimensional variability. Another item on our shopping list is improvement in the design of anti-centrifugal chucks. Possibly some holding device other than a chuck will evolve thru the tooling or machine tool manufacturers. We are seeking automatic tool changing devices for application on transfer lines, as non-cutting time becomes a larger percentage of the total. Hole manufacture including drilling, boring, etc., is one of our costliest manufacturing operations and we're looking for improvements in that area. We will be looking at machining after heat treat. The new ceramic tools should be capable of turning parts over 50 R_c, which are now ground. In general, we are looking for cleaner, quieter, less physically demanding machining operations in accordance with OSHA and good common sense. We want to improve our heavy milling rates for cast iron and steel into the 200-300 ipm range. We feel that with the help of tool producers it can be accomplished.

PANEL QUESTIONS AND ANSWERS

Dr. K.M. Kulkarni (Consolidated Metallurgical Industries)

Mr. Hansen, did you or your suppliers analyize the spread in life of end mills? Was there any correlation between price and performance?

Panelist: Mr. Harold Hansen.

The more expensive tools do seem to be the better performers. Calculations of "using cost" reveal that initial tool cost is not nearly as significant as the performance. We did very little exploratory work to determine why a particular tool did not perform. We did some evaluation of hardness and microstructure but no in-depth studies. Our main concern is to buy performance.

Dr. Norihiko Narutaki (Hiroshima University)

Dr. Ekemar, how did you measure the cutting temperature of the coated tools, and how accurate was the method?

Panelist: Dr. Sven Ekemar

I'm sorry I do not have the details as the work was done years ago by another researcher.

Dr. Vijay Tipnis (Tipnis Associates)

Let me ask a question of the various cutting tool producers. How do you control the quality of your coated carbide tools?

Panelist: Mr. Dave McCray

The Q.C. of the substrate is a fairly straight forward process involving control of properties mentioned by Mr. Ekemar. Referring to the coating, our internal processes are quite proprietary at this time. Each of us probably uses different methods for coating and the interface between substrate and coating is something we are careful not to talk about. For the final coating we control thickness by one of several methods, and the structure of the coating. We direct our attention to impurities or inconsistancies in the coating.

Panelist: Dr. J. Hazra

I'm not at liberty to discuss our Q.C. procedures, but I can tell you that, because of the complexity of the system, the Q.C. that is employed is much more complicated than just looking at the substrate itself. Obviously we have our set techniques for doing it and we evaluate our product to meet them. Other than that, I cannot comment.

Professor R.F. Bunshah (UCLA)

Is the scatter in performance of the coated tools less than that of the uncoated tools?

Panelist: Mr. Howard Kline

There is considerable difference in some products, particularly if it has just been introduced. Others are very consistant. Now the Aluminum-Oxide coated tools are quite consistant and we don't have any problems with them.

Mr. Gregory Bouse (General Electric - M&P Lab)

With the availability of Tungsten questionable, it would seem advantageous to introduce more coated H.S.S. tools ie. TiC, TiN, Al_2O_3. What is being done in this area?

Panelist: Arthur Bahmiller

We do make some throw-aways, but at the present they are not coated. However, as a result of stimulation by this conference, the idea will be pursued further.

Mr. Yoshihiko Doi (Sumitomo Electric Industries Ltd.)

We have produced a few coated H.S.S. inserts, but unfortunately we have not found a substantial market.

Moderator: Dr. Vijay Tipnis

How about form tools for screw machines, broaches and the like?

Mr. Yoshihiko Doi (Sumitomo Electric Industries Ltd.)

Yes, we find a market for these expensive tools, but the

coating process is also costly.

Mr. Ronald Mills (Tungsten Carbide Manufacturing)

Mr. Kline, addressing your comment about the need for holding mechanisms compatible with higher speeds, what is the position of the chuck manufacturers on this? Why should the cutting tool business be involved with it?

Panelist: Howard Kline

Our experience is that the chuck manufacturers are settling on the counter-centrifical design as being the answer to the problem. If you are not familiar with it, an internal weight on a lever is supposed to stabilize the jaw force. Unfortunately it's only designed for a certain diameter and weight of jaw. Our operations require versatility due to casting sizes and non-round surfaces. Therefore, we consider the counter-centrifical chuck a stop-gap which will have to be improved upon if even higher-speed tool materials are to be utilized. If the tool manufacturers want to market these products they will have to come up with something to hold the parts with.

Panelist: J. Hazra

I'd like to ask a machine tool builder if he could throw some light on this area.

Mr. Mike Schmenk (Cincinnati Milacron)

Not really, because we buy our chucks. Several years ago when employed at the GM Technical Center, when we saw that 2000 ft/min. was going to be a practical cutting speed, we recognized the chuck problem. People have been struggling with it ever since and there have been countless designs, some of which have been successful on a development basis. I think this is a major problem, hence a major opportunity, one which, I think, the chuck manufacturers haven't realized. I think what Howard is saying is that if they don't want the business, how about someone else picking it up because there appears to be a strong need.

Another area being addressed at Milacron is Ultra High Speed Machining. Even machining aluminum within the current capabilities of our machine tools (3500-4000 RPM) you have extreme problems with safty. Chucks tend to fly apart, an in the case of machining centers, rotating tools must be dynamically balanced. Our position as a machine tool builder is "hold on, we are going to take a very conservative look because we don't want to injure anybody".

I have two questions. On Howard's wish list he has optimization of the cutting tool and machine. Do you want some scheme of picking out the right cutting tool, or do you want to take the existing cutting tool and do more with it? Similarly with the machine tool. Are you looking for better machine tools, or do you want to apply it better? My second question, to the cutting tool manufacturers, is what are you doing to increase strength while maintaining some ductility?

Panelist: Howard Kline

I was referring to the optimization of new cutting tool materials on existing machine tools, although as we get improved machine tools we will look for further increases in productivity.

Panelist: J. Hazra

In answer to Mr. Schmenk's question on tool material properties, we are trying to balance two sets of properties which are at either end of the spectrum (toughness vs. hardness-at-temperature). One or the other can be achieved with a basic material. That is why the coating approach has been successful, promising the best of both worlds. Machining applications have shifted because of the demands on the machine tool and the way the applications are tailored. Consequently the kinds of materials we are looking into are coated systems where the coating and substrate play an important role. We are approaching the goal of that unique conbination of properties thru the systems approach.

Panelist: David McCray

In answer to the question "what are we doing" I would like to point out recent developments in ceramics. Each material is an evolutionary process. The coatings are one manifistation of hybridizing, of combining properties. That is apparantly the way we are going to go from here on. We've reached some of the basic limits of, for example, aluminum oxide. We can't go much faster with that, although we can increase its strength. Strength is probably limited by

500,000 T.R. I can't help thinking it's going to be an evolutionary process. Perhaps one of us here will come out with the next advance.

Dr. Kishor Kulkarni (Consolidated Metallurgical Ind.)

Mr. McCray was just commenting about evolutionary changes. Would any of the panel members comment on revolutionary new techniques like laser machining which could have a major effect on machining technology as we know it now?

Panelist: Frank Gorsler

I expect that Laser-Assisted Machining is quite a way into the future. The area in which I'm involved is high tempterature alloys, an area of high potential payoff. There is an investigation underway utilizing ceramic tools. With carbide tools the addition of external heat seems to shorten tool life. In theory, with the proper thermal gradient, this can be overcome. I do see potential, but hesitate to guess how many years it will take to solve the technical and economic problems. Certainly when that happens, new tool material systems will be required.

Panel Moderator: Dr. Vijay Tipnis

I'd like to make a few comments on this since I am involved in evaluating the economics of Laser-Assisted Machining. It is too early to make conclusions about the economic feasability of LAM. We have evaluated large 15 KW lasers, but there are other types such as pulse lasers which may turn out to be economically feasable. We must first make this process technically feasable. Perhaps in a year and a half we'll be in a better position to make a judgement.

Mr. Mike Schmenk (Cincinnati Milacron)

I, too, would like to make one comment on that. Years ago we looked at LAM. The CO_2 laser was only effectively imputting about 10% of its energy to the workpiece. Even through available in massive quantity, it is not well absorbed. However at the recent machine tool show AVCO-EVERETT announced a new process for dramatically increasing the beam absorption without using coatings on the part. There could be a breakthru in this area.

Panel Moderator: Dr. Vijay Tipnis

To pursue the question which Frank Gorsler asked, is there any new cutting tool material on the horizon for increasing the speed at which Titanium can be machined?

Panelist: David McCray

Most of the materials we're using now in metal cutting have an affinity for Titanium. It is certainly a significant problem that I'm sure all of us here are working on. I suspect that, due to the potential business we will come up with something, which may not be just a material, but a material/tooling system. It has to be done more efficiently than it is now.

Panelist: Dr. Jay Hazra

I cannot help but feel that this development will not be overnight. Obviously considerable resources are being spent and my personal opinion is that it will have to come via a materials/geometry system rather than just a material development.

Panel Moderator: Dr. Vijay Tipnis

On this note I think we will end our panel discussion. Here is a challenge for all the researchers in the audience - come up with a better cutting tool material for Titanium. Thank you.

SESSION V

DIAMOND AND CBN TOOL MATERIALS

ASPECTS OF MACHINABILITY USING POLYCRYSTALLINE DIAMOND AND CBN COMPACT TOOLING

S.K. Bhattacharyya
D.K. Aspinwall
University of Birmingham, U.K.

SUMMARY

Extensive industrial studies undertaken on high volume Silicon Aluminium automotive components highlight the many advantages possible when using polycrystalline diamond tooling as an alternative to Tungsten Carbide, for both single point and milling applications. Improvements in machine utilization and component quality coupled with reduced tool costs and fewer machine operations are characteristic of this type of tooling.

Increasing cutting speed when machining Incoloy 901, a nickel based superalloy with CBN compact tooling is shown to significantly reduce depth of cut notch or groove wear. Cutting speeds of 244-305m/min are possible compared with 24-30m/min using premium grades of carbide, effectively providing a 10 fold increase in metal removal rate. In addition, comparisons with cold pressed and sintered Alumina inserts show the CBN tool to have a longer and more predictable life and the facility for multi stop/start cycling.

INTRODUCTION

As with cemented carbide and the more conventional ceramic tool materials, the range of polycrystalline compacts based on diamond and cubic boron nitride (CBN) has increased dramatically since their commercial introduction in the early 1970's. Current tooling has a cosmopolitan air with products from major manufacturers in the USA, Japan, South Africa and the USSR. Both synthetic and natural aggregates are used with a number of manufacturers offering an infrastructure of blank/insert shapes and sizes of various grades, based on grain size.

The development of synthetic polycrystalline agglomerates, notably diamond, formed as a natural progression from single stone synthesis, the technique evolved in the mid 1950's. The conversion of graphite to diamond and hexagonal boron nitride to the harder more durable cubic form, requires extremely high temperatures and pressures, in the region of 2000K and 115Kbar,(1). The exact transition is governed by the presence of additional catalytic materials which act to lower the combined pressure/temperature transition ceiling and are, therefore, a commercial necessity, (2).

Both the diamond and CBN compacts, used for the machining tests, are outwardly similar, comprising a dense layer approximately 0.5mm thick of either synthetic polycrystalline diamond or CBN, bonded to a stiff cemented carbide substrate. The blanks or inserts have, therefore, a laminate appearance, the base carbide layer serving as a dual purpose support and toughening agent, while in addition, providing flexibility in both handling and fabrication. Structurally the compacts rely on a combination of intermediate and direct bonding and are more akin to the naturally forming polycrystalline materials such as Carbonado, than the early bonded aggregates where both strength and toughness were functions of the metal bond, (3). In line with similar products, diamond grain size is quoted as being in the region of 20-100μm in diameter. The CBN layer, however, has a somewhat finer crystal structure with a particle size of between 3-5μm.

Application Areas

Like natural diamond, the synthetic diamond compact is susceptible to temperature and it is this feature that to a large extent limits the field of application to non-ferrous/non metallic materials. In contrast, the CBN tool offers the capability to operate with tool interface temperatures of up to 1200°C with minimal oxidation effect (4). The retention of edge strength at high temperature is a primary requisite for the machining of nickel based super alloys and related Cobalt and Iron based materials, it is in this area that the CBN tool finds major application. As a general rule, with exceptions for super alloys, materials having a hardness greater than

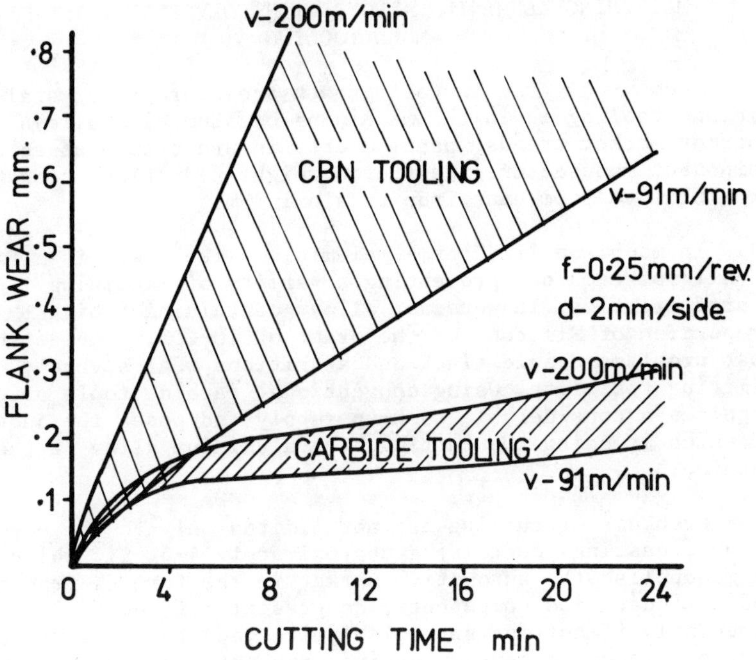

Figure 1. Flank Wear to Cutting Time for CBN and Carbide Tools.

55R_c are also suitable candidates for CBN tooling. Hardened tool and die steels together with chilled cast iron, fall into this category.

While it is technically feasible in most cases to machine the more conventional medium carbon and free machining steels with CBN tools, in purely cost terms the exercise tends to be uneconomical. Surprisingly, however, a number of these softer ferrous alloys do present machining problems over and above the cost aspect. Figure 1 shows flank wear results for a CBN tool and a steel cutting grade of carbide. The work material is a 3.25% Ni, 1% Cr steel used for gear and camshaft applications. Whether cutting wet or dry, the CBN tool shows a rate of wear appreciably higher than that of the carbide tool.

MACHINING ALUMINIUM ALLOYS WITH POLYCRYSTALLINE DIAMOND TOOLING

Undoubtedly the largest growth area for polycrystalline diamond tooling is in the machining of aluminium alloys. Current market trends, particularly in areas such as automotive component manufacture, centre on weight reduction, hence the growing move from cast iron to aluminium.

In machinability terms, aluminium like steel is a family of alloys, each one presenting a variety of machining characteristics. In general, alloys containing a high proportion of Silicon, of the order of 10-20%, provide the most problems, rapid flank and attritious wear severely limiting tool life. Using conventional carbide tools often requires a considerable back-up supply and poses the question of which is doing the cutting, the aluminium alloy or the carbide.

Problems of cutting are not limited only to the eutectic alloys, castings containing approximately 4-6% Si, which tend to monopolise the automotive market in the form of transmission and cylinder head components, do present difficulties especially if cutting speed is limited and the quality of castings inconsistent. Excessive work material build-up on the rake face of the cutting tool is often a major feature, its effect on geometrical size and component surface finish having a far greater consequence than the abrasive effects of Silicon.

Industrial Machining

Evaluation studies using polycrystalline diamond tooling were performed under actual production conditions in a number of U.K. Midlands Manufacturing Companies. Recorded observations relate to volume production of aluminium automotive components, notably transmission equipment, i.e. flywheel housings. Mention is also made of current trials on an aluminium cylinder head and cam cover.

Initial tooling studies were confined to a limited number of operations on the component shown in Figure 2 and used extensively on the Mini. The housing was selected as a consequence of the variety of tooling configurations, throughput level and the relatively poor condition of many of the machines on which it is manufactured. Actual production of several thousand components per week is achieved through a double shift and two working lines, incorporating a variety of

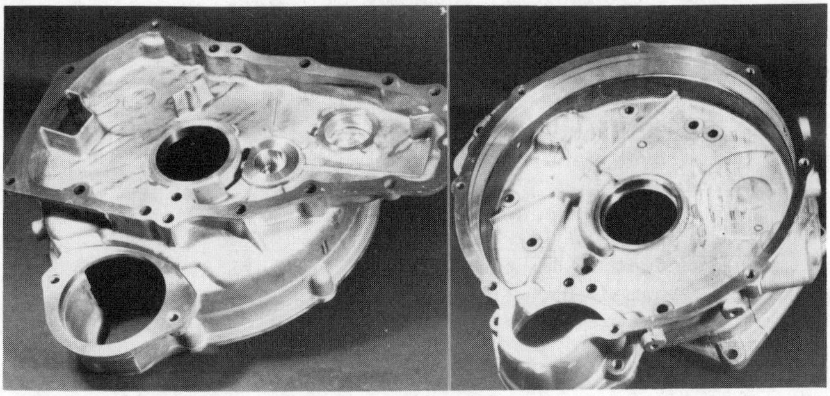

Figure 2. Aluminium Flywheel Housing.

vertical rotary and horizontal machines.

Experimental Results

Results from feasibility tests were encouraging and enabled a full scale changeover to polycrystalline diamond for both single and multi point cutting operations (5). In all twenty six tools were installed in addition to an eight tooth fixed blade milling cutter.

At the outset of the exercise, it was anticipated that not all the diamond tools would be justified on a permanent basis and that carbide would prove more economical for applications such as chamferring, where metal removal rate was low, surface finish unimportant and existing tool life acceptable. For operations involving fine boring, facing and milling, the reverse was true, the carbide tools requiring to be changed on average every 600-800 components, as a result of built up edge and flank wear.

With a large proportion of the carbide tools in shank rather than in insert form, tool change was slow, a factor that when allied with the tool wear problems, resulted in low machine utilization and high overhead costs. Table 1 summarises the main results from the changeover.

Table 1. Summary of Results

- Reduction in direct tool costs.
- Improved tool life and reduced setting time (60,000-80,000 components/edge compared with 600-800 for carbide).
- Improved machine utilization and increased output.
- Improved component quality and a reduction in scrap rate.
- A reduction in the number of operations performed.
- Simplification of tool stock holding.

In addition to the more obvious advantages relating to tool life, a further factor was the reduction in the total number of machining operations. Whereas with carbide tools, operations such as roughing, semi-finishing and finishing were performed separately, using polycrystalline diamond enabled operations to be combined.

Figure 3 shows details of a fixed blade milling cutter used on the flywheel housing to machine the starter motor face. The head incorporates eight blades each employing either a 90° or 180° diamond segment. Installed about thirty months ago, the cutter has yet to be reground and has currently machined

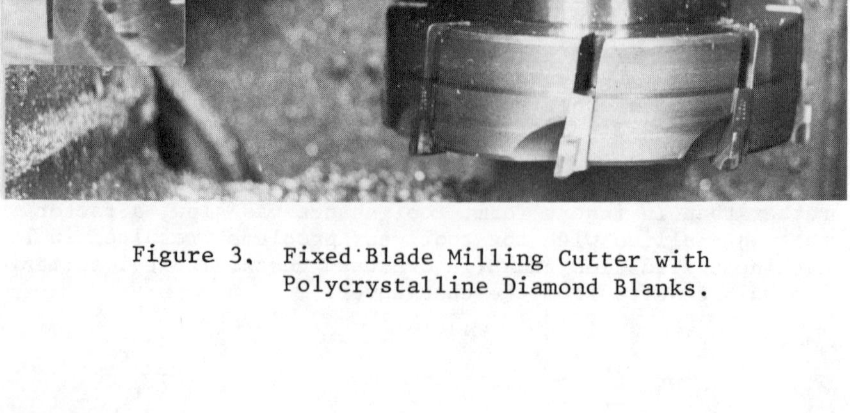

Figure 3. Fixed Blade Milling Cutter with Polycrystalline Diamond Blanks.

500,000 components (similar carbide milling heads are currently reground every 1000-1500 components). Surface finish produced by the diamond blades has varied between Ra 0.42μm and 1μm, depth of cut being in the region of 9mm. In addition to its normal cutting load, the head has been through a number of machine malfunctions, including cutting on fast traverse and machining unclamped components.

Tool Design. One of the chief advantages of using brazable polycrystalline segments is that the existing tool configurations, whether insert or shank, can be maintained. The concept is simple and provides minimal shop floor upheavel. It does however, pose the question of which is the more suitable parent, the shank or the insert. As tool life tends to be measured in tens of thousands of components rather than just hundreds, the argument in favour of using indexable inserts in order to limit setting time, is no longer valid, at least as far as the majority of single point tools are concerned (milling presents a slightly different picture).

The shank format offers a number of benefits in that it provides a more resilient mounting for the blank and enables a greater number of regrinds to be made, identification with the term "throw away" is also avoided. Conversely, repeat use of the insert combination in the same tool holder necessitates either transfer of the worn blank to a new insert after regrind, or modification of the existing one. Neither procedure is ideal, indeed repeat brazing of the blank increases the chance of thermal damage. Figure 4 shows details of a variety of shank tools positioned in a multi tool boring head.

Figure 4. Boring Head with Polycrystalline Diamond Shank Tools.

The fact that rake face build up was not a significant problem, even with cutting speeds as low as 41m/min, greatly simplified the design function and backed up the original philosophy of designing primarily for edge strength. Where possible a standard +7° top rake and 7° side relief was used. For small diameter boring applications, where additional side relief was required, the use of secondary angles or indeed above centre cutting provided acceptable results.

Design considerations relative to chip breaking are currently under investigation on cylinder head castings, a number of "clamp on" and "ground in" designs having so far been tested. Though simple to fabricate, the "clamp on" chip breaker has a number of drawbacks, not least in that swarf has a habit of eroding the clamp or indeed, lodging between it and the tool face.

To safeguard against problems of quality, tool drawings detailed specific requirements both with regard to edge finish, blank type and manufacturer etc. and general tool construction. Tool inspection procedures were also implemented. While edge finish is often a major consideration in determining whether to accept or reject a particular tool, factors such as blank size and shape, together with general tool construction can be of equal importance, particularly if initial cost estimates are based on a specific number of regrinds.

Training. Without overemphasising the fact that the tools were made of diamond, for obvious reasons, damage through chipping did occur during both handling and setting, even after a cautionary note. Application usage placed the tools firmly in the sphere occupied by carbide as opposed to natural diamond, with the result that setting and handling procedures acceptable to carbide and judged valid for the polycrystalline tool, were too robust. Training periods for Supervisory, Setting and Operating Personnel, were therefore planned prior to the introduction of the main batch of tools. The Company were quick to appreciate the importance of these preliminary sessions and eager to co-operate at all levels, so much so that an in-Company training film was produced.

Conclusions

As a cutting tool material, synthetic polycrystalline diamond has a large part to play in the future machining of non ferrous alloys, particularly at high volume production, where automatic and transfer machines predominate and where operator involvement is at a minimum.

MACHINING OF INCOLOY 901

Incoloy 901 is a nickel based, high temperature alloy used extensively in aerospace applications, notably engine discs. Table 2 gives typical composition details. From a machining viewpoint, relative to both carbide and Al_2O_3 ceramic cutting tools, the material is best described as "difficult", poor chip control, relatively low cutting speed and rapid notch or groove formation playing decisive roles. Using premium grades of carbide, cutting speeds for a 20 minute tool life are in the region of 30m/min with a feed rate of 0.25mm/rev.

Although cutting speeds of 244m/min are feasible using cold pressed and sintered Al_2O_3 ceramics, tool wear is not as predictable as when using carbide and tool life is substantially lower, in the region of 3 minutes. Rapid notch growth, particularly at the depth of cut position, is a primary cause of tool failure, re-entry of the cutting edge where grooving has occurred often leading to spontaneous collapse.

Discussion of the mechanisms of groove formation are outlined in (6), (7), (8) and (9). As with flank and crater wear, grooving appears to result from a combination of factors notably the chemical relationship between tool and work material in the presence of oxygen, cutting temperature and the effects of localised stress concentrations. Development of trailing edge or radius grooves, which are similar in appearance to those at the depth of cut position, are generally attributed to feed marks from the previous cut. Similar explanations for leading edge grooving are not as conclusive however, and it appears reasonable to assume that although visually similar, the mechanisms of wear are to some extent

Table 2. Composition of Incoloy 901.

	Ni	C	Mn	Si	Cr	Fe	Ti	Al	Mo	B
%	42.73	.05	.45	.40	13.5	34.0	2.5	.25	6.1	.015

unrelated. Conditions that produce trailing edge grooving do not always effect the same result on the leading edge and vice-versa.

Experimental Set Up

In order to further quantify the mechanism of grooving, machining tests using square CBN compact inserts were undertaken on a cylindrical billet of Incoloy 901, initial diameter 177mm and length 610mm. The lathe used was a Dean Smith and Grace 1910, incorporating continuously variable speed control between 5 and 2000rpm and 15kw drive motor.

The compacts were mounted in a standard insert tool-holder providing 6° negative rake geometry and a 45° approach angle. All tests were conducted "wet" with flood application of a water soluble emulsion type fluid to the cutting zone, flow rate being maintained throughout at $4\,\ell\,min^{-1}$. Prior to use, all cutting edges were chamfered using a diamond hand lap (45° x 0.05mm). A 2mm/side depth of cut together with a fresh edge was used for each test. In addition, tool entry format was held constant with 45° chamfering of the workpiece before each test cut. Insert specification - SNGN 120416.

Experimental Results

Figures 5 to 11 indicate the magnitude of flank and groove wear at varying cutting speeds and feed rates. At speeds of 122 and 153m/min the rate of notch growth exceeds that of the flank, test ceasure occurring after only 3.7 minutes of cutting. Coincident with notch growth was burr formation on the leading edge of the work material. As the cutting speed is increased, however, the groove dimension reduces to that of the flank. Increasing the feed rate when operating at 244m/min had minimal effect in terms of groove formation, indeed clearance face photographs, see Figure 11, show a complete absence of notching. The uniform nature of flank growth is an additional feature at this cutting speed and is clearly evident from the comparison of Figures 10 and 7. Doubling the feed rate consolidated this effect while in addition producing a lower rate of flank growth. Measurement of notching on the rake face produced similar dimensions to that on the flank, when flank notching diminished so did its counterpart on the rake face.

Build up of work material on the flank face and cutting

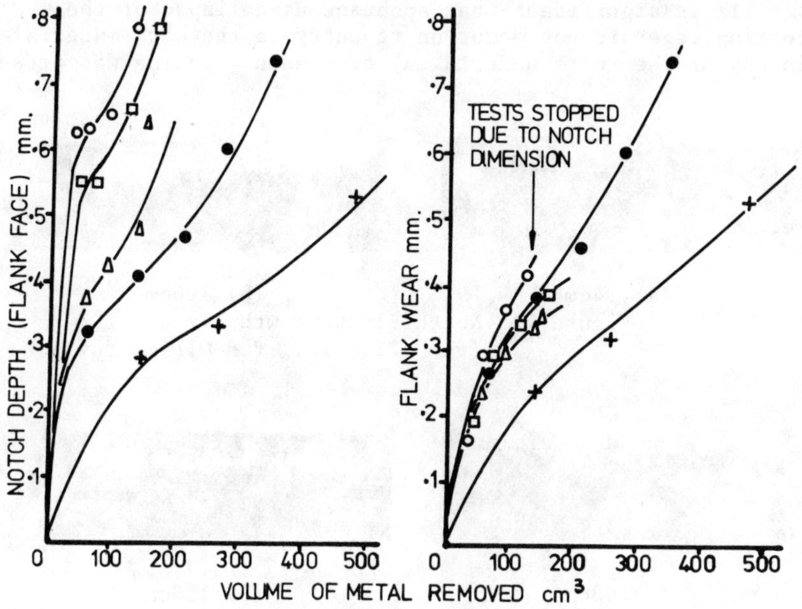

SYMBOL	v m/min	f mm/rev	d/side mm
○	122	0·14	2
□	153	0·14	2
Δ	183	0·14	2
●	244	0·14	2
+	244	0·28	2

Figures 5 and 6. Notch Depth and Flank Wear to Volume of Metal Removed. CBN/Incoloy 901.

edge as shown in Figure 9b occurred at all speeds and feeds. Estimates of tool/workpiece interface temperatures at maximum metal removal rate were of the order of 1000°C, a red/yellow glow being visible through the coolant. As a consequence, problems with surface finish occurred due to swarf pick-up and related directly to chip flow. Positioning of the coolant supply to prevent chip fall back and spatter onto the freshly generated surface required a fair degree of accuracy. The flow rate of 4 $\ell\,min^{-1}$ was judged totally inadequate for the requirements.

It is significant that spontaneous collapse of the cutting edge did not occur on re-entry to the work material in any of the tests undertaken, even when grooving was present.

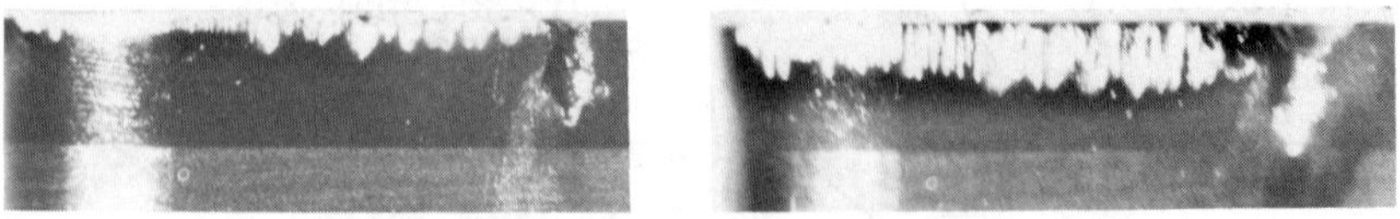

(a) $34cm^3$ (b) $126cm^3$

Figure 7. Notch/Flank Growth 17x
(v = 122m/min., f = 0.14mm/rev.)

(a) $43cm^3$ (b) $158cm^3$

Figure 8. Notch/Flank Growth 17x
(v = 153m/min., f = 0.14mm/rev.)

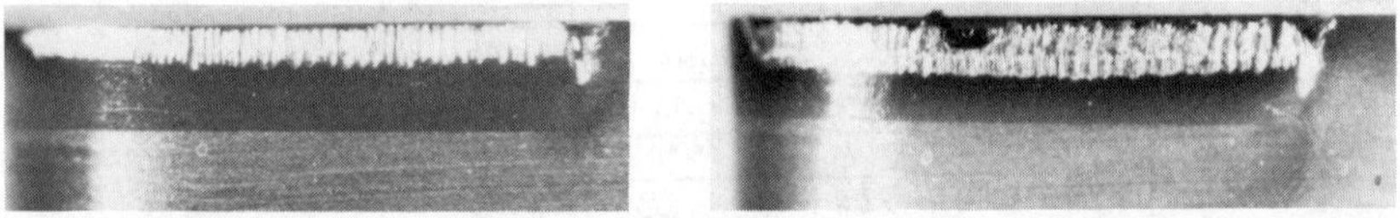

(a) $51cm^3$ (b) $138cm^3$

Figure 9. Notch/Flank Growth 17x
(v = 183m/min., f = 0.14mm/rev.)

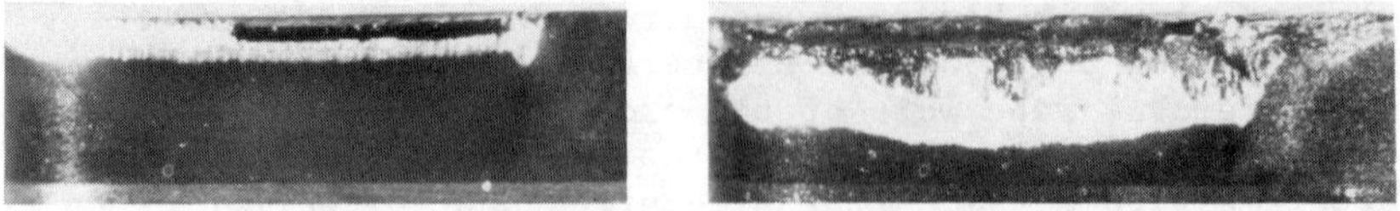

(a) $69cm^3$ (b) $341cm^3$

Figure 10. Notch/Flank Growth 17x
(v = 244m/min., f = 0.14mm/rev.)

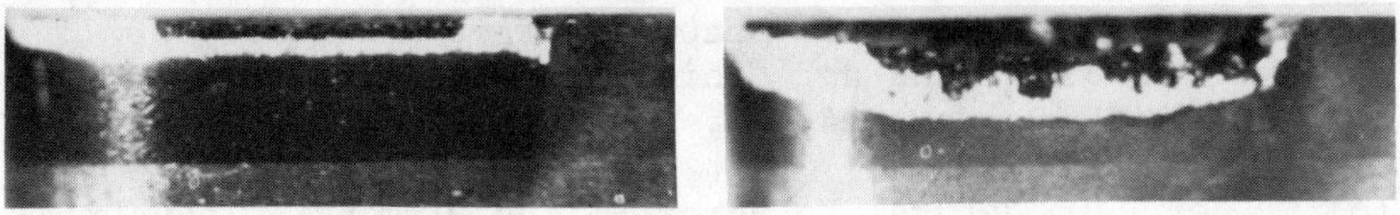

(a) $137cm^3$ (b) $573cm^3$

Figure 11. Notch/Flank Growth 17x
(v = 244m/min., f = 0.28mm/rev.)

Conclusions

Consideration of the wear at high cutting speed, points to temperature as a key factor in groove formation when using CBN tooling. The results, to a certain extent, contradict Shaw (9) in that higher temperatures would be expected to produce increased notching. The data merely consolidates the view previously outlined that grooving results from not one factor but a combination.

Analysis of estimated yield strength curves for a variety of nickel based alloys, suggests that effective machining will occur at temperatures above 850°C, the level at which they begin to loose their strength, in effect soften. The CBN compact appears capable of operating at or above this level without experiencing any undue problems of oxidation.

ACKNOWLEDGEMENTS

We would like to thank Professor N.A. Dudley CBE, Head of the Department of Engineering Production, for his encouragement and for providing facilities for this work to be carried out. Additional thanks go to the General Electric Company USA for funding this investigation together with Austin Morris Ltd. for their help and co-operation.

REFERENCES

(1) F. P. Bundy and R. H. Wentorf Jr, "Direct Transformation of Hexagonal Boron Nitride to Denser Forms", Journal of Chemical Physics, Vol. 38, No. 5, March 1963.

(2) R. Komanduri and M. C. Shaw, "Surface Morphology of Synthetic Diamonds and Cubic Boron Nitride", International Journal MTDR, Vol. 14, No. 1, March 1974, p. 63-84.

(3) C. Jeynes, "Natural Polycrystalline Diamond", Industrial Diamond Review, January 1978.

(4) R. E. Hanneman, "Borazon Compact Cutting Tools", SME Technical Paper, MR73 - 143, 1973.

(5) S. K. Bhattacharyya and D. K. Aspinwall, "Diamond Compacts Pay Dividends on Aluminium Alloys", Metalworking Production, February 1978.

(6) E. M. Trent,"Metal Cutting", Butterworth Publishers, 1977.

(7) V. Solaja, "Wear of Carbide Tools and Surface Finish Generated in Finish Turning of Steel", Wear, Vol. 2, 1958/59, p. 40-58.

(8) C. T. Ansell and J. Taylor, "The Surface Finishing Properties of a Carbide and Ceramic Cutting Tool", MTDR, September 1962, p. 225-243.

(9) M. C. Shaw, A. L. Thurman and H. J. Ahlgren, "A Plasticity Problem Involving Plane Strain and Plane Stress Simultaneously: Groove Formation in the Machining of High Temperature Alloys", Transactions of ASME, No. 142, May 1966, p. 142-146.

DISCUSSION

Thomas Skingle (Acme Cleveland)

Were speeds and/or feeds modified on the 1958-installed rotary machine when the tooling was changed from carbide to diamond?

David Aspinwall

It was impractical in our line to change feeds and speeds although some tool geometry changes were made to accommodate diamond.

Professor Hoshi (Toyohashi University)

Are you concerned with vibration of the machine and setup when using diamond and CBN tooling? If so, what are your criteria for accepting a machine with respect to vibration?

David Aspinwall

We've done very little vibration work. On diamond its not a significant parameter because we're machining aluminum alloys. On BZN applications it is. It's noticeable that when machining Incoloy 901 at high speeds or not vibration will produce a notch.

Charles Tobin (Caterpillar Tractor)

You used the term "Compax" in your discussion. That is a trade name of a General Electric product. Did you use GE's Compax?

David Aspinwall

Yes. This work was started 4 or 5 years ago, and although there are other suppliers in the U.K. there were adverse reports about some of their products. The information we had at that time was that the GE Compax was the most consistent.

Charles Tobin (Caterpillar Tractor)

In rough milling you used the term "Stratapax". Whose product is this?

David Aspinwall

Stratapax is the "rock drilling" equivalent of "Compax" and is also by General Electric. It has very coarse grain size, about 100-170 microns in diameter.

WEAR RESISTANCE AND CUTTING ABILITY OF A NEWLY DEVELOPED CUTTING TOOL

Hiroshi Eda*
Kozo Kishi*
Hiroshi Hashimoto**

* Faculty of Engineering, Utsunomiya University
** Research and Development Dept., Nippon Kogaku K.K.

1. Introduction

Borazon (BZN), produced by the General Electric Co. and used to make cutting tools, is made of sintered CBN which contains cobalt alloy as a bonding material. The newly developed bonding material for cutting tools, SUMIBORON BN200, is a particular ceramic which is heat resistant, wear resistant and has a lower affinity for the workpiece material. The toughness of BN200 was improved by adjusting its composition. As a result of fracture tests, which were carried out in order to investigate bonding strength, it was found that almost all the grains were fractured in a cleavage plane. This denotes that the bonding strength of BN200 grains and the particular ceramics is greater than that of a conventional CBN. In addition, the toughness of BN200 was improved by making the bond ing layer grains in the form of fine crystals below 1 μm. At room temperature, the hardness of grains of both conventional CBN and BN200 is, Hv=4200 and Hv=3200 respectively. However, at 1000°C, BN200 shows a Vicker's hardness of 1800 which is almost the same as that of a conventional CBN. This results in a lower incidence of heat crack, due to thermal shock and thermal damage due to the low thermal dependence of material hardness.

In this paper, development details of the

BN200 cutting tool, and experiments comparing the BN200 cutting tool with the CBN(BZN) tool are described. Furthermore, the relationship of the wear mechanism of the cutting edge in relation to cutting temperature is investigated from the thermal elastic-plastic mechanics view point.

2. Characteristics of the newly developed BN200 cutting tool

Table 1 compares the characteristics of BN200 with those of CBN(BZN).

Table 1. Characteristics of BN200 cutting tools

	BN200	CBN (BZN)
Specific gravity	4.2	3.48
hardness (Hv)	3000 - 3500	4500
Heat conductivity (cal/cm·sec·°c)	0.09	3.1
thermal expansion (1/°C)	$4.7X10^{-6}$	$5.6X10^{-6}$

In particular, the heat conductivity is 0.09 cal/cm.sec.°C for both BN200 and Al_2O_3, at room temperature. At 600°C, however, the heat conductivity of BN200 increases markedly to 0.13 cal/cm·sec ·°c, while that of Al_2O_3 decreases to 0.02 cal/cm·sec ·°c. This implies that the cutting heat energy which is produced around the cutting edge of the high termal conductivity of BN200. The heat diffusion of the tool lead to a decrease in tool wear.

It is quite difficult to sinter CBN grains for the following reason: CBN grains have the same crystal structure as a diamond, therefore, they have a high hardness, and resistance to deformations and atomic diffusions. For these reasons, BN200 is sintered at a temperature of more than 1500°C and a pressure of more than 40 Kbar (400kg/cm^2). Fig.1 shows a belt-type high pressure generator used for sintering.

Super high pressure is generated through a pair of spindles which are made from a high hardness alloy, WC-Co. This compresses a pressure medium and belt in a cylinder. A high temperature

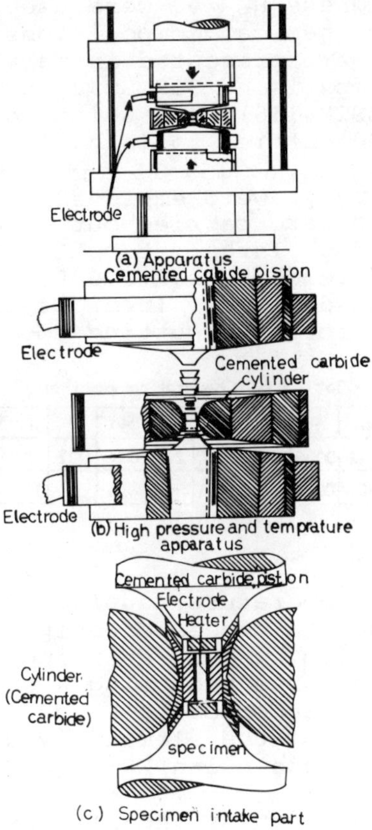

Fig. 1 A belt-type super high pressure generator

is obtained from the heater which is a graphite cylinder and is placed inside a testpiece chamber. Piroferrite which is a natural mineral, is used as a pressure medium. The BN200 grains and the particular ceramic are mixed and, with a WC-Co high hardness alloy basemetal, are inserted into the apparatus shown in Fig.1. During sintering, the liquid phase compound, which is mainly composed of Co in WC-Co super alloy, permeates into the BN200 powder and particular ceramic. Then BN200, a compound of three materials, is formed.

3. Experimental method and apparatus

High strength rolled steel alloys which have hardnesses H_{RC}=66,45,32,13, and maraging steel which has a hardness H_{RC}=60 were used as workpiece materials. The chemical compositions of the workpiece materials are presented in Table 2. For an experimental comparison, JIS SUJ2, SNCM9(ASTM 5046) and JIS SCM21(ASTM4118) were selected. CBN(BZN), BN200 and JISP20 cutting tools having the tool shape of -5, -6, 5, 6, 15, 15 and the nose radius 0.8 mm respectively, were employed.

Experiments were carried out on a shown-caznube HB-500 type lathe, at cutting speeds V=30-170m/min., feed rates f=0,075, 0.15 mm/rev, and depth of cut d=0.5mm. In order to obtain machinabilities for each cutting tool, cutting

Table2 Chemical composition

C	Si	Mn	P	S	Ni	Cr	Mo	V	Ti	Al	Co
0.85	0.31	0.42	0.008	0.005	0.017	2.43	0.24	0.07	—	—	—
0.0029	0.03	0.01	0.006	0.0025	17.66	—	4.37	—	1.28	0.068	11.80

force, cutting temperature, surface roughness of the workpiece and thickness of chip were measured. To assess wear also, crater wear and flank wear were measured and observed using a scanning electron microscope. In order to elucidate diffusion wear, an analysis of the elementary density of the wear surface was made using EPMA.

4. Experimental results and Discussions

4.1 The machinability of BN200 tools

In cutting with conventional tools, wear on the tools increases as the hardness of the workpiece material increases. Wear on a CBN tool decreases, however, when the hardness of the workpiece material increases although, when the hardness of the workpiece increases to a greater extent than described previously, the wear is found to be increased. Cutting force and cutting temperature in relation to the material's hardness are shown in Fig.2 and Fig.3. As shown in Fig.2, the tangential force and the traversing force decrease up to a hardness H_{RC}=45, but increase when hardness is greater than H_{RC}=45. Such a change is also found in the cutting temperature, i.e, when the hardness of the workpiece material increases to H_{RC}=60 - 65,

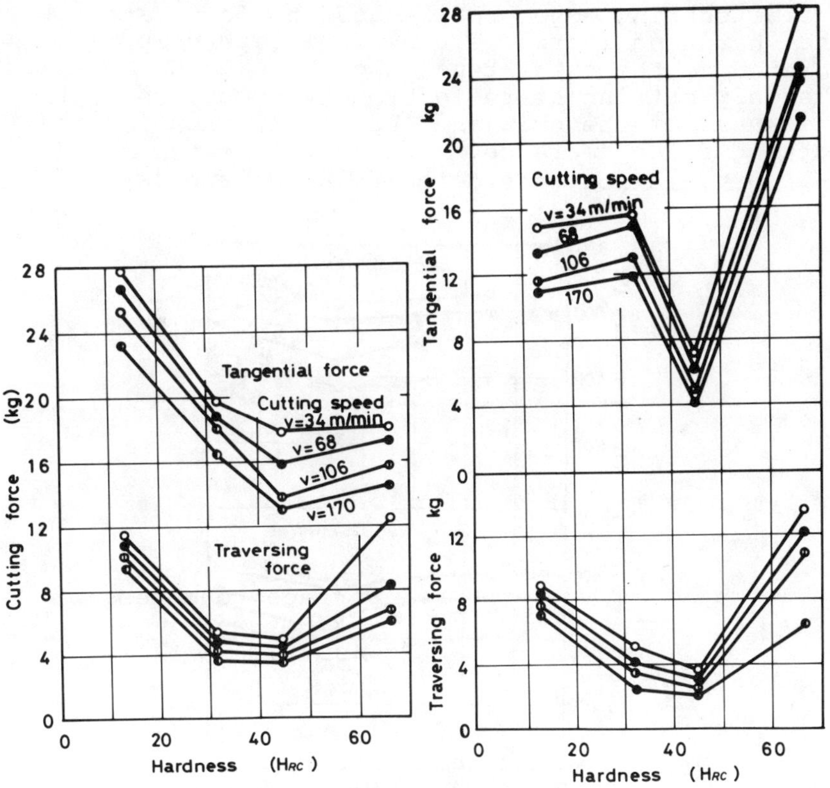

Fig.2 Cutting force in relation to material's hardness

marked wear begins to occur. This is considered not to be wear on the CBN grain itself but to be macroscopic wear produced by the plasticity of the cutting tool.

Fig.4 and Fig.5 show the cutting ratio in oblique cutting and the surface roughness. The surface roughness (Rmax) shows a marked difference from the geometrical surface roughness (Rth) theoretically obtained, providing that the hardness of the work-piece material (H_{RC}) is small, but it gradually approaches Rth when the H_{RC} value increases.

The change of chip profiles are shown in Fig.6. The changes in surface roughness are elucidated not only from the results of the recorded surface roughness but also from the regularity of the geometrical profile of the surface, which was formed by the cutting edge profile and the feed rate. When hardness increases, the flow-type chip is changed to the shear-type chip. At the same time, a steady cutting state is brought about, while the thickness of the chip and the fluctuations of the cutting force decrease. The cutting ratio shown in Fig.5, is not considered to be an accurate

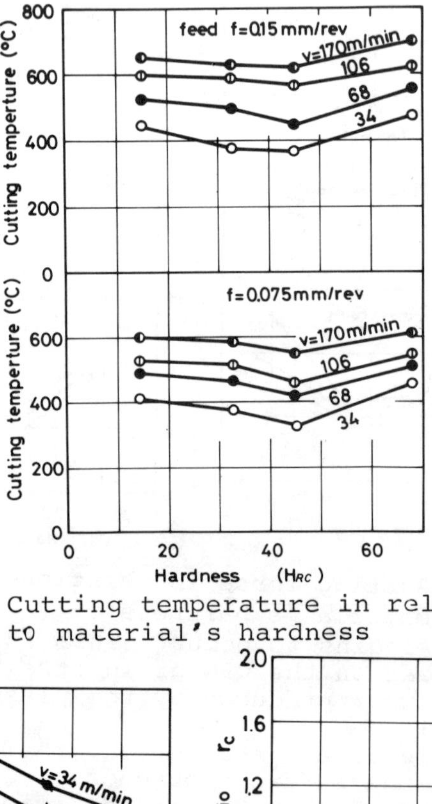

Fig. 3 Cutting temperature in relation to material's hardness

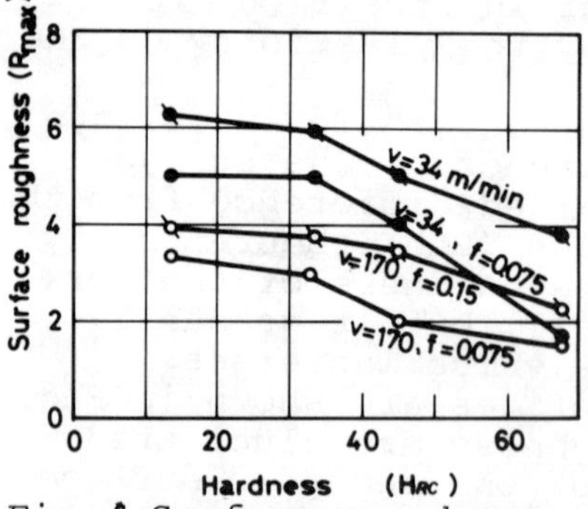

Fig.4 Surface roughness in relation to material's hardness

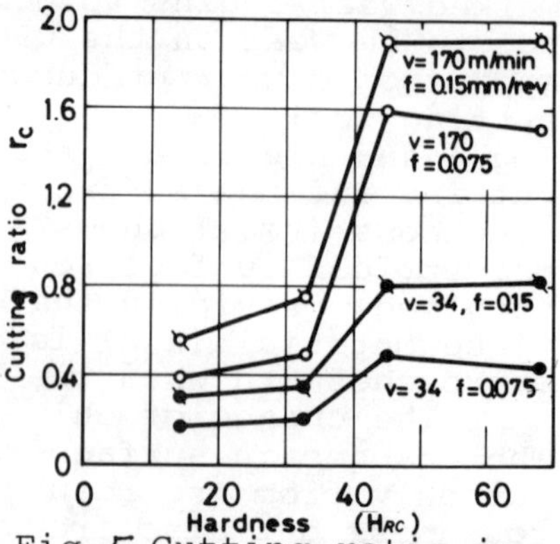

Fig.5 Cutting ratio in relation to material's hardness

value, but it is possible to use it to judge the cutting ability of the tools. It is previously described that, when the hardness of the workpiece material increases the cutting temperature increases. This increase in the cutting ratio is illustrated by the marked increase in cutting ratio which is shown in Fig.5. When the hardness of the workpiece material increases to more than $H_{RC}50$, the cutting ratio and the temperature of the rake surface increase. In the range of hardness H_{RC}= 13-45, fluctuations in the thickness of chip are not observed. In the range of hardness above H_{RC}=45 cutting ability is improved because of the increase in the cutting ratio, the decrease in surface roughness, and the increase in cutting temperature and cutting force. Thus, the more plastic the workpiece material the more adhesive it is for the cutting edge. The developed BN200 tool is appropriate for the workpiece material which shows a low rate of elongation. However, the BN200 tool is not expected to be useful for a workpiece material which shows a high rate of deformation, relative to the hardness of the BN200 tool. The greater the hardness of the tool therefore, the harder the workpiece materials which can be cut. In this case, the upper limit of hardness for workpiece materials, which allows the workpiece material to be cut, is determined. The thermal stability of the tool, however, must be sufficiently satisfied. For example, although the hardness of a diamond tool is sufficiently high, the thermal stability is restricted "to temperatures below 600°C". Therefore, if the cutting temperature is above 600°C, cutting ability is decreased. Conversely, for the workpiece material which has a soft matrix and is sufficiently plastic, the hardness of the tools is not closely related to the cutting ability, but the sharpness of the cutting edge and the surface roughness of the tool do show a relationship to cutting ability. In the low hardness range, surface roughness of the workpiece is comparable with that of the cutting tools because the surface roughness of the matrix is comparable with the surface roughness described above. The reason why hard cutting tools are used to cut high hardness materials is that if the thermal stability is sufficient at the cutting point, the stress at the contact point of the workpiece with the tool, is concentrated and cracks are easily brought about because the stress is in a wedgetype distribution. In other words, cutting ability is decreased for the following two reasons. 1) the lower the hardness of the tool the larger the area of the contact zone at the cutting point.

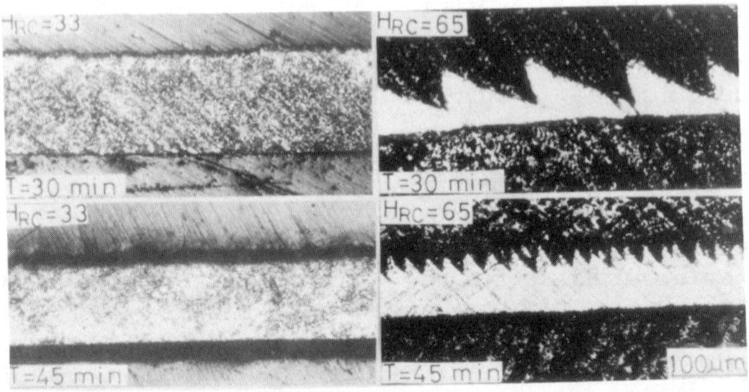

Fig. 6 The change of chip profile

2) Moreover, the stress is distributed over a large area and the distribution pattern is changed from spheroidal to ellipsoidal, the latter having a long axis in the cutting direction.

4.2 Comparison of wear on the various tools

The wear on conventional CBN tools increases when the hardness of the workpiece material is low, whilst when cutting with the newly developed BN200 tool, the wear is almost constant for a workpiece which has a hardness below H_{RC}=45. At around H_{RC}= 55, a CBN (B2N) tool shows slightly less wear than the BN200 tool. At around H_{RC}=60, the wear on a CBN (BZN) tool becomes markedly greater than on a BN200 tool as shown in Fig.7. The profile of wear on a CBN tool is observed as attritious wear accompanied by plastic flow as shown in Fig.8. Wear of the CBN grain itself is not observed. Therefore, wear is considered to be dependent upon the thermal strength of the base compound such as bonding materials, high hardness powders (WC-Co) and particular ceramics. The low wear shown by BN200 tools in the range of H_R=13 - 45, is considered to be dependent on the low wear resulting from plastic flow because of the high bonding strength of WC-Co bonding materials and CBN grains. In diffusion experiments of

BN200 grains and electrolytic iron, and of sintered BN200 and a workpiece material, for investigating cutting tool wear mechanism, it was elucidated that BN200 grain are stable with Fe but boron and cobalt, which are contained in sintered BN200, are partly dispersed into Fe. As shown in Fig.7, the cutting temperature decreases at around H_{RC}=60, where a BZN tool shows lower wear than a BN200 tool, because of the decrease of shearing heat and friction heat in the rake surface, which is dependant upon transformation of the chips from the flow type to the shear type. The hardness of the CBN (BZN) tool, which is highly sensitive to temperature change, increases because the temperature of the cutting edge decreases. Therefore, wear on a CBN (BZN) tool is reduced. In Fig.9, the relation of tool life T (min.) to the cutting speed V(m/min) is shown, for JIS SUJ2, SCM21, SNCM9 and SKD11. It is found that a BBN(BZN) tool shows a longer tool life and less wear than a BN200 tool, but the differences between them are reduced when the hardness is increased. The above observations are considered to be related to the temperature depend-

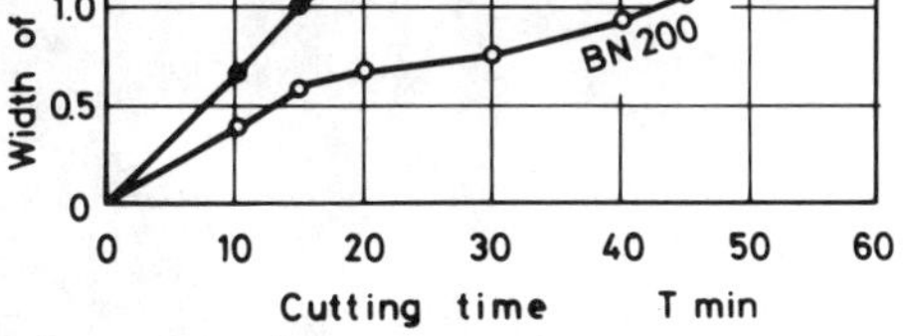

Fig.7 Flank wear in relation to cutting time

ences of CBN(BZN) and BN200 tools. In order to supplement the above experiments, the high speed steel JIS SKH57 (H_{RC}60), the chilled steel (Hs90), the cast iron FC25 (H_B250) and Maraging steel (H_{RC}60) were cut. It was found that the BN200 tool has a longer tool life than the CBN (BZN) tool. In particular, a marked difference in tool life is shown in a workpiece which has a hardness below H_{RC}=50[2]. Moreover, in wet cutting experiments on

Fig.8 The profile of attritious wear on a CBN tool, accompanied by elastic-plastic flow
Cutting speed = 170 mm/min,
workpiece feed rate = 0.075 mm/rev.,
workpiece material = SUJ2 (H_{RC}65)

JIS SNCM29 using solution-type cutting fluid, at a cutting speed of V=60m/min. the tool life increased from T=35min which is obtained in dry cutting, to T=45min. At V=150m/min, the tool life increased from T=20min in dry cutting to T-30min in wet cutting regardless of the type of workpiece material.

4.3 Investigation of tool wear from the thermal elastic-plastic mechanics viewpoint

Cutting heat can be obtained from the three component force as shown by M.C. Show in the orthogonal cutting theory 3), where it is supposed that, (1) there is a heat source which has a constant energy respectively on the flank surface and rake surface. (2) the heat sources move at constant speeds. The temperature distribution in the vicinity of the cutting edge can then be calculateed.

Firstly, the transtitional temperature caused by a heat source which moves on the flank surface and on the rake surface is calculated using FEM (Finite Element Method).

In cartesian co-ordinates, the following partial differential equation which is related to the heat conduction in a solid, is obtained,

$$\frac{\partial}{\partial xi}(\lambda i\frac{\partial T}{\partial xi}) - \rho c\frac{\partial T}{\partial t} + \bar{Q} = 0 \qquad (1)$$

where, T is temperature, is heat conductivty, is density, and C is specific heat. The differential equation of temperature T at the joint is expressed as follows, using the functional stationary conditions for the equation (1) and keeping the time constant:

$$[\mathbf{H}]\ \{\mathbf{T}\} + [\mathbf{C}]\frac{\partial}{\partial t}\{\mathbf{T}\} + \{\bar{\mathbf{F}}\} = 0 \qquad (2)$$

where,

$$[\mathbf{h}^e] = \int_v [\mathbf{B}]^T[\mathbf{K}]\,[\mathbf{B}]dv+\int_{s_2}\alpha\,[\mathbf{N}]^T[\mathbf{N}]d\mathbf{S}$$

$$[\mathbf{C}^e] = \int_v \rho c\,[\mathbf{N}]^T[\mathbf{N}]dv$$

$$[f^e] = -\int_v \mathbf{Q}[\mathbf{N}]^T dv+\int_{s_1} q\,[\mathbf{N}]^T ds-\int_{s_2} h\mathbf{T}\infty\,[\mathbf{N}]^T d\mathbf{S}$$

and h is heat conductivity, q is heat flux (function of time), and **T** is ambient temperature, furthermore, the following approximation equation can be obtained using the weighted differential method proposed by Galerkin for each time interval.

From equation (3), the joint temperature **T** n after t can be obtained with an initial value **T** n-1 and in-flow heat quantity during t.

Next, the stress thermal elastic-plastic analysis is carried out by the method proposed by YAMADA[4]), i.e., calculation of the increment that adds incremental load temperature which makes each triangular element yield in order. In every calculation step, it must be judged whether each triangular element is in an elastic or plastic state. For an elastic or plastic state, the stress-strain matrix $[\mathbf{D}^p]$ for the plastic state is used. The equivalent load increment $\{d\mathbf{F}\}_{thermal}$ which corresponds to the thermal increment $\{d\mathbf{T}\}$ is expressed as follows replacing the stress-strain matrix in the elastic state $[\mathbf{D}^e]$ with $[\mathbf{D}^p]$ described above:

$$\{d\mathbf{F}\}thermal=\int_V [\mathbf{B}][\mathbf{D}^p]\{\alpha\}d\mathbf{T}d\mathbf{V} \qquad (4)$$

where, α is coefficient of linear expansion and dT is thermal increment of triangular element.

Calculation results are shown in Fig.10a, Fig. 10b. It is found that BN200 has a gentle gradient of thermal stress as a whole, in comparison with a CBN (BZN) tool. In particular, a marked difference is shown in the vicinity of the cutting edge. Comparing with the attritious wear caused by plastic flow in the vicinity of the cutting edge, which is shown in Fig.8, it is found that a high tensile stress of 30-40kg/mm^2 is brought about locally around the ragged flank surface.

Comparatively high tensile stress is found in the out-flow direction of chips from the cutting edge. However, the stress value is lower and the stress grandient is more gentle than that brought about on the flank surface. Although this numerical calculation is a rather qualitative assessment of tool wear, it is considered that plastic

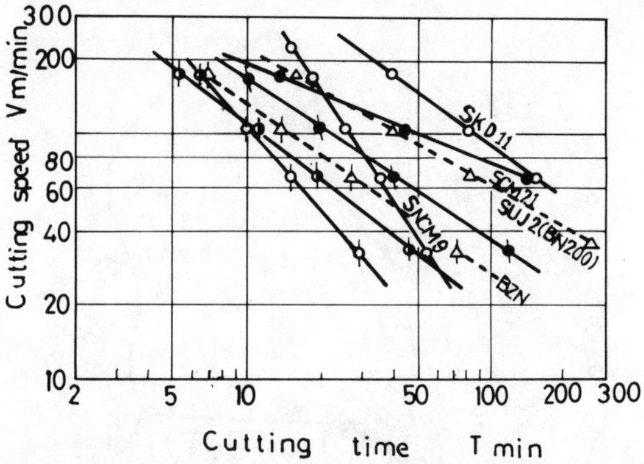

Fig.9 The relation of tool life (cutting time) T to cutting speed

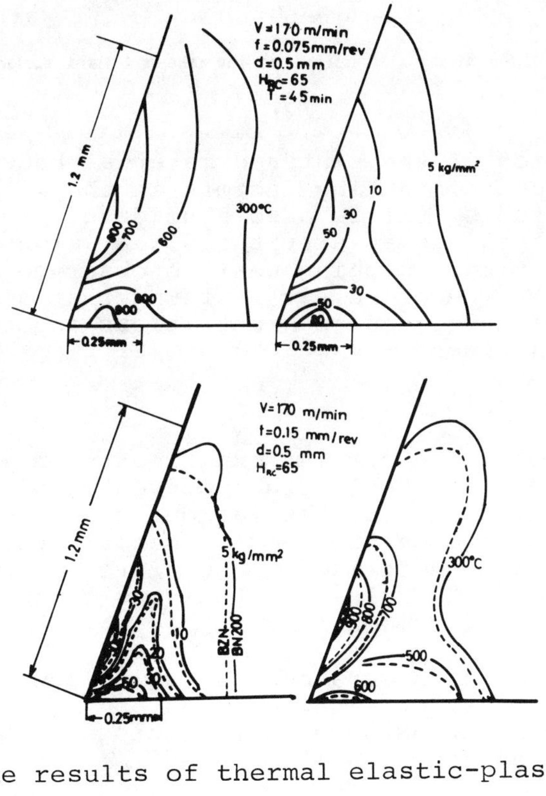

Fig.10a The results of thermal elastic-plastic analyses

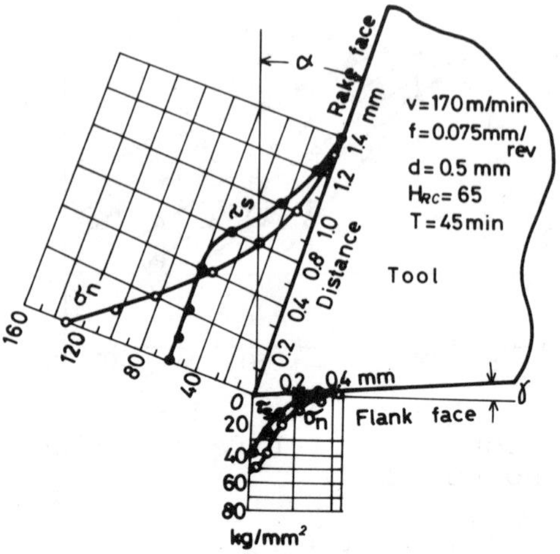

Fig.10b Stress distribution on the rake and flank surface

deformation of the sintered material base is one of the most dominating factors in the wear of CBN cutting tools. It is also elucidated that a CBN cutting tool has an attritious wear mechanism. At the same time, chipping wear is observed as shown in Fig.8. Therefore, it is found that a CBN cutting tool undergoes both attritious and chipping wear mechanisms.

5. Conclusion

Comparative cutting experiments on a newly developed BN200 tool with a conventional CBN (BZN) were carried out for the various workpiece materials. Based upon these experimental results, FEM analysis was performed to elucidate the wear mechanism of CNB tools from the view point of thermal elastic-plastic mechanics.

The results obtained are as follows:

1) Grains of BN200 show attritious wear in a wide range of cutting conditions.

2) Comparing the tool life T_{BN} (with a flank wear V_B=0.15mm) of a BN200 cutting tool with T_{BZN} of

CBN (BZN) tool, in dry cutting of JIS SCM21 (ASTM 4118) which is H_{RC}=65, it was found that at a cutting speed of V=60m/min, T_{BN} was 200min., while T_{BZN} was 50min., at V=170m/min T_{BN} was 17min, while T_{BZN} was 10min. Thus, the tool life of the BN200 cutting tool was markedly increased.

3) In the wet cutting of JIS SNCM9 (ASTM5046) which is H_{RC}=55, using water soluble cutting fluid, at a speed of V=60m/min, it was found that the life of a BN200 cutting tool T_{BN} was 45min., T_{BZN} was 35min. At V-150m/min, T_{BN} was 40min., while T_{BZN} was 20min. Thus, a BN200 cutting tool has a longer tool life than a conventional CBN cutting tool in conditions of wet cutting as well.

4) It was found from our experiments that a BN200 cutting tool has a long tool life in the cutting of high hardness steel, high hardness cast iron (chilled cast iron which has H_S=80), cast iron F35, heat resistant steel and a high temperature alloy such as a maraging steel.

5) In the cutting of Rolled material which is H_{RC}=13-65, at workpiece speed f=0.075-0.15mm/rev., depth of cut d=0.5mm and cutting speed V=30-200m/min, a BN200 cutting tool produced a cutting temperature 300-700°C. A CBN cutting tool, however, produced a cutting temperature which was 50-100°C higher.

6) In diffusion experiments of BN200 grains and electrolytic iron, and of sintered BN200 and a workpiece material, for investigating cutting tool wear mechanism, it was elucidated that BN200 grain are stable with Fe but boron and cobalt, which are contained in sintered BN200, are partly dispersed into Fe.

Acknowledgement

We should like to conclude by expressing our sincere gratitude to Sumitomo Elastic Co., Ltd. We should also like to express our gratitude to Mrs. K. Skerrett for her assistance in completing this manuscript.

References

(1) N. Tabuchi: Performance of "SUMIBORON BN200", Sumitomo Electoric, Technical Review, No. 18 12 (1978) 57

(2) T. Uchida: Cutting Operation in CBN Sintered Cutting Tool Japan Machinist, No. 4(1979)1

(3) M.C. Show: Metal Cutting Principles, MIT Press (1965) 12-1

(4) Y. Yamada: Stiffness Matrix in Plastic Elestic Problems of Continua, Seisan-Kenkyu (Monthly Jour. of the Institute of Industrial Science, University of Tokyo), 19, 3(1967) 75.

CUTTING PERFORMANCE OF SINTERED CBN TOOLS

YUICHIRO KONO, AKIO HARA
SHUJI YAZU, TAKASHI UCHIDA
YOSHIKATSU MORI

SUMITOMO ELECTRIC INDUSTRIES,LTD

1. INTRODUCTION

Cubic boron nitride is a superior tool material since it is the hardest substance next to diamond and shows less reaction to ferrous metals under elevated temperature.

Recently a few types of sintered CBN cutting tools have developed and begun to be saled commercially. These sintered CBNtools are now used for machining hard ferrous metals-quench hardened steels and hard cast irons-,super alloys and sintered ferrous metals.

The main advantages of using the sintered CBN tools in machining above mentioned materials are the improvements in tool life and in productivity (material removal rate) compared with other cutting tools. But in the case of machining hardened steels , it is needed for the sintered CBN tools not only to get the longer tool life and the higher productivity but also a good surface integrity of the machined work. Since the purpose of this case is to change the finish machining process of hardened steel parts from grinding to cutting.

In this paper first the tool life test results of a few types of the sintered CBN tools in machining of quench hardened steels and Ni base super alloy are described. Second, the surface integrities of the work which is machined by the sintered CBN tool are described.

2. EXPERIMENTAL PROCEDURE

2-1. Tool Life Test

Turning test were conducted under the conditions shown in Table 1. Tools tested were three types of the sintered CBN tools marked CBN1(CBN-TiN composite, Sumiboron BN200),CBN2(CBN-TiN composite,Sumiboron BN100,CBN content is higher than CBN1) and CBN3 (CBN-Co composite) and carbide K10. The micro-structure of the sintered CBN tools are shown in Fig 1.

The work materials were case hardened steel SCM21,hardened die steel SKD11 and Ni base super alloy INCONEL718. The die steel SKD11 were machined to the shape as shown in Fig 2 before cutting test and then using this work interrupted turning test were conducted.

The chemical compositions of the work materials are listed in Table 2.

Table 1. Turning Test Conditions

Test 1	Tool	: CBN 1,CBN2,CBN3,K10
	Work Material	: SCM21 $H_{RC}65$
	Cutting Speed	: 60-200 m/min
	Depth of Cut	: 0.2mm
	Feed	: 0.1mm/rev
	Coolant	: Dry
Test 2	Tool	: CBN1,CBN2,CBN3
	Work Material	: SKD11 $H_{RC}63$
	Cutting Speed	: 80m/min
	Depth of Cut	: 0.3mm
	Feed	: 0.25mm/rev
	Coolant	: Dry
Test 3	Tool	: CBN1,CBN2,CBN3
	Work Material	: INCONEL 718
	Cutting Speed	: 100m/min
	Depth of Cut	: 0.2mm
	Feed	: 0.1mm/rev
	Coolant	: Emulsion Type Water Soluble Oil
	Tool Geometries (-5,-6,5,6,15,15,0.8mm)	

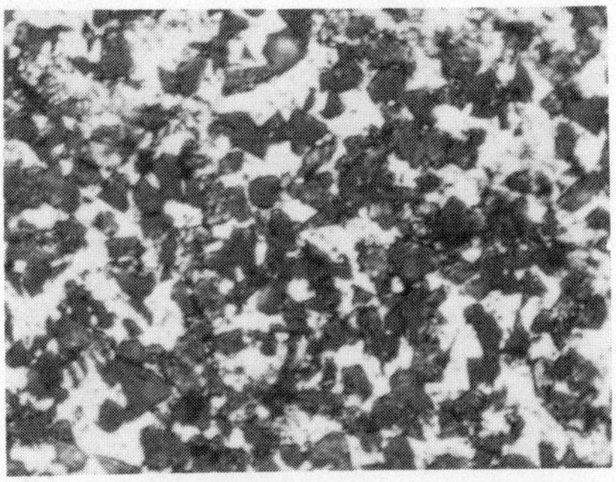

CBN1

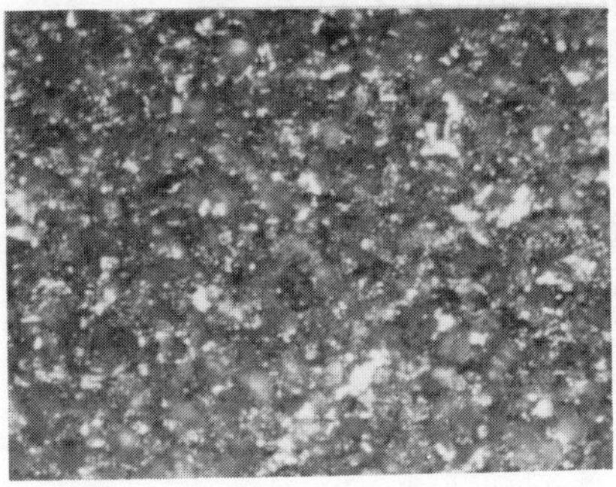

CBN2

CBN3

Fig 1 Microstructures of the Sintered CBN Tools

10μm

Table 2. Chemical Compositions Of Work Materials

Work	C	Si	Mn	P	S	Ni	Cr	Mo	V	Ti	Al	Ta+Nb	Fe
SCM21	0.24	0.31	0.8	0.034	0.020	--	1.04	0.27	--	--	--	--	bal.
SKD11	1.47	0.31	o.41	0.019	0.007	--	11.82	0.83	0.24	--	--	--	bal.
INCONEL718	0.05	--	--	--	--	53.12	19.05	3.21	--	0.82	0.49	5.03	bal.

Chemical Compositions (%)

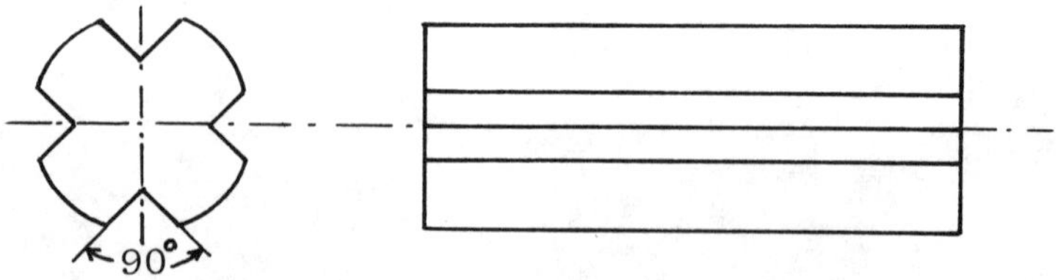

Fig 2 Work shape for Interrupted Turning Test (Work material SKD11)

2-2 Surface Integrities of the Machined Work

In this experiments hardened SNCM8 and SKD11 were used for work material. The chemical compositions of SNCM8 are shown in Table 3. (SKD11 is the same as tool life test.)

First the work materials were machined by CBN1 tool and by WA60K grinding wheel under conditions shown in Table 4 respectively.

Table 3 Chemical Compositions of SNCM8 (%)

C	Si	Mn	P	S	Ni	Cr
0.38	0.29	0.70	0.007	0.011	1.86	0.88
Mo						
0.23	Heat Treated H_{RC} 50					

Table 4 Turning and Grinding Conditions

Turning	Tool	: CBN1, SNG432 type
	Tool Geometries	: (-5,-6,5,6,15,15,0.8mm)
	Cutting Speed	: 50 - 300 m/min
	Depth of Cut	: 0.025 - 1.0 mm
	Feed	: 0.025 - 0.3 mm/rev
	Coolant	: dry and emulsion type water soluble oil
Grinding	Wheel	: WA60K
	Wheel Speed	: 1450 m/min
	Work Rotation Speed	: 30 m/min
	Depth of Cut	: 0.015 mm
	Coolant	: dry and solution type water soluble oil

CBN2 and CBN3 tools were not used because these two tools showed less wear resistivity than CBN1 tool in machining hardened steels. Then the surface integrities of the machined work were examined. Table 5 shows the examined surface integrities and method of measurement.

Table 5 Examined Surface Integrities and Method of Measurement

Surface Integrities	Method of Measurement
Hardness	Micro Vickers Hardness Tester Load 0.98N x 15s
Surface Roughness	Needle-contact type Roughness Tester ("TALYSURF")
Microstructure	Microscope specimens were polished and etched by Nital
Residual Stress	Xray $\sin^2\psi$ Method Measuring Conditions Characteriatic Xray : Cr-Kα Bulb Voltage : 30 KV Bulb Current : 10 mA Count Range : $2x10^3$ c/s Time Constant : 8s Divergent Angle of Slit : 1.02° Goniometer Speed : 2°/min Chart Speed : 40 mm/min in the case of measurement of stress distribution in surfase layer, the specimen was prepared by electrolysis-machining

3 RESULT AND DISCUSSION

3-1 Tool Life

Fig 3 shows the tool wear in turning SCM21. CBN2 and CBN3 produced considerably more wear than CBN1. Tool life curve obtained in turning SCM21 are shown in Fig 4. CBN1 has the longer tool life than the other tools at wide range of cutting speed.

Fig 5 shows the result of interrupted turning test of hardened die steel. CBN1 is more tougher than the other two CBN tools.

Fig 6 and Fig 7 show the tool wear and the worn tool edges in turning INCONEL718 respectively. In this case CBN1 produced more wear than the other two.

To consider the difference of wear in turning SCN21 between three CBN tools, the worn flank surfaces of CBN1 and CBN3 were observed in scanning electron microscope. Fig 8 shows the SEM image of the flank surfaces. Many micro-cracks and striations are observed in both CBN tools. The width of a single striation corresponds to several CBN grains. From this observation the wear mechanism of the CBN tool in machining hardened steels was considered as following.

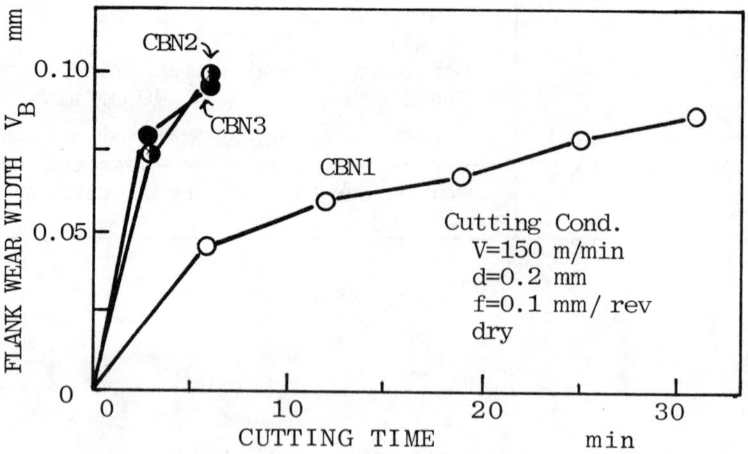

Fig 3 Tool Wear in Turning SCM21(H_{RC} 65)

The falling out of CBN grains as a small lump from flank surface due to rubbing off by work takes place , then the falled out CBN grains rubb the flank surface and the striations are formed on the flank surface. As shown in Fig 8, the quantity of micro-cracks and the depth of striation of CBN3 are more than those of CBN1. The CBN grains of CBN1 are apert from each other and hold by the matrix, on the other hand the CBN grains of CBN3 are contact with each other as shown in Fig 1.

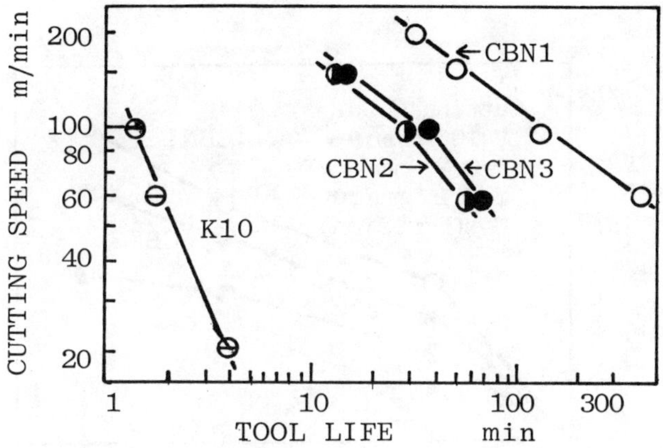

Fig 4 Tool Life Curve in Turning SCM21 (H_{RC} 65) V_B=0.15 mm

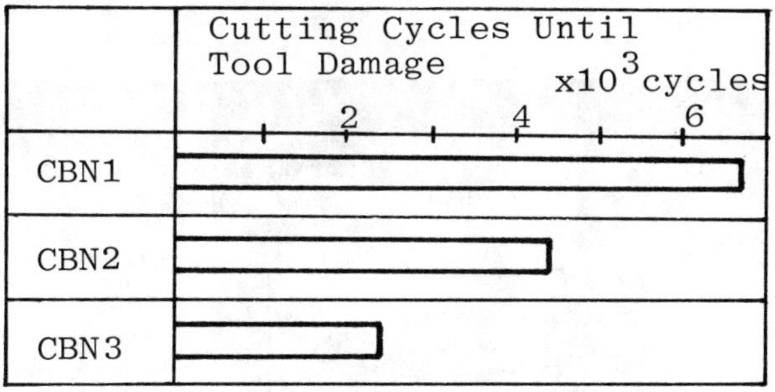

Fig 5 Interrupted Turning Test Result
Work Material : SKD11 H_{RC} 63
Cutting Cond.
V=80 m/min,d=0.3 mm,f=0.25 mm/rev dry

From these considerations and experimental facts, it was considered that the CBN grains in CBN1 are held firmly by matrix and hard to fall out from the flank surface but the CBN grains in CBN3 are less firmly bonded with each other and easy to fall out. This is the reason of difference in wear resistance between CBN1 and CBN3 in machining SCM21.

The toughness of CBN tools shown in Fig 5 was also considered to be explained from above mentioned discussions.

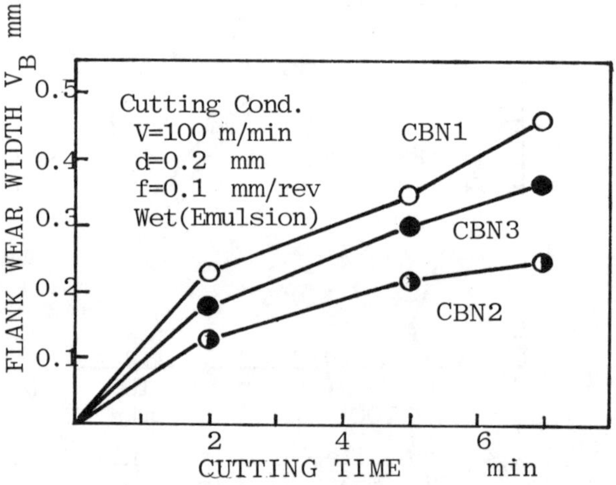

Fig 6 Tool Wear in Turning INCONEL718

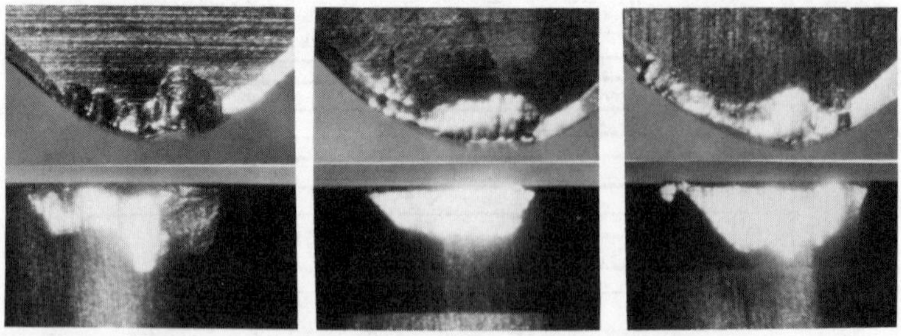

CBN1 CBN2 CBN3

Fig 7 Worn Tool Edge in Turning INCONEL718
Cutting Cond.: V=100m/min, d=0.2mm
f=0.1mm/rev, wet(emulsion type)

Quench hardened steels are very hard but brittle, so the cutting temperature and stress during machining hardened steels are not so high. Super alloys are commonly very strong and tough even at elevated temperature, so the stress and temperature of the cutting edge are considerably intense during maching. Therfore high hardness at elevated temperatureis needed for cutting tools in machining super alloys. According to this point of view, it is natural that CBN2 and CBN3 which contain more CBN than CBN1 showed less wear in machining INCONEL718.

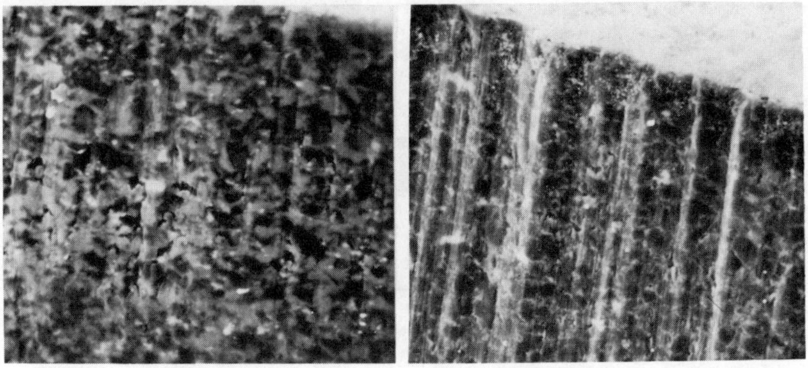

CBN1 CBN3

10μm

Fig 8 SEM Image of Flank Surface in Turning SCM21(H_{RC} 65)
Cutting Cond.: V=100m/min, d=0.2mm f=0.1mm/rev, dry

3-2 Surface Integrities

Fig 9 and Fig 10 show the hardness of surface layer of the machined work by cutting and by grinding respectively. The work machined by grinding shows a little drop down in hardness near the surface but the work machined by cutting shows almost no changes.

Fig 11 shows the roughness curves of the machined surface. The surface machined by grinding is irregular shaped but that machined by cutting is regular.

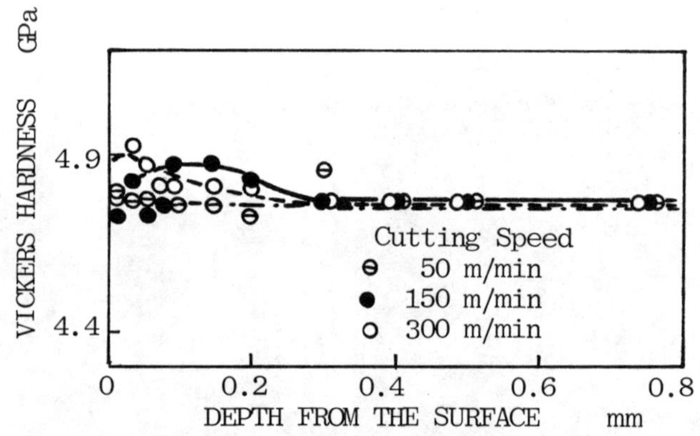

Fig 9 Hardness at Surface Layer of the Work Machined by CBN1 (SNCM8)
Cutting Cond.: d=0.2mm, f=0.1mm/rev

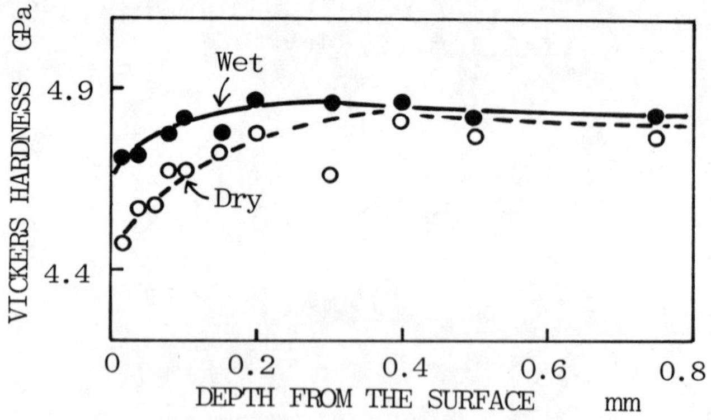

Fig 10 Hardness at Surface Layer of the Work Machined by Grinding (SNCM8)
Wheel : WA60K, Wheel Speed : 1450m/min

Fig 12 shows the surface roughness of the work machined by cutting in relation to feed rate. Roughness close to the theoretical value are generated by cutting.

Fig 13 shows the microstructures of the machined work surface and the chip. The chip is deformed strongly during cutting operation, so the breakage of the carbide prticles (white particles in Fig) are observed. On the other hand the carbide particles near the machined surface show no breakage. So it is concluded that the remarkable changes in microstructure did not occur by machining with CBN1 tool.

Fig 14 shows the residual stress of the work surface machined by CBN1 in relation to the cutting conditions. The stress level shows remarkable changes according to the cutting conditions.

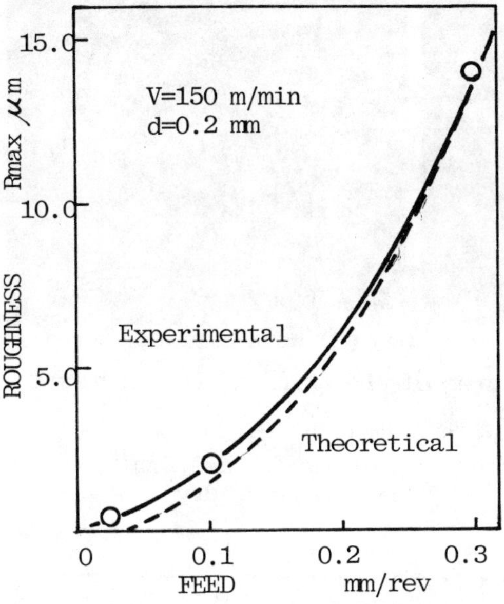

Fig 12 Surface Roughness of Work Machined by CBN1

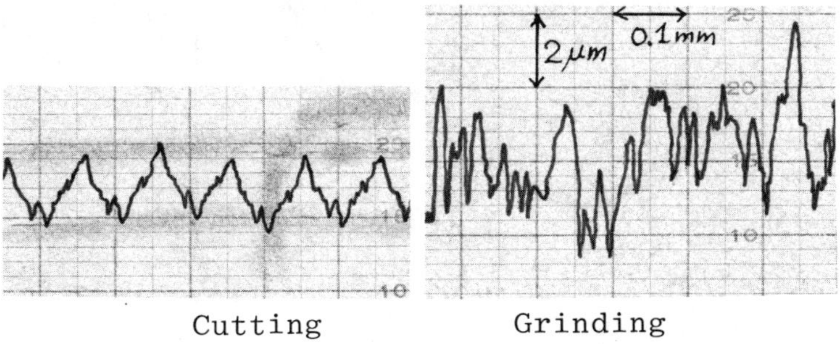

Fig11 Roughness Curves of Machined Work

Fig 15 shows the residual stress distributions in the surface layer of the machined work.

The machining mechanisms of the cutting and the grinding processes are the same essentialy. Namely in both processes the cutting edge engages into the work material and scrapes off the chips from the work. But the quatity or the shape of cutting edges

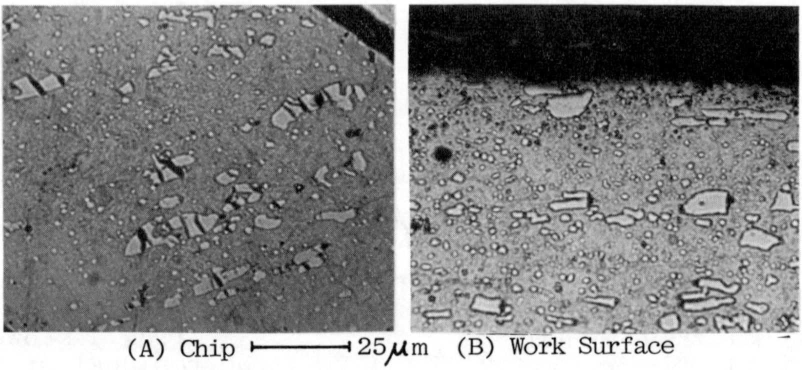

Fig 13 Microstructures of the Chip and the Machined Work Surface
Work Material : SKD11, H_{RC}63
Cutting Cond. : V=100m/min, d=0.2mm
f=0.1mm/rev, dry

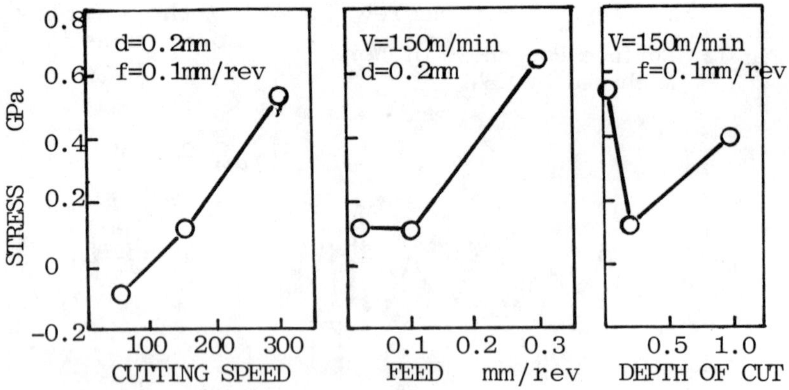

Fig 14 Circumferential Residual Stress of Work Surface Machined by CBN1 Tool

are different from each other.

Grinding wheel is consist of a lot of abrasive grains and the cutting edges of abrasive grains are distributed randomly on the grinding wheel surface. So the surface machined by grinding becomes irregular as shown in Fig 11. On the other hand, in case of cutting, the work is machined by the single point cutting edge which is ground to regular shape. So the machined surface by cutting naturally becomes rgular.

Moreover the hardened steels are brittle and shows less reaction to the tool materials than soft steels, so in case of hardened steels the deformation zone of the work in front of the cutting edge and the exsessive plastic deformation of the work are smaller than in soft steels.

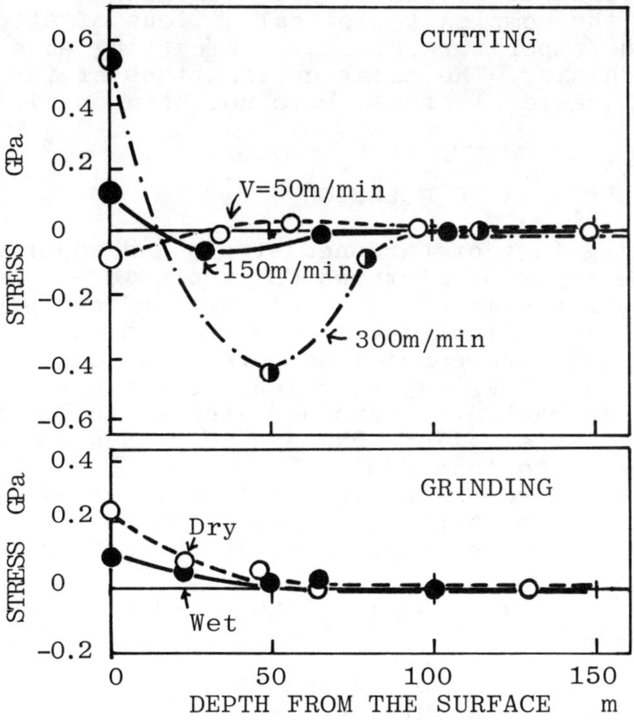

Fig 15 Residual Stress Distribution in the Surface Layer of Machined Work (Circumferential Stress)

These points are the reason why the very regilar shaped surface is generated by CBN tool in machining hardened steels.

The tool geometries are also quite different between grinding wheels and single point cutting edges. The rake angle of the single point cutting tool is commonly in the range of $-6^{\circ}\sim+10^{\circ}$ in machining hardened steels. But the rake angle of the cutting edges of the grinding wheel is remarkably large negative angle. Therefore the deformation of the work during machining by grinding is larger than in cutting and the cutting temperature is also high than in cutting.

These facts seem to explain the difference of the hardness distribution in surface layer shown in Fig 9 and 10.

The residual stress due to machining are the result of the complex reciplocal actions of stress, strain and temperature, at near the cutting edge during machining. The clear explanations of the results of residual stress have not been obtained.

4 CONCLUSION

Turning test of hardened steels and super alloy with three types of sintered CBN tools were done.

The surface integrities of the hardened steel machined by a CBN cutting tool were studied.

The following conditions were obtained.

(1) CBN1 shows superior wear resistance and toughness in machining hardened steels. While in machining a super alloy CBN2 and CBN3 show better wear resistances than CBN1.

(2) The hardness at the surface layer of the hardened steel machined by grinding shows some drop down but that of the work machined by cutting shows little change.

(3) The very regular shaped surface was generated inmachining a hardened steel by a CBN cutting tool.

(4) No remarkable changes of the microstructure was observed in the surface layer of the work machined by a CBN tool.

(5) The residual stress at the surface layer of the work machined by a CBN tool shows fairly changes according to cutting conditions.

DISCUSSION

Paul Hartung (M.I.T.)

Many of the cutting tests described were at 100 m/min. What criterion was used in choosing this speed?

Yuichiro Kono

A range of 100-150 m/min. was used to balance machine time with tool life.

Dr. Oswald Bergmann (E.I. du Pont De Nemours)

How do the properties of your CBN compare with WC/6 C_o materials?

Yuichiro Kono

(The properties were written on the blackboard and not recorded for the proceedings - Editor)

George Wilson (General Electric - SMBD)

Why did you select a feed of .1 mm/rev. on steel? Did you try higher rates, and if so, what were the results?

Yuichiro Kono

When machining heat treated steel the roughness is important since it becomes a final surface. In this case the customer requires a surface roughness of 2-4 microns so this feed rate was chosen.

MACHINING OF SELECTED CAST ALUMINUM ALLOYS WITH COMPAX* BLANK TOOLS

George F. Wilson
Dieter K. Bruschek
General Electric Company

ABSTRACT

Machining of sand mold cast and permanent mold cast AA390 silicon-aluminum alloys (16-18% Si) and permanent mold cast AA319 aluminum alloy (5.5-6.5% Si) was investigated in General Electric Company's Specialty Materials Department Technical Service Centers in Worthington, Ohio and in Dreieich-Sprendlingen, West Germany.

The aluminum alloys were machined with cutting tools tipped with COMPAX industrial diamond blanks. These tools consist of a layer of polycrystalline Man-Made* diamond on a tungsten carbide substrate. The COMPAX blank is brazed onto a tool shank or indexable insert and then ground to a specified geometry.

A 13 inch swing Sheldon lathe, with infinitely variable spindle speed up to 4,000 rpm was used for machining these aluminum alloys in the laboratory in Worthington, Ohio. A Heidenreich and Harbeck 540mm swing lathe with infinitely variable spindle speed up to 5,600 rpm was used for this program in the laboratory in Dreieich-Sprendlingen, West Germany.

Cutting speeds between 600 m/min. (2,000 fpm) and 1,370 m/min. (4,000 fpm) were used to evaluate the performance of COMPAX blank tools machining these aluminum alloys.

A three force component piezoelectric crystal tool post dynamometer was used for measuring cutting tool forces.

*Trademark of General Electric Company, U.S.A.

Surface finish in Ra and Rt was measured with suitable surface texture measuring instruments on the machined workpieces.

Tool life determination for two aluminum alloys are presented in tabular form and as a plot of the Taylor tool life equation. Cutting forces on sharp and worn tools are presented along with the resulting workpiece surface texture. Tool life and surface texture are presented for both dry cutting and for machining with a water soluable cutting fluid.

The result of this study shows what can be achieved using COMPAX blank tooling for the machining of these cast aluminum alloys.

INTRODUCTION

Aluminum alloys have received much attention recently because of their increasing use in automotive components in the United States [1,2] and elsewhere. Initially many components made from these aluminum alloy castings were machined on transfer machines designed for cast iron workpieces. Often cutting speeds and feed rates were not the most efficient and also cutting fluid systems were required for equipment not designed for such systems. There is a vast difference in the ease or degree of difficulty in machining the various cast aluminum alloys.[3] An alloy will machine differently if it is sand mold cast, permanent mold cast, or die cast. A recent alloy development is the hypereutectic silicon-aluminum alloy with the aluminum association designation 390. This alloy has a silicon content of 16-18%. Above 11.7% silicon in an aluminum alloy there will be primary silicon present. This primary silicon is desirable in wear parts but affects cutting tool life drastically when trying to machine this material. The 390 Si-Al alloy machines differently than other cast aluminum alloys, such as 319 (5.5-6.5 Si) and 380 (7.5-9.5 Si) in that the primary silicon causes rapid tool wear. There have been laboratory studies on machining these alloys with various tool materials [4,5] and subsequent applications in manufacturing cast aluminum components with polycrystalline diamond tooling.[6,7,8]

Polycrystalline diamond, which has proven quite effective in machining both hypereutectic and hypoeutectic cast Si-Al alloys, consists of a layer of polycrystalline diamond on a tungsten carbide substrate. This cutting tool material is produced utilizing high pressure - high temperature technology with many small diamond crystals randomly orientated and with bonding by growth of the crystals to each other. The layer of diamonds, which is typically 0.5mm thick, is bonded to a tungsten carbide substrate. The polycrystalline diamond is produced in a cylindrical shape, typically 13.5mm and 8.2mm diameter, and then cut into segments that can be brazed onto a tool shank or insert. The polycrystalline diamond layer produced by the General Electric Company are made with Man-Made Diamond and the resulting cutting elements are known as COMPAX blanks.

An investigation in the machining of both sand mold cast and permanent mold cast Si-Al alloys was conducted in the laboratories of the Specialty Materials Department of General Electric Company, both in Worthington, Ohio U.S.A. and Dreieich-Sprendlingen, West Germany. Some of the results of this activity are reported in this paper.

PROCEDURES

Specialty Materials Department - Worthington, Ohio

All machining done in the laboratory in Worthington, Ohio was accomplished on a Sheldon 13 in. (330mm) tool room lathe equipped with a 5hp (3.7kW) spindle drive motor. Spindle speed on this machine is infinitely variable from 0 to 4000 rpm.

Workpiece materials used in this investigation were:

1. Sand mold cast A390 Si-Al alloy tube - 100mm (4 in.) od x 76mm (3 in.) id x 406mm (16 in.) long.
2. Continuous, permanent mold type 390 Si-Al alloy - 152mm (6 in.) diameter cut to 457mm (18 in.) lengths.
3. Continuous, permanent mold type 319 aluminum alloy - 152mm (6 in.) diameter cut to 457mm (18 in.) lengths.

The AA319 alloy specification for silicon is 5.5-6.5%. Chemical and physical specifications for these materials can be found in the literature.[2,3,5,9]

The cast workpiece materials were trimmed to avoid the as-cast surface and avoid problems with run-out. The sand cast tubes were chamfered on the I.D. of one end for a bull nose live center and chucked with a three jaw universal chuck on the other end. The solid, permanent mold castings were driven with a face driver with the workpiece being held to the spindle nose end through a hydraulically loaded live center.

A BR-12 style tool with a 19mm square shank tipped with 6.6mm length per side x 3.2mm thick COMPAX blank was used for these tests. Tool geometry for these tools was:[10,11]

Back Rake, γ_p	- 0°
Side Rake, γ_f	- 5°
End (Back) Clearance, α_p	- 5°
Side Clearance, α_f	- 5°
Minor Cutting Edge Angle, κ_f	- 15°
Lead Angle, ψ_r	- 15°
Corner Radius, r_ε	- 0.8mm (1/32 in.)

Care was taken to produce a good quality cutting edge on the polycrystalline diamond material with the COMPAX blank being polished on the rake face prior to brazing. The final grind on the relief surfaces was done with a fine diamond polishing wheel. Final inspection of the cutting edge was done at 40X with a toolmaker's microscope.

Cutting forces were continuously monitored with a Kistler 3-force component Model 9257A piezoelectric crystal dynamometer. A suitable tool holder for holding the BR-12 tool was mounted on the dynamometer's platform. This set up is shown in Figure 1.

FIGURE 1

Set-Up on Sheldon Lathe for Machining Si-Al Alloy Sand Cast Tube With Tool Mounted on Dynamometer

Machining conditions were feed 0.127mm/rev. (0.005 ipr) and depth of cut 1.27mm (0.050 in.). Cutting speed was variable for the determining of the Taylor tool life relationship.[10]

$$VT^n = C$$

Where:

V = Cutting Speed, m/min
T = Tool Life, min
n = An exponent
C = A Constant

End point criteria was a uniform flank wear scar, V_B, of 0.40mm (0.015 in.). The wear was measured after each pass of approximately 356mm (14 in.) for the sand mold cast AA390 Si-Al alloy and 457mm (18 in.) for the 390 Si-Al alloy permanent mold workpieces. A plot of flank wear, V_B, versus time was made for each tool. The time, T, to obtain the 0.40mm (0.015 in.) flank wear was determined from this plot. Each tool was photographed through the toolmaker's microscope, at 40X, prior to the test and upon completion of each test.

Surface finish was measured, on a limited number of tests, with a Talysurf 10 surface texture measuring instrument. Surface finish was measured in both Ra and Rt and a profile trace was made on the surface.

European Technical Service Center - Dreieich-Sprendlingen, W. Germany

Comparison tests were run on the permanent mold AA390 and AA319 alloys at the European Technical Service Center. A Heidenreich and Harbeck 540mm swing lathe with a 30kW SCR drive giving infinitely variable spindle speeds between 90 and 5600 rpm was used for these tests.

Tool geometry was the same as used on the 13" Sheldon lathe except 25mm square shanks were used. All tool life tests on the AA390 Si-Al alloy were run using flood applied water soluable cutting fluid. Surface finish evaluation on the AA319 alloy was run both with and without cutting fluid with 1.2mm and 1.6mm nose radius. Surface finish values in both Ra and Rt were obtained on a Perth-o-Meter surface texture measuring instrument.

The lathe with an aluminum alloy workpiece set up on the machine is shown in Figure 2.

FIGURE 2. Heidenreich and Harbeck Lathe With Aluminum Alloy Workpiece Set-Up With Workdriver on Spindle Nose

RESULTS AND DISCUSSION OF RESULTS

Cutting speeds for obtaining tool life values ranged between 457 and 914 m/min (1500 and 3000 fpm). Most tool life tests consist of at least four (4) and usually five (5) tests at different cutting speeds. The tool life curves were "curve fit" using the method of least squares.

Tool life results are given in Table I.

TABLE I

Tool Life Values Machining AA390 Aluminum Alloys

Work-piece Material	Cutting Tool Material	C	n	V_{20} m/min	V_{30} m/min	V_{60} m/min
Sand Cast	COMPAX Blank	1906	0.32	731	642	514
Permanent Mold	COMPAX Blank	1588	0.33	591	517	411
Permanent Mold**	COMPAX Blank	1211	0.23	608	554	472
Sand Cast	Tungsten Carbide K-20	329	0.21	175	161	139

**Note: With flood applied cutting fluid on Heidenreich and Harbeck lathe.

The values in Table I are from the Taylor tool life equation:

$$VT^n = C \quad (1)$$

The tool life results are illustrated graphically in Figure 3. This is the log-log coordinate plot of tool life, T, versus cutting speed, V.

These results differ somewhat from other published literature.[3,5] The sand mold cast Si-Al alloy shows better tool life values over the permanent mold cast material. For these semi-finishing operations the surface scale from each workpiece was completely removed. The machining characteristics of sand mold cast AA390.1 Si-Al alloy will improve when the as-cast surface is removed.[3] Some voids (or blow holes)

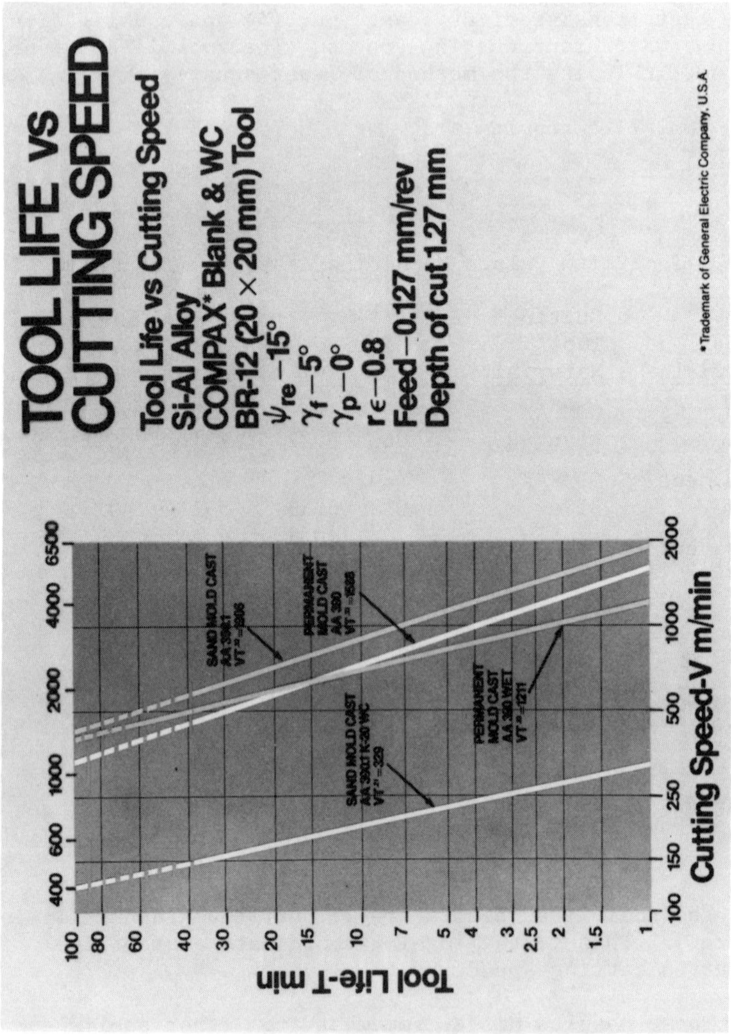
TOOL LIFE VS CUTTING SPEED
Tool Life vs Cutting Speed
Si-Al Alloy
COMPAX* Blank & WC
BR-12 (20 × 20 mm) Tool
ψ_{re}—15°
γ_f—5°
γ_p—0°
$r\epsilon$—0.8
Feed—0.127 mm/rev
Depth of cut 1.27 mm
*Trademark of General Electric Company, U.S.A.
Tool Life-T min
Cutting Speed-V m/min
100 80 60 40 30 20 15 10 7 5 4 3 2.5 2 1.5 1
400 600 1000 2000 4000 6500
100 150 250 500 1000 2000

FIGURE 3.

were encountered on some of the workpieces but these did not appear to have significant effect on the tool performance. The dispersion of silicon was not completely uniform on some of the permanent mold AA390 Si-Al alloy. When machining across a silicon deficient area built-up-edge or workpiece smearing on the rake surface of the tool could become a problem. Figure 4 shows the worst case of build-up with workpiece material smearing on the relief surface.

FIGURE 4.

Workpiece Material Build-Up on Tool Machining Permanent Mold Si-Al Alloy

Build-up of workpiece material was not a problem when cutting fluid was applied. It was somewhat surprising that improved tool life was not achieved when machining with cutting fluid on a much more rigid lathe.

Using the relationship for determining cutting speed for least cost[12] a further comparison with tungsten carbide and polycrystalline cutting tools machining this Si-Al alloy can be made. The relationship for the cutting speed for least cost, Vc, is:

$$Vc = C \left(\frac{n}{1-n}\right)^n \left(\frac{K_1}{K_1\ TCT + K_2}\right)^n \tag{2}$$

and for tool life at that cutting speed:

$$Tc = \left(\frac{1}{n} - 1\right) \left(\frac{K_1\ TCT + K_2}{K_1}\right) \tag{3}$$

where:

K_1 = Direct labor rate plus overhead rate
K_2 = Tool cost per grind, includes original and regrinding costs
TCT = Tool changing time

The values used for overhead rate and tool cost, while realistic in U.S. dollars, are for comparative purposes and may not reflect actual costs under specific circumstances.

K_1 = \$25.00/Hour = \$0.42/min

K_2 = (polycrystalline diamond) =

$$\frac{\$95.00 + (6)\ \text{Regrinds}\ (\$15/\text{Regrind})}{7}$$

= \$26.43/Grind

K_2 = (tungsten carbide) =

$$\frac{\$11.76 + (6)\ \text{Regrinds}\ (\$4.00/\text{Regrind})}{7}$$

= \$3.67/Grind

TCT = 6 Minutes

Table II tabulates the results of these calculations for cutting speed for minimum cost, Vc, and tool life for minimum cost, Tc, for machining the AA390 Si-Al alloys with polycrystalline diamond and tungsten carbide tools.

TABLE II

Cutting Speed and Tool Life for Minimum Cost

Tool Material	Workpiece Material - Cast	Vc - m/min Minimum Cost	Tc -min Minimum Cost
COMPAX Blank	Permanent Mold (Dry)	311	140
COMPAX Blank	Permanent Mold (Wet)	346	231
COMPAX Blank	Sand Mold (Dry)	387	146
Tungsten Carbide	Sand Mold (Dry)	142	55
WC,SPGN-120308 (insert)	Sand Mold (Dry)	186	15

Cutting speed and tool life for minimum cost were considered for both a milled and brazed carbide tool and a SPGN-120308, square tungsten carbide insert. Insert cost would be $4.97 for four cutting edges with $0.05 added for the tool holder and one minute allowance for indexing the insert. The cutting speed is higher with subsequent lower tool life for the most economic cutting condition because of less time to index the insert (tool changing time) and eliminating regrinding. A polycrystalline diamond insert was not considered in the evaluation because of only one cutting edge per insert and problems of holding the insert dimensions when regrinding.

An increase in productivity on a high production operation can be significant with the increase in cutting speed of 2.5 to 3X and tool life 4 to 5X for polycrystalline diamond over tungsten carbide. In actual production, increases in tool life of 40X over carbide when cutting speed used for tungsten carbide tooling was not increased have been achieved.[1] However, when using the polycrystalline diamond tool at carbide cutting speeds it is not being utilized efficiently.

A perfunctory look at surface finishes produced both cutting with and without flood applied cutting fluid was made. Cutting speeds were 610 and 457 m/min, feed 0.13mm/rev, and depth of cut 1.3mm. Nose radius of the polycrystalline tools was 1.2 and 1.6mm. Table III presents the results of surface finish measurements.

TABLE III

Surface Finish Results
Machining AA319 Si-Al Alloy

Nose Radius mm	Cutting Speed m/min	Cutting Fluid	Surface Finish μm-Ra	Surface Finish μm-Rt
1.2	610	Yes	1.45	7.7
1.2	457	Yes	1.40	7.5
1.2	610	No	1.55	8.2
1.2	457	No	1.20	8.5
1.6	610	Yes	1.10	5.8
1.6	457	Yes	1.15	5.8
1.6	610	No	1.20	7.5
1.6	457	No	1.20	6.7

All of the surface finish values were measured at a cut-off of 0.80mm. A smoother surface finish is usually expected when machining aluminum alloys with polycrystalline diamond tools. An improved finish was obtained on this AA319 Si-Al alloy when the water soluable cutting fluid was applied. The machining conditions, particularly depth-of-cut, were those for semi-finishing and not for fine finishing. Dynamic stability of cutting is affected by feed-depth of cut product; in this case depth-of-cut.[13] Large depth-of-cut can cause large cutting forces which increase deflection of the entire machining system. Because the initial tool life test parameters were selected to machine at a depth greater than the 0.8mm nose radius, the same depth-of-cut was selected when surface finish was measured. A further study of surface finishes obtainable with polycrystalline cutting tools will involve smaller depth-of-cuts.

In order to evaluate three dimensional cutting forces, it is necessary to determine some further angles involved in the cutting tool's geometry.[10,11,14,15,16,17]

Inclination angle, λ_s,

$$\tan \lambda_s = \tan \gamma_p \cos \psi_r - \tan \gamma_f \sin \psi_r \tag{4}$$

Tool orthogonal rake angle, γ_o,

$$\tan \gamma_o = \tan \gamma_f \cos \psi_r + \tan \gamma_p \sin \psi_r \tag{5}$$

Tool normal rake angle, γ_n,

$$\tan \gamma_n = \tan \gamma_o \cos \lambda_s \tag{6}$$

Referring back to the basic tool geometry as stated under Procedures, the resulting values for inclination, orthogonal rake, and normal rake angles are:

$$\lambda_s = -1.3^\circ$$
$$\gamma_o = 1.3^\circ$$
$$\gamma_n = -0.03^\circ$$

It is to be noted that the inclination angle, λ_s, is negative. This would tend to direct the chip back towards the machined surface of the workpiece effecting the resultant surface finish. The values obtained were with a large nose radius which in effect is an increasing lead angle causing the inclination angle to increase negatively and further directing the chip towards the newly generated workpiece surface.

Three mutually perpendicular force components were measured with the tool post dynamometer. These are:

$F_c = F_y$ = Cutting force acting in direction of tool travel relative to workpiece, N.

$F_t = F_z$ = Thrust force acting in direction of feed, N.

$F_r = F_x$ = Radial force component normal to workpiece axis, N.

Measured values from the cutting tests for both sand mold cast AA390.1 Si-Al alloy and permanent mold cast AA390 Si-Al alloy are given in Table IV:

TABLE IV

Measured Cutting Force Values

Material	Cutting Speed m/min	Tool Condition	F_y Newton	F_z Newton	F_x Newton
Sand Mold	762	Sharp	126.8	62.3	57.8
Sand Mold	762	Dull	266.9	404.8	409.2
Permanent Mold	670	Sharp	97.9	35.6	35.6
Permanent Mold	670	Dull	195.7	384.6	462.6

There was considerable increases in the cutting forces from a sharp tool to a dull tool with a much greater increase in the thrust force, F_t or F_z, and the radial force, F_r or F_x. Only the static force is being considered in this paper. It is to be noted, however, that the dynamic forces were much greater for a dull tool than a sharp tool.

It is necessary to resolve the measured cutting forces, F_x-F_y-F_z, into X'Y'Z' axis in order to find the resultant force in the plane of the tool face. The three component forces are F'_x acting along the cutting edge, F'_y acting normal to the rake face, and F'_z acting normal to the cutting edge in the plane of the rake face. These component forces are determined from the relationships:

$$F_x' = F_x \cos \lambda_s - F_y \sin \lambda_s \quad (7)$$

$$F_y' = F_y \cos \gamma_n \sin \lambda_s + F_y \cos \gamma_n \cos \lambda_s + F_2 \sin \gamma_n \quad (8)$$

$$F_z' = F_y \sin \gamma_n \sin \lambda_s - F_y \sin \gamma_n \cos \lambda_s + F_z \cos \gamma_n \quad (9)$$

The resultant force, R, acting on the rake face of the tool and the resultant force acting on the shear plan, R', are equal in magnitude and may be determined from the relationship:

$$R^2 = (F_x')^2 + (F_y')^2 + (F_z')^2 \quad (10)$$

The results of these calculations are presented in Table V.

TABLE V

Forces Acting in Plane of Rake Face and Resultant Force

Material and Tool	Cutting Speed m/min	F_y' Newton	F_z' Newton	F_x' Newton	R Newton
Sand Mold - Sharp	762	129.6	62.2	60.7	156.0
Sand Mold - Dull	762	272.9	404.7	415.2	640.8
Permanent Mold-Sharp	670	100.1	35.5	37.8	112.7
Permanent Mold-Dull	670	199.9	384.5	467.0	637.1

The cutting force (F_c or F_y) and its component, F_y', are approximately doubled for the dull tool over the sharp tool. The thrust force (F_t or F_z) and radial force (F_r or F_y) show increases between 6 and 12 times for the sharp tool over the dull tool. This large increase not only causes an increase in the resultant force but also in the direction in which it acts.

Also of interest is the power and specific energy required for machining these cast Si-Al alloys with polycrystalline tools. The net power, P_n, (in kilowatts) can be determined from the expression:

$$P_n = \frac{F_c V}{60\ 000} \qquad (11)$$

The rate of material removal, W_n, in kilowatts per cubic centimeter per minute can be determined from the expression:

$$W_n = \frac{P_n}{U} \qquad (12)$$

Where:

U = cubic centimeters work material removed per minute.

Table VI presents the net power and specific power consumption for these examples of Si-Al alloys:

TABLE VI

Net Power and Specific Power Consumption

Material	Cutting Speed m/min	Tool Condition	P_n kW	W_{n_3} kW/(cm/min)
Sand Mold	762	Sharp	1.61	0.013
Sand Mold	762	Dull	3.39	0.028
Permanent Mold	670	Sharp	1.09	0.011
Permanent Mold	670	Dull	2.18	0.020

The specific power consumption has approximately doubled over the sharp to the dull tool, being directly proportional, the same as the cutting force, F_c.

CONCLUSIONS

Polycrystalline diamond tools can machine the hypereutectic cast Si-Al alloys at cutting speeds three times that of tungsten carbide with tool life, at the higher cutting speed, approximately three times that of the tungsten carbide cutting tool.

This is a cutting tool material that should be considered when machining the hypereutectic aluminum alloys, particularly as their use becomes more common, in high production operations.

Some of the operating parameters for most efficient use of this cutting tool material are presented in this paper.

ACKNOWLEDGEMENTS

The laboratory work in Worthington, Ohio was done by Robert L. Payne and in Dreieich-Sprendlingen, West Germany by Reiner E. Rudolf, technicians of the Specialty Materials Department of General Electric Company. Their contribution is gratefully acknowledged.

Advice and assistance regarding use and metallurgy of cast silicon-aluminum alloys was given by John C. Miller and John L. Jorstad of Reynolds Metals Company. Their assistance is greatly appreciated.

LITERATURE REFERENCE

1. E. W. Krumrei, "Major U.S. Automotive Plant Improves Productivity by Using COMPAX Blank Tools". Proceedings of the Vth Industrial Diamond Seminar, Industrial Diamond Association of Japan. May 1978.

2. J. L. Jorstad, "The Hypereutectic Aluminum-Silicon Alloy Used to Cast the Vega Engine Block". Casting Engineering, September/October 1971.

3. J. L. Jorstad, "Influence of Aluminum Casting Alloy Metallurgical Factors on Machinability", Society of Automotive Engineers, Paper 800486 presented at Cobo Hall, Detroit, MI. February 1980.

4. M. D. Dennis and J. D. Christopher, "The Characterization and Performance of COMPAX Diamond Tools in Machining Applications". Presented at 17th Annual Seminar, Society of Manufacturing Engineers, Kansas City, MO, May 1975. SME Paper MR75-986.

5. J. C. Miller and R. E. DeVor, "An Experimental Approach to Machinability Testing", Society of Automotive Engineers, Paper 800487, presented at Cobo Hall, Detroit, MI. February 1980.

6. E. W. Krumrei, "Diamond Tooling Cuts Into Traditional Carbide Use". Tool and Production, February 1980.

7. S. K. Bhattacharyya and D. K. Aspinwall, "The Application of Polycrystalline Tooling", Proceedings of the Vth Industrial Diamond Seminar, Industrial Diamond Association of Japan, May 1978.

8. S. K. Bhattacharyya and D. K. Aspinwall, "Diamond Compacts Pay Dividends on Aluminum Alloys", Metalworking Production, February 1978.

9. J. L. Jorstad, "-390 Alloy Technology - Foundry Characteristics", Reynolds Metals Company Report, June 1975.

10. International Standard, ISO3002/1-1977(E), "Geometry of the Active Part of Cutting Tools - Part 1: General Terms, Reference Systems, Tool and Working Angles".

11. American National Standard, ANSI B94.50-1975, "Basic Nomenclature and Definitions for Single-Point Cutting Tools".

12. W. W. Gilbert, "Economics of Machining", Machining - Theory Practice, PP. 465-485, American Society for Metals, 1950.

13. B. F. von Turkovich, "Influence of Very High Cutting Speeds on Chip Formation Mechanics," Appendix C, Advanced Machining Research Program, Report No. SRD-80-018, 15 February 1980.

14. M. C. Shaw, "Metal Cutting Principles", 3rd Edition, 2nd Printing, The Technology Press, Massachusetts Institute of Technology, Cambridge, MA, 1960.

15. K. J. Trigger, "Metal Cutting Principles", Machining With Carbide and Oxides, Chapter 2, printed for ASTME by McGraw-Hill, 1962.

16. A. O. Schmidt, I. Ham, and G. F. Wilson, "Comparative Turning Performance Study of Different Tool Geometries", Proceedings of the 7th International M.T.D.R. Conference, 1966, Perganon Press 1967.

17. G. F. Wilson, "Geometry of 3-Dimensional Chip Formation", ASTME Paper MR68-611, presented at Metal Cutting Symposium, Cincinnati, OH, April 1968.

DISCUSSION

Dr. J. Hazra (General Electric - Carboloy)

What is the economic speed-life relationship of carbide on 390 type material corresponding to COMPAX* blanks?

George Wilson

This relationship, for a sample turning operation, is presented in Table II of the paper. Vc for a COMPAX blank tool is 387 m/min. with a tool life of 146 minutes, while Carboloy grade 883 WC brazed BR-12 tool has a Vc of 142 m/min. with a tool life of 55 minutes. This does not take into account surface finish requirements or other considerations that would be found on automatic equipment such as transfer lines. These other considerations include the effect of other operating stations and the rework of parts after tool change because of adjustments for obtaining size. The source for determining most economic cutting speed is Literature Reference Number 12 in the paper.

Dr. O.R. Beramann (E.I. duPont de Nemours & Co.)

What is the grain size in the different types of COMPAX blanks, and the market outlook for COMPAX blanks in cutting tools in the future?

George Wilson

The specific grain size used in COMPAX blanks is proprietary. The 1500 series product, which is the COMPAX blank used on the tools reported on in this paper, has a coarser particle size than does 1300 size products (which have been used successfully machining the hypereutectic Si-Al alloys). A fine grain COMPAX blank has recently been introduced for fine finishing operations. This 1600 series product has a closely controlled distribution of fine particle size diamonds. It is suspected that this question comes from David Aspinwall's comments regarding use of STRATAPAX drill blanks for milling of aluminum castings. The STRATAPAX blank product has a relatively large particle size and is designed for use in oil well drilling bits and mining tools where toughness is required. They are not generally recommended for machining applications.

* Trademark of General Electric Company, U.S.A.

The Market outlook for COMPAX blanks is very good. They have solved and will continue to solve many machining problems in the automotive industries for machining both the hyper and hypo-eutectic Si-Al alloys. They have replaced or supplemented mined diamond tools for machining electric motor commutators. They are being used in increasing numbers machining non-metallic materials such as graphite, wood products, and both green and fired ceramics.

Dhioaj Shah (Valeron Corporation)

Do you see any impurities in your polycrystalline diamond grades? If there are any, what are those and do they play an important role in cutting Si-Al alloy.

George Wilson

There would be little, if any, impurities found in COMPAX blanks. There are very rigorous quality control procedures in their manufacture. There may be some sintering agent(s) present in very small amounts in the polycrystalline diamond mass. These are in such small amounts that they would not be a factor in machining Si-Al alloys.

SESSION VI

CERAMIC TOOL MATERIALS

CUTTING PERFORMANCE OF CERMET TOOLS

N. NARUTAKI* and Y. YAMANE**

*Professor, Hiroshima University, 3 Senda Machi, Hirshima 730, JAPAN
**Lecturer, Hiroshima Denki Institute of Technology, 595 Nakano, Senogawa, Hiroshima 739-03

1. INTRODUCTION

Conventional cermet tools (carbide type) sometimes show an excellent cutting performance under a certain situation of metal cutting. But the application field has been quite limited because of their low toughness and large plastic deformation at an elevated temperature.

In order to overcome the above shortness, another carbides such as WC, TaC, etc. and nitrides such as TiN, TaN, etc. have been added to the tool recently (nitride type). It has been said that the addition of these materials made the tool grains small and has improved the cutting performance of the tool largely. But the basic data of the cutting performance of these tools have hardly been reported so far.

In this paper therefore, the results of cutting tests and static diffusion tests were introduced to understand the wear characteristics of cermet tools. Tool life and cutting temperature were measured to evaluate the cutting performance macroscopically and the diffusion between cermet tools and work material or non-metallic inclusions were examined to discuss the metallurgical behviour of the tool.

2. CUTTING TEST

2-1 TEST PROCEDURE

Dry turning tests were conducted under the conditions shown in Table 1. Chemical compositions of tested cermet tools analyzed by EPMA are shown in Table 2. The prefix CT means a carbide type cermet tool and NT means a nitride type cermet tool. Steel cutting grade carbide tool P10 and pure alumina ceramic tool C1 were also used as comparisons. According to the manufacturer's catalogue, NT 1, 2 and 3 contain TiN, while NT 4 and 5 contain TaN. Referring to Table 2, nitride type cermets contain more Ta and W than carbide type cermets.

The work material machined was Al-deoxidized medium carbon steel S45C and the chemical compositions were 0.46%C, 0.26%Si, 0.75%Mn, 0.02%P, 0.02%S, and 0.12%Cr. The average hardness of the workpiece was HB 167.

Cutting temperature was measured by tool-work thermocouple method using the cermet tools and carbide tool P10. The calibration of temperature was operated up to 1300°C using the tested tool inserts.

Table 1. TURNING TEST CONDITIONS.

WORK MATERIAL	S45C
CUTTING SPEED	50 ~ 500 m/min
DEPTH OF CUT	1.0 mm
FEED	0.21 mm/rev
TOOL GEOMETRY	-5, -6, 5, 6, 15, 15, 0.8 mm

Table 2. CHEMICAL COMPOSITIONS OF THE TESTED CERMET TOOLS ANALYZED BY EPMA.

Tool	Chemical compositions (%)							
	Ti	Ta	W	Mo	Ni	Co	C	Total
CT1	44	7	3	11	16	-	8	89
CT2	54	-	3	7	19	-	8	91
CT3	44	10	2	7	8	4	7	82
CT4	57	1	1	10	12	-	9	90
CT5	49	4	3	9	11	-	9	85
NT1	38	7	12	8	16	-	8	89
NT2	40	8	14	4	6	6	6	84
NT3	36	-	18	12	8	4	7	85
NT4	36	10	14	9	5	9	7	90
NT5	44	4	2	14	20	-	7	90

2-2 TEST RESULTS

Fig. 1 shows the V-T curves of the tested tools. The tool life was determined by the flank wear (VB = 0.3 mm). Fig. 1a) is the case of carbide type cermets and there are considerable difference in tool life among the tools at low cutting speed range. In addition, a change in the inclination of V-T curve is recognized around the cutting speed of 300 m/min.

On the other hand, the V-T curves of nitride type cermets [Fig. 1b)] are straight lines within the cutting speed range of this experiment and the tool life curves are almost the same except NT 5. But, the tool lives of nitride type cermets are a little shorter than those of carbide type

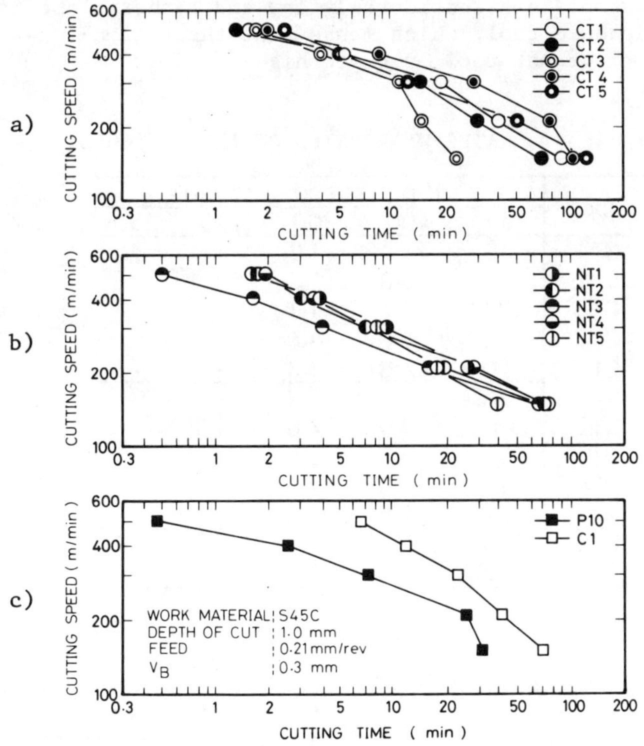

Fig. 1 TOOL LIFE CURVES OF TESTED TOOLS.

cermets between the cutting speed of 150 and 400 m/min. This means that the wear resistance of nitride type cermet is a little lower than that of carbide type cermet.

Comparing with the other tool materials, both cermets have longer tool life than carbide tool P10 within the cutting speed range of this experiment, and carbide type cermet has almost the same tool life as ceramic tool C1 at the cutting speed range lower than 300 m/min. At higher cutting speeds, ceramic tool is superior to the other tools.

Table 3 shows the crater wear rates of the tools for each cutting speed. According to this table, the crater wear of cermet tool occurs above the cutting speed of 300 m/min. and the wear rates show no significant difference between the two types of cermet tools.

Comparing with the wear rate of carbide tool P10, those of cermet tools are considerably low and rather close to that of ceramic tool, which means that the thermal wear resistance of cermet tool is quite high.

Table 3 CRATER WEAR RATES OF TESTED TOOLS.

Tool	Cutting speed (m/min)				
	500	400	300	210	150
CT1	35	11	1.7	-	-
CT2	20	6.0	1.7	-	-
CT3	23	4.5	1.1	-	-
CT4	19	5.8	1.0	-	-
CT5	20	4.3	1.3	-	-
NT1	38	11	2.2	-	-
NT2	18	7.0	2.0	1.3	-
NT3	33	8.1	0.8	-	-
NT4	23	5.3	2.1	-	-
NT5	21	5.9	1.6	-	-
P10	480	105	20	3.1	1.0
C1	3.0	2.0	1.1	-	-

Fig. 2 shows the average cutting temperature of each tool plotted with regard to cutting speed. When the cutting speed is less than 210 m/min., the temperature difference among the cermet tools is very little. But at higher cutting speeds, the difference becomes larger with the increase

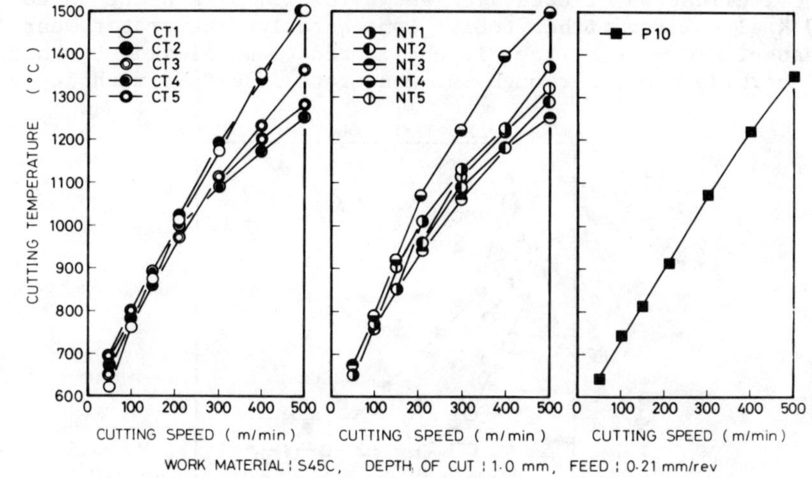

Fig. 2 AVERAGE CUTTING TEMPERATURE OF TESTED TOOLS.

of cutting speed. This is probably due to the difference in thermal conductivity of tools at elevated temperatures, and nitride type cermets show a little lower temperature than carbide type cermets. Compared with carbide tool P10, the cutting temperature of cermet tool is a little higher regardless of its type.

The wear of cutting tool is closely related with the cutting temperature and it has been said that the wear of carbide tool is describable as a rate process (1) to (4). In this experiment, the relation between cutting temperature and crater wear rate or flank wear rate was examined with the two types of cermet tools from the above viewpoint.

Fig. 3 shows the relation of the flank wear depth per unit cutting length $dVB\sin\alpha/dL$ versus the reciprocal of cutting temperature 1/°K. As can be seen in this figure, the flank wear of the tested cermet tools CT1 and NT1 show characteristics of rate process above the temperature around 850 to 900°C. Namely, thermal flank wear becomes remarkable above this temperature on these tools.

Fig. 4 is the case of crater wear and in this figure it is found that the crater wear of cermet tools occur above 1100°C. The activation energy of the tools calculated from the inclination of each line shown in Fig. 4 is about 50 to 60 Kcal/mol for either tool. Consequently, the crater wear mechanism of cermet tool is considered to be similar to that of carbide tool, although the wear rate itself is much less.

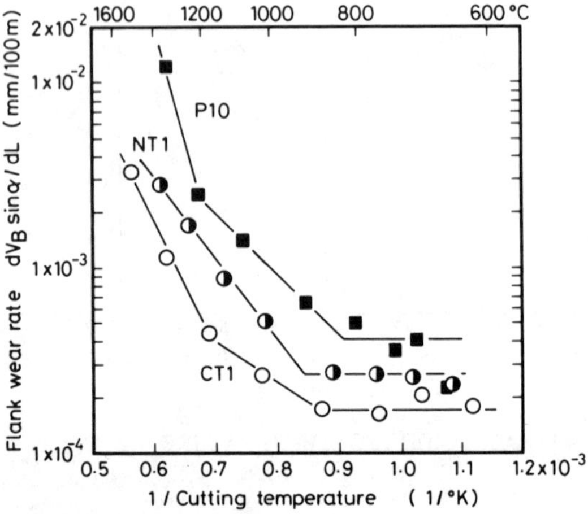

Fig. 3 RELATION BETWEEN THE FLANK WEAR RATE AND THE RECIPROCAL OF CUTTING TEMP.

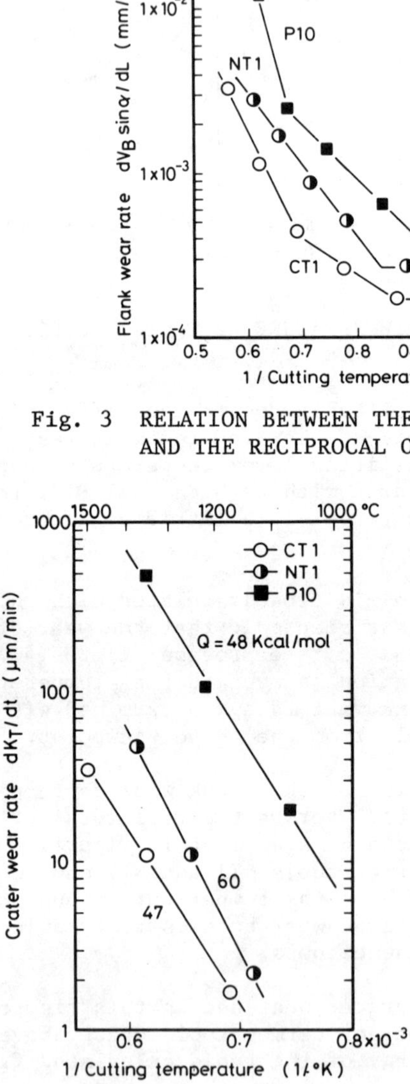

Fig. 4 RELATION BETWEEN THE CRATER WEAR RATE AND THE RECIPROCAL OF CUTTING TEMP.

3. DIFFUSION TESTS

3-1 TEST PROCEDURE

Static diffusion tests were conducted using a universal testing machine equipped with an electric oven. Couples of tools and work materials were set in a graphite vessel (samples were insulated by ceramic plates from the vessel) and held for 30 minutes under the programmed temperature and pressure.

Four kinds of tests were designed to examine the effect of the following variables on the diffusion.

1) temperature
2) compositions of the tool
3) pressure
4) non-metallic inclusions

After the tests, samples were cut vertical to the tool-work interface and polished to serve microscopic observation and EPMA analysis.

3-2 TEST RESULTS

Fig. 5 shows the change in microstructure of cermet tool CT1 and 0.15%C steel S15C after the reaction test under various temperature. As can be seen in this figure, the width of the reaction zone near the interface and the diffusion of carbon (judged from pearlite area in the steel) increase with the raise of reaction temperature. Especially at 1300°C, the reaction zone becomes very wide and cannot be distinguished from the tool surface, and the diffusion of carbon from the tool is also very drastic.

Fig. 6 is an example of the result of line analysis by EPMA across the tool-work interface. As shown in this figure, the reaction zone mainly consists of Ni and Fe, and another elements such as Ti, Ta, W and Mo are also detected slightly. The behviour of Ni is very much alike as that of Co in carbide tools (5).

Fig. 7 shows the compositions in the reaction zone of each tool reacted with S15C under the conditions of 1200°C, 62 kg/cm^2 and 30 min. In this figure, it is recognized that more W and Mo diffuse from nitride type cermets than carbide type cermets. Referring to Table 2, nitride type cermets

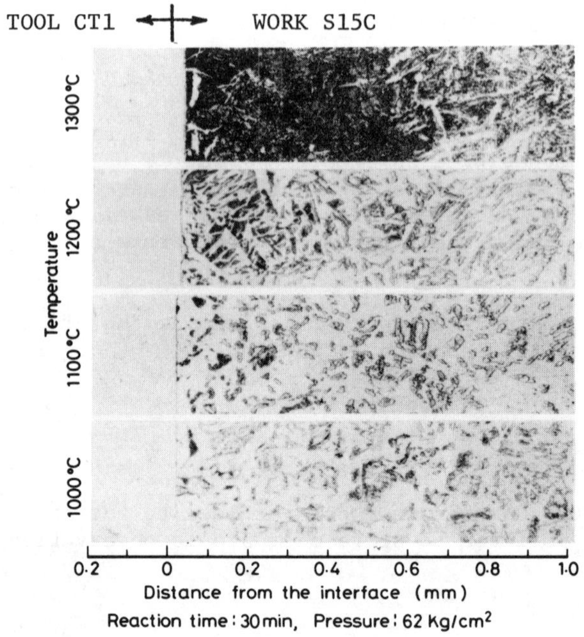

Fig. 5
CHANGE IN MICRO-STRUCTURE OF THE REACTED COUPLE.

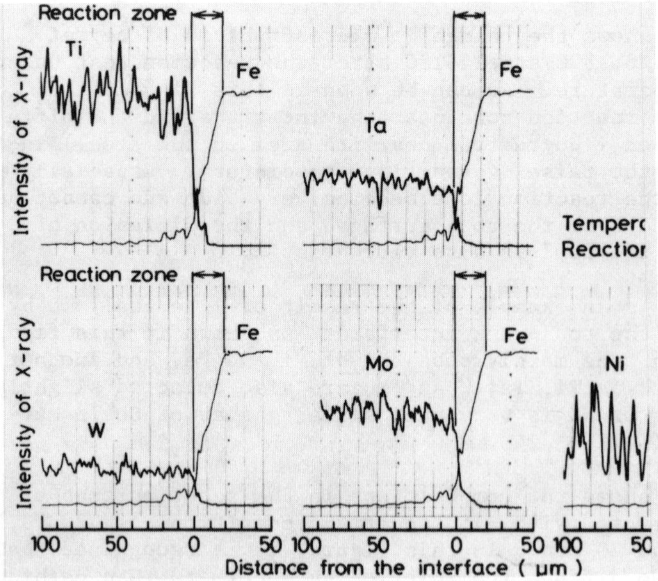

Fig. 6 EPMA LINE ANALYSIS DATA ACROSS THE REACTION ZONE.

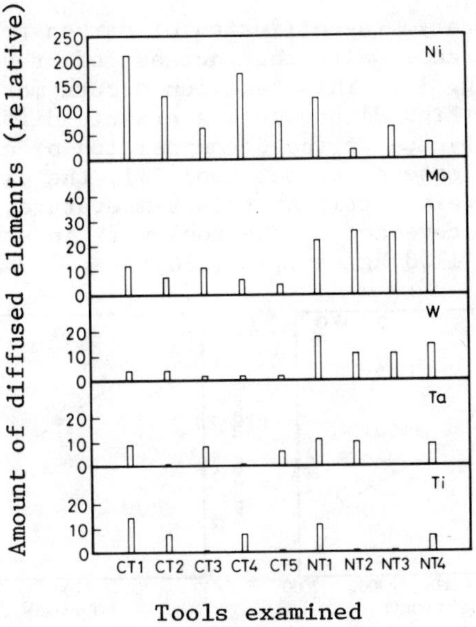

Fig. 7 COMPOSITIONS OF THE REACTION ZONE OF EACH TOOL (trace Fe)

contain more W and this might be the reason of the above phenomenon with W. Concerning with Mo, the content in the tool is almost the same with the two types of cermets. But, according to Suzuki et. al. (6), (7), Mo in carbide type cermets exists as a solid solution of (Ti-Mo)C around the grain of TiC, while in nitride type cermets the binder phase become to contain more Mo since Mo_2C does not be occluded in TiN. Consequently, the difference in Mo content in the binder phase is considered to be the reason that nitride type cermet generates more diffusion of Mo than carbide type tool. The diffusion of the other elements of the tool such as Ti, Ta and Ni changes rather randomly and does not depend on the types of the tool.

Fig. 8a) and 8b) illustrate the effect of pressure on the diffusion of the tool binder and Fe. In the case of carbide tool P10 [Fig. 8a)], the diffusion of Co decreases with the increase of pressure, while that of Fe is independent from the pressure. In the case of cermet tool CT1 [Fig. 8b)], the diffusion of Fe and the tool binder Ni show quite a similar tendency as that of Fe and Co in carbide tool.

On the contrary, the diffusion of carbon from the tool to the steel increases with the increase of pressure as shown in Fig. 9 and Fig. 10. This reaction occurs more remarkably on carbide tool [Fig. 9] and such a drastic diffusion of carbon is probably due to the decomposition of carbides in the tool. In the case of cermet tool CT1, the diffusion of carbon is relatively little at this temperature, but plastic deformation is recognized at the tool-work interface when the pressure exceeds 1100 kg/cm^2 [Fig. 10].

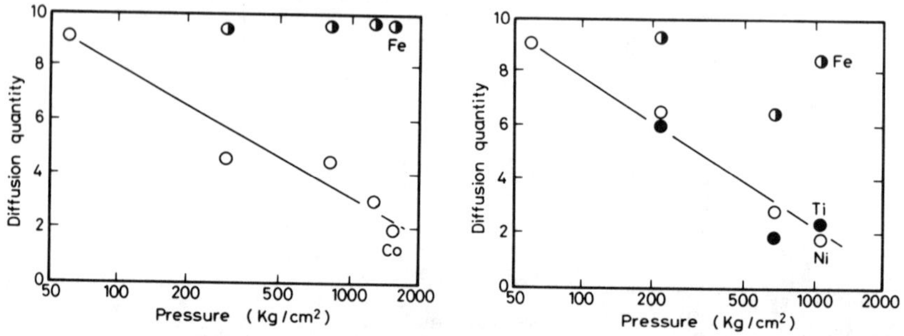

a)
Carbide tool P10

b)
Cermet tool CT1

Fig. 8 EFFECT OF PRESSURE ON THE DIFFUSION OF TOOL BINDER AND Fe.

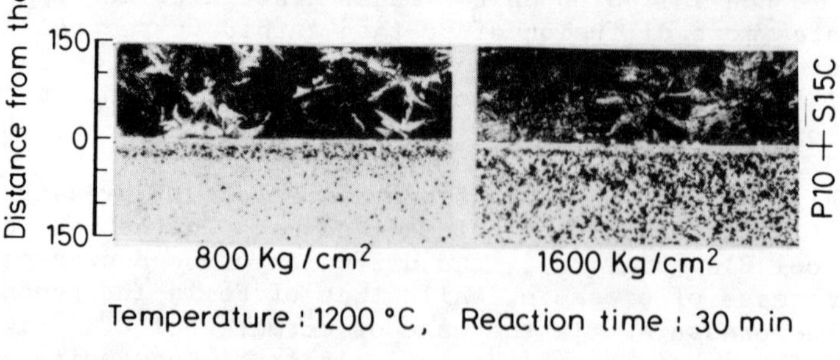

Fig. 9 EFFECT OF PRESSURE ON THE DIFFUSION OF CARBON FROM CARBIDE TOOL P10.

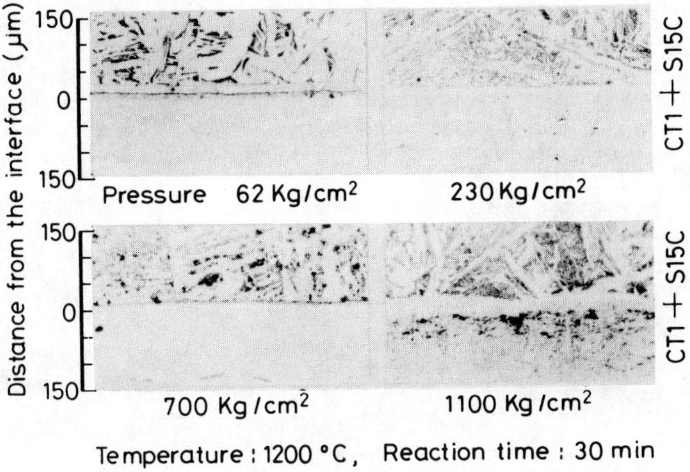

Fig. 10 EFFECT OF PRESSURE ON THE DIFFUSION OF CARBON FROM CERMET TOOL CT1.

Since the pressure on the tool surface during actual metal cutting is much higher, the diffusion of carbon or the decomposition of tool carbides is considered to have more important meanings for the tool wear than the diffusion of tool binder. Considering the above results, the reason that the thermal wear resistance of cermet tool is higher than that of carbide tool is due to the less diffusivity of carbon from the tool.

The reaction of cermet tool with non-metallic inclusions were examined for various couples, and examples with MnO and MnS are shown in Fig. 11 and Fig. 12 respectively. In Fig.11, a reaction zone where Ti, Mn and O are detected with a certain ratio is recognized between cermet tool CT3 and MnO. Neither Ni nor C was detected in this zone. Therefore, this zone is considered to be made of the complex oxide of (Ti-Mn) O, and the Ti in this zone is supplied from the tool through the decomposition of TiC. Consequently, MnO is considered to have harmful effect on TiC containing tools such as cermet tools and steel cutting grade carbide tools.

Concerning the reaction with MnS, many carbide grains are detached from the tool surface and flow into the MnS zone as shown in Fig. 12. According to the phase diagram of Mn-Ni and Ni-S (8), the eutectic point of these mixtures are very

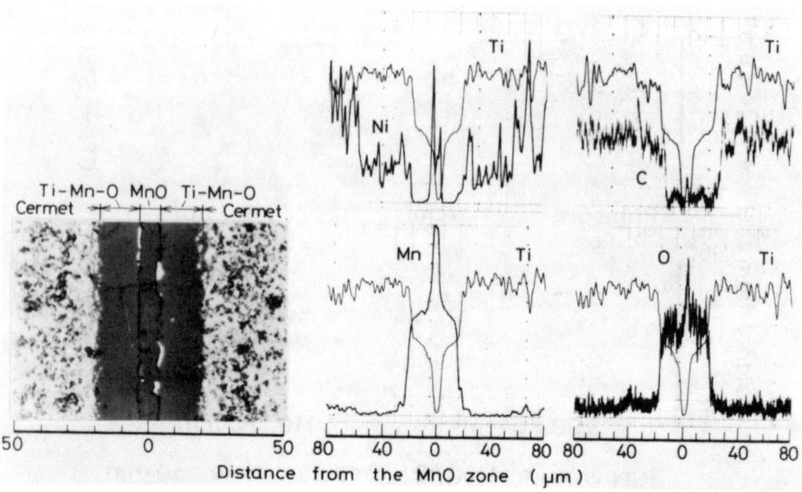

Fig. 11 DIFFUSION BETWEEN CERMET TOOL CT3 AND MnO.

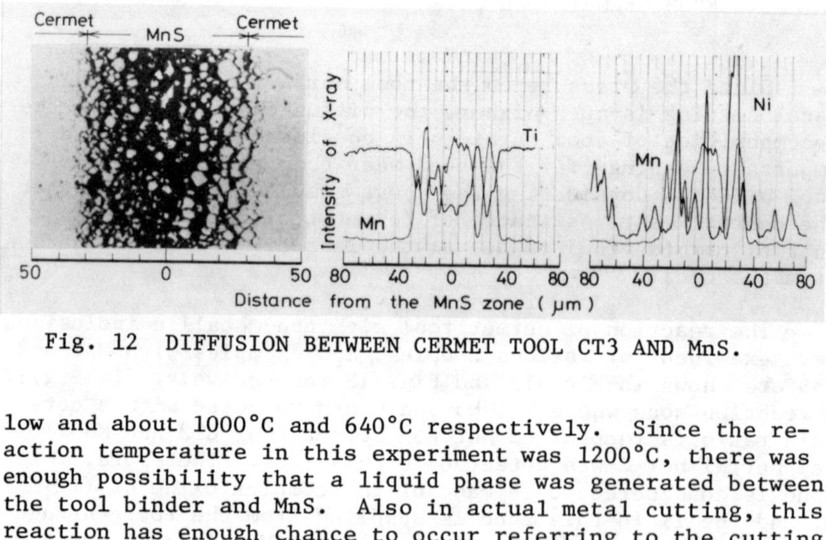

Fig. 12 DIFFUSION BETWEEN CERMET TOOL CT3 AND MnS.

low and about 1000°C and 640°C respectively. Since the reaction temperature in this experiment was 1200°C, there was enough possibility that a liquid phase was generated between the tool binder and MnS. Also in actual metal cutting, this reaction has enough chance to occur referring to the cutting temperature shown in Fig. 2. Therefore, it is considered to be unfavourable to machine high MnS content steels with cermet tools from the viewpoint of thermal reaction.

4. CONCLUSIONS

Commercially available 5 kinds of carbide type cermet tools and 5 kinds of nitride type cermet tools together with a carbide tool and a ceramic tool were tested to evaluate their cutting performance in machining of carbon steels. In order to clarify the thermal wear of cermet tools, diffusion between the tool and steels or non-metallic inclusions in steels were examined.

Main results obtained are as follows;

1] Both types of cermet tools showed longer tool life than carbide tool P10 in turning operation, and the wear resistance of carbide type cermet seemed to be slightly higher than that of nitride type cermet.

2] Cutting temperature of cermet tools were a little higher than that of carbide tool P10 at high cutting speed range.

3] Thermal wear of cermet tools tend to become remarkable above 900°C on the flank surface and above 1100°C on the rake surface.

4] Diffusion of Fe and Ni were recognized above 1000°C on cermet tools. When the temperature became 1300°C, a thick reaction zone which contained Fe, Ni, Mo, W and Ti was generated near the tool surface and drastic diffusion of carbon occurred from the tool to steel.

5] The diffusion of carbon from the tool tends to increase with the increase of pressure on the tool surface, while that of tool binder tends to decrease. Consequently, in actual metal cutting operation the diffusion of carbon is more important for the tool wear than the reaction of tool binder.

6] Cermet tools react with non-metallic inclusions such as MnO and MnS quite actively. MnO reacts with the tool grain TiC and makes (Ti-Mn)O solid solution, while MnS reacts with the tool binder Ni and lowers the melting point of the binder.

REFERENCES

(1) K.J.Trigger and B.T.Chao, "The Mechanism of Crater Wear of Cemented Carbide Tools.", Trans. ASME, Vol. 78, 1956, p. 1119.

(2) T.Kitagawa, T.Shirakashi and E.Usui, "Characteristic Equation of Crater Wear - Study on Analytical Prediction of Cutting Tool Life (1st Report).", J. of JSPE, Vol. 42 No. 12, 1976, p. 1178.

(3) N.H.Cook, "Tool Wear and Tool Life." Trans. ASME, Vol. 95, 1973, p. 931.

(4) H.Takeyama and R.Murata, "Temperature Dependence of Tool Wear.", J. of JSPE, Vol. 27, No. 1, 1961, p. 33.

(5) N.Narutaki and Y.Yamane, "Wear Mechanism of Carbide Tool Based on the Reaction between Tool and Work Material. (Part 1 Reaction Test)", Bull. of JSPE, Vol. 10, No. 3, 1976, p. 95.

(6) H.Suzuki, K.Hayashi and O.Terada, "Mechanisms of Surrounding Structure Formation in Sintered $TiC-Mo_2C-Ni$ Alloy.", J. of JSM, Vol. 35, 1971, p. 936.

(7) H.Suzuki, K.Hayashi, T.Yamamoto and W.J.Lee, "Effects of Additional Titanium Nitride on Strength of $TiC-Mo_2C-Ni$ Alloy.", J. of J. Soc. of Powder and Powdwe Metallurgy, Vol. 23, No. 7, 1976, p. 224.

(8) ASM, Metals Handbook, Vol. 8, Metallography, Structures and Phase Diagrams, 1973, p. 317 and p. 325.

DISCUSSION

Tom Skingle (Acme-Cleveland Development Co.)

How did you measure tool/chip interface temperatures?

Dr. Narutaki

We used a tool holder thermocouple.

Dr. R. Wertheim (Iscar, Israel)

What edge preparation did you use on the cermet tools? Have any experiments been done to evaluate the effect of edge preparation on the life of cermet tools?

Dr. Narutaki

All tools were received with a .02 mm by 20 degree chamfer. We did not test different edge geometries, but have not had any trouble machining carbon steels with the above-mentioned.

MACHINING PERFORMANCE OF CERAMIC TOOLS

S.T. Buljan
V.K. Sarin
GTE Laboratories, Inc.
Waltham, Massachusetts 02254

INTRODUCTION

In one form or another, ceramic cutting tools have been available for forty years. During this period, a better understanding of microstructure-property relationships and their influence on performance, in conjunction with advancements in processing techniques has led to a gradual enhancement in the properties of these tools. By combining these materials with improved machine tools, an exponential increase in productivity capability has been realized, as measured by increased cutting speed. It is imminent that this trend will continue through new developments due to economic desires to minimize machining costs and maximize productivity.

The choice of cutting tool material for a specific application is governed by the material being cut and the nature of the machining operation, its performance being dependent on several of the material's intrinsic properties, i.e. high temperature stability and oxidation resistance, chemical reactivity, elastic modulus, expansion coefficient, ambient and elevated temperature strength and wear resistance. In spite of the capability of Al_2O_3 based ceramic tools to be operated at cutting speeds well above those used for conventional cemented carbides, they have not gained widespread acceptance. Their major limitation can be principally attributed to low shock resistance and poor fracture toughness.

Development work in the ceramic cutting tool area has been geared towards improvements in fracture toughness. Fracture toughness of hot pressed Al_2O_3 + TiC composites, today, is significantly higher than that of sintered Al_2O_3 ceramics.

Silicon nitride (Si_3N_4) has long been recognized as one of the toughest ceramic materials available. Its potential as a cutting tool material has not been fully realized. This in part is due to intense interest directed at its use as structural components in turbine related areas, which has overshadowed any other developments.

Si_3N_4 is a compound of highly covalent bond character and as such requires sintering aids to achieve theoretical densities. For this purpose, a wide variety of compounds (mainly oxides and nitrides) have been investigated, yielding a number of materials with a wide spectrum of properties. Except for some silicon-aluminum oxinitride (SiAlON) compositions (1,2) the performance of a majority of these materials has not been evaluated in cutting applications. The limited testing of SiAlON cutting tools has indicated a definite potential.

Based on the growing usage of Al_2O_3 based ceramic tools and the indicated potential of SiAlON's, an investigation to study the machinability characteristics of some ceramic based cutting tools was conducted. The prime goal of this study was to evaluate and compare the machining performance of state of the art Al_2O_3 based ceramic tools to some Si_3N_4 based tools developed at GTE Laboratories.

EXPERIMENTAL PROCEDURE

Materials studies in this investigation can be divided into three groups.

(1) commercial sintered Al_2O_3 tools
(2) commercial hot pressed Al_2O_3 + TiC tools
(3) GTE hot pressed Si_3N_4 based tools

Tool performance was evaluated on SNG 432 inserts with equivalent edge preparation on a 40 hp N/C lathe equipped with a direct tool wear monitoring system. This system consists of a device placed adjacent to the tool which measures and registers workpiece radius change (ΔR) due only to the tool wear and deformation during machining. In all real life machining operations, it is the dimensional tolerance of the workpiece that determines the useful life span of the tool. Workpiece radius change (ΔR) is geometrically related to the traditionally measured tool wear indicators (i.e. the width of the wear land developed at the tool nose). Cutting time vs. ΔR, which was monitored dynamically in the process of machining, provided data for characterizing tool performance. In each comparative test, five specimens of each material were run. In addition, tool performance was also evaluated using the conventional techniques of measuring flank, nose and crater wear at several intervals during some of the tests. These measurements were then related to workpiece radius changes for control purposes.

In order to establish a baseline for comparison of these materials, several properties (i.e. fracture toughness, hardness, and abrasive wear resistance) were also measured. Relative fracture toughness values were obtained from an indentation fracture test utilizing a Vickers diamond pyramid indenter. In this test, the length of cracks developed at the corners of the indentation and the indentation size are used to obtain fracture toughness (K_H) values from a relationship proposed by Evans and Charles (3).

$$K_H = 0.113\ HD^{1/2} \left[1 + \frac{C_L}{2D}\right]^{-3/2}$$

where

K_H = indentation fracture toughness

H = hardness in GN/m^2

D = indentation diagonal in microns

C_L = total crack length in microns

Abrasion wear resistance was evaluated utilizing a pin and wheel method. The same samples (SNG 432) which had been used for turning, served as the pin and were abraded under constant load by a rotating metal bonded diamond wheel. All test specimens were run concurrently with a standard in order to minimize secondary effects (i.e. degradation of abrasive media cutting action).

RESULTS AND DISCUSSION

The complexity of tool wear mechanisms prevents meaningful extrapolation of cutting performance from structure and room temperature mechanical properties. Nevertheless, it should be noted that some of the materials intrinsic properties can be used to indicate or define certain relative trends with respect to performance in specific applications.

The two parameters most often considered are wear or abrasion resistance and toughness. The lack of widespread acceptance of ceramic cutting tools is believed to be primarily due to their low shock resistance, which increases the probability of catastrophic failure. It is therefore evident that one of the primary factors to be considered for the development of improved ceramic cutting tool materials is to increase their fracture toughness without considerable loss in their wear resistance.

Table I lists hardness and indentation fracture toughness (K_H) values of several commercial Al_2O_3 and Al_2O_3 - TiC tools obtained from four major suppliers (A,B,C,D) and a series of experimental silicon nitride based compositions. One member of the SiAlON family of materials was also included for comparison. The SiAlON (silicon aluminum oxinitride family of ceramics) encompasses a wide range of compositions of which the selected SiAlON composition represents only a small part. Three distinct groups differing significantly in both hardness and K_H are evident. As can be seen, Al_2O_3 + TiC tool materials exhibit the highest hardness (17.0 to 17.4 GN/m^2) and have significantly improved fracture toughness (3.1 to 3.5 $MN/m^{3/2}$) over Al_2O_3 tools (2.2 to 2.5 $MN/m^{3/2}$), while Si_3N_4 based materials have the highest fracture toughness (3.6 to 5.2 $MN/^{3/2}$) and the lowest hardness (12.2 to 15.2 $MN/m^{3/2}$). While it is safe to assume that this increase in fracture toughness of the Si_3N_4 based materials could bring about a substantial improvement in the reliability of the cutting tool, the question arises as to how much its wear resistance is compromised due to its lower hardness.

In order to examine the relationship among hardness, fracture toughness and abrasion wear resistance, several specimens from each group were tested. Relative wear rates were obtained by calculating volume change calibrated to the standard which was run concurrently with each specimen. Relative wear vs. respective hardness and fracture toughness are represented in Figure 1.

The harder Al_2O_3 + TiC materials, as would be expected, were found to have superior resistance to abrasion wear than the sintered Al_2O_3 specimens. Although lower in hardness, the Si_3N_4 based materials exhibited resistance to abrasion wear which was similar to the Al_2O_3 based materials. In fact, systematic changes (from compositions of 5 to 4 to 3) resulted in a material (composition 3) which was more resistant to abrasion wear than all the samples tested.

Table 1. Hardness and Relative Fracture Toughness of Investigated Tool Materials

SOURCE	HARDNESS		FRACTURE TOUGHNESS	MATERIAL
	R_A	KHN GN/m^2	K_H $MN/m^{3/2}$	
A_1	92.6	15.6	2.2	Al_2O_3
B_1	93.6	15.9	2.5	
C_1	92.5	15.8	2.2	
D_1	93.2	15.3	2.4	
A	94.4	17.0	3.1	Al_2O_3 + TiC
B	94.0	17.4	3.5	
C	94.3	17.0	3.5	
D	94.4	17.4	3.3	
1	92.6	13.5	4.8	Silicon Nitride Based
2	93.8	14.0	4.2	
3	93.5	15.2	4.3	
4	92.9	13.7	4.4	
5	92.9	14.3	4.5	
6	92.7	14.4	5.2	
7	92.1	13.8	4.9	
8	92.6	12.9	3.6	
9	90.8	12.2	4.0	SiALON

Considering the nature of abrasive wear, it would stand to reason that both hardness and fracture toughness combine to determine the overall abrasion wear resistance. Thus, materials with higher toughness can exhibit wear resistance comparable to harder and less tough material.

Relative performance in machining was determined via turning gray cast iron (ASTM A48 class 30) using a wide range of feed rates. These ranges were selected in order to investigate both the wear and shock resistance of these materials. The superiority of ceramic cutting tools over CVD TiC coated WC-Co cemented carbide at high speeds (1000 sfpm) becomes clearly evident (Figure 2) in a plot of running average of bar radius increase due to tool wear as a function of cutting time.

Although specimen A (Al_2O_3 + TiC) shows slightly superior wear characteristics, it can be easily grouped with samples 4 and 5 (Si_3N_4 based) in comparison to the relatively inferior performance of specimen 9 (SiAlON) in machining cast iron. It should also be noted (Table I) that specimen 9 was

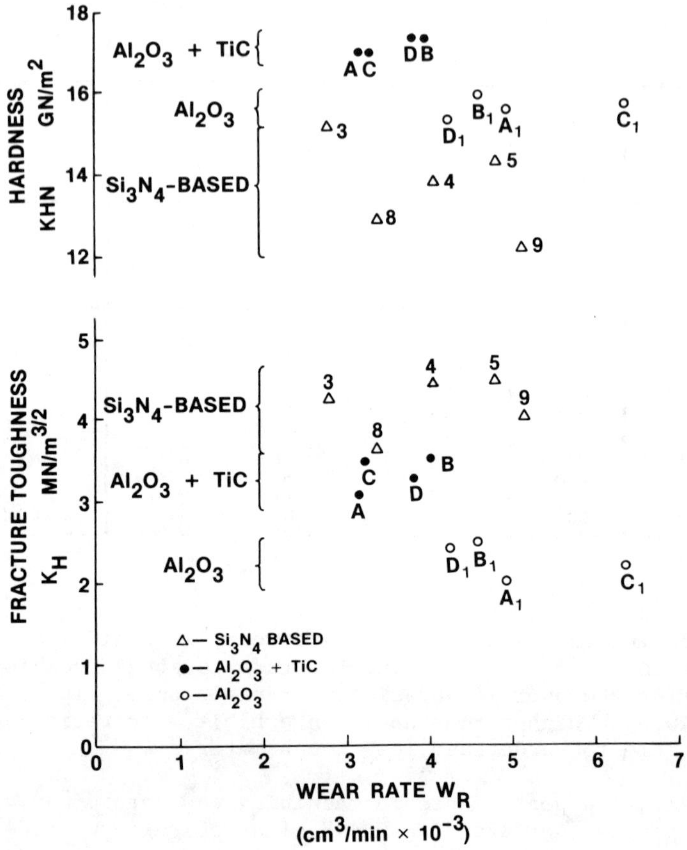

Figure 1. Wear rate, hardness and fracture toughness of alumina and silicon nitride based ceramics

found to exhibit the lowest hardness (12.1 GN/m^2) and abrasion wear resistance (Figure 1) among the Si_3N_4 based group of materials. Improved tool reliability stemming from the higher fracture toughness of the Si_3N_4 based materials is evident in Figure 3. Although the Al_2O_3 + TiC (sample A) materials demonstrate slightly better wear resistance than the Si_3N_4 based materials (sample 3) their lower fracture toughness resulted in a much higher probability of catastrophic failure. In this particular test, four out of five tools either failed on contact or after a relatively short time into the cut, whereas none of the five Si_3N_4 based tools failed.

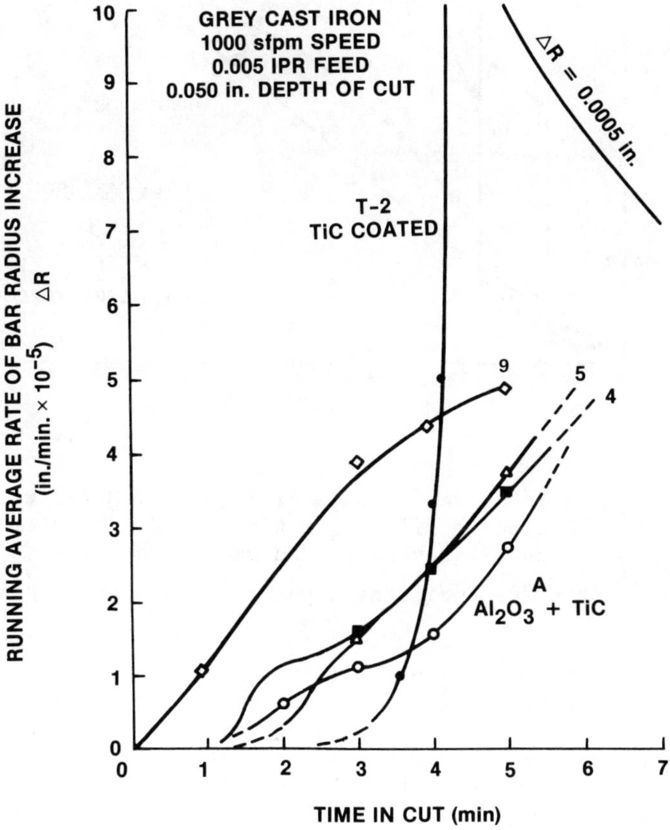

Figure 2. Performance of selected ceramic cutting tools in gray cast iron machining compared to TiC coated WC-Co tool

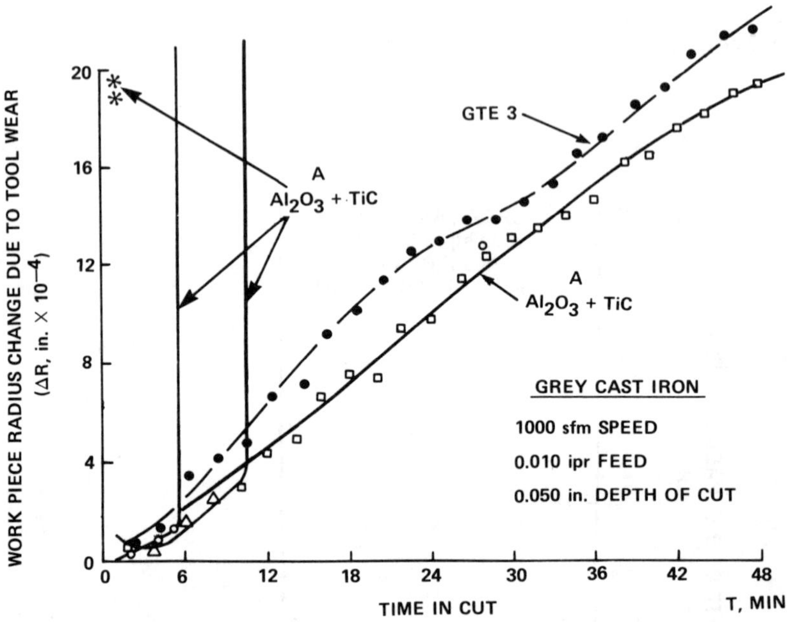

Figure 3. Machining Performance of Al_2O_3 - TiC (A) tools compared to silicon nitride based composition 3 (5 runs), gray cast iron machining

Star (*) indicates a point of catastrophic failure

Additional testing on cast iron machining which included tools from another selected manufacturer further indicated that even Al_2O_3 + TiC with improved fracture toughness (sample D; 3.3 $MN/m^{3/2}$ as opposed to sample A; 3.1 $MN/m^{3/2}$) could also not be successfully used under severe cutting conditions. The tests were conducted at the speed of a 1000 sfpm with a 0.050 in. depth of cut, while the feed rate was varied from 0.010 ipr to 0.030 ipr. Figures 4 and 5 show the performance of tested materials under these conditions. It is again evident that although at the lower feed rate (0.010 ipr) and the Al_2O_3 + TiC tools demonstrate better wear resistance, the use of higher feed rates (0.020 ipr) resulted in their failing catastrophically in most cases. Although the difference in fracture toughness between samples A and D is not marked, tools of material D were found to have considerably lower tendency to fail catastrophically. At increased feed rates of 0.030 ipr, all tested silicon nitride tools (compositions 3 and 6) survived for the duration of the test (5 min time in cut) while all the Al_2O_3 + TiC tools failed catastrophically in the first few seconds of cut, and therefore a valid comparison could not be made.

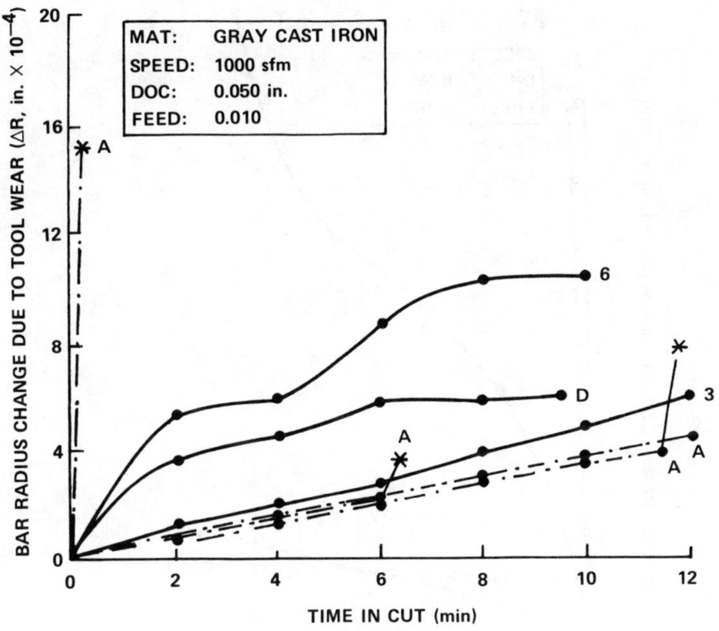

Figure 4. Machining performance of Al_2O_3 - TiC (A,D) and silicon nitride based cutting tools (compositions 3 and 6)

*Star (*) indicates point of catastrophic failure

It is fairly well recognized that wear modes of specific cutting tool materials can vary, depending on the workpiece material and cutting conditions. In the case of Si_3N_4 based materials, improved wear resistance rather than toughness seems to be the predominant characteristic required for the machining of steel, where long chip formation is obtained.

Systematic changes in the composition of the Si_3N_4 based cutting tools brought about substantial improvements in their wear rate in the machining of AISI 1020 steel (Figure 6). Specimen 3 with a hardness of 15.2 GN/m^2 resulted in improved performance over samples 5 (14.3 MN/m^2) and 4 (13.7 GN/m^2). Work is in progress to further optimize a Si_3N_4 composition for steel cutting.

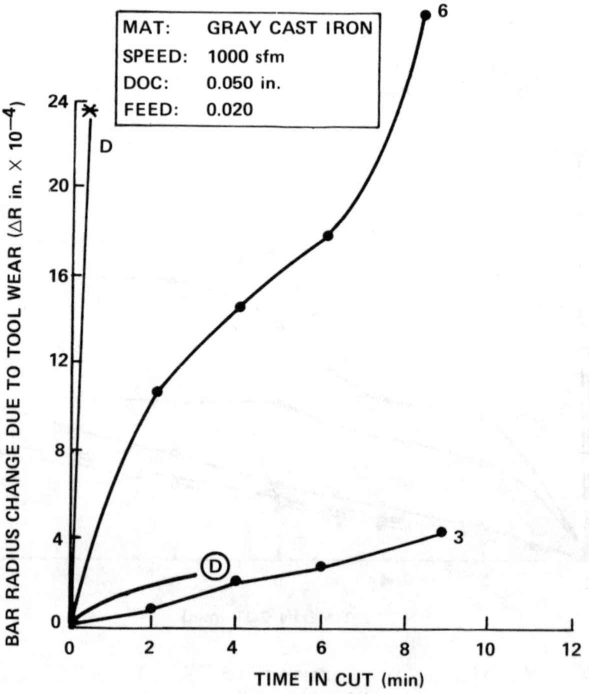

Figure 5. Machining performance of Al_2O_3 - TiC (D) and silicon nitride cutting tools (compositions 3 and 6). Feed rate 0.020 ipr

Star (*) indicates point of catastrophic failure

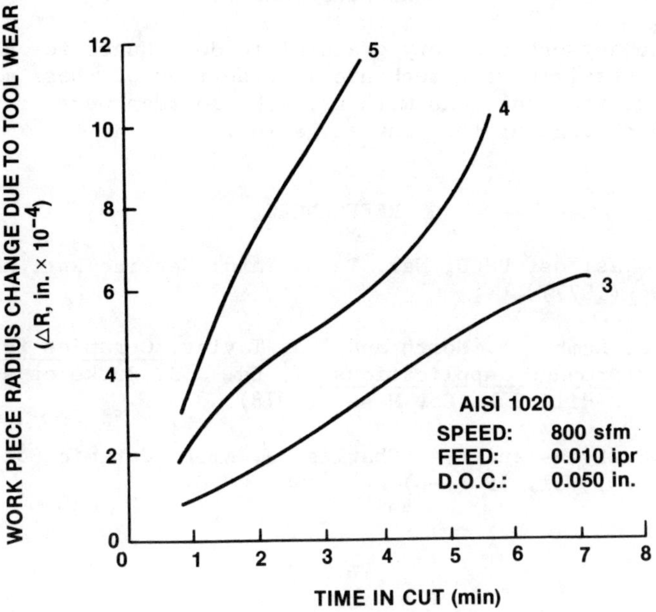

Figure 6. Performance of silicon nitride based cutting tools in the machining of AISI 1020 steel

CONCLUSIONS

At present composite ceramic cutting tools occupy a 1% to 2% share of the cutting tool market. Our results support the contention that improvements in toughness of Al_2O_3 + TiC tools and the introduction of new ceramic materials such as Si_3N_4 create a strong potential for their increased usage.

It should be pointed out that the Si_3N_4 based materials discussed in this investigation are first generation. Improvements through further development of their structure and properties are anticipated. Nevertheless, the present materials have indicated satisfactory wear resistance (in cast iron machining) and much improved reliability due to their superior toughness. Furthermore, due to their excellent high temperature stability and oxidation resistance, higher cutting speeds and therefore improved productivity can easily be visualized.

ACKNOWLEDGEMENTS

The authors are very grateful to John Nunes for his help in characterizing the mechanical properties of these materials. Stimulating discussions with Dr. D.M. Koffman were extremely helpful throughout this investigation.

REFERENCES

(1) T. Vasilos, USCD, Nat. Tech. Info. Service Rpt, AD 4044-594 (1977).

(2) R.J. Lumby, B. North and A.J. Taylor, Ceramics for High Performance Applications-II. Ed. J.J. Burke et al, Brook Hill Pub. Co. Mass. (1978).

(3) A.G. Evans and E.A. Charles, J. Amer. Ceramic Soc. (1976) Vol. 59 (7-8).

DISCUSSION

Paul Hartung (MIT)

Material #3 seemed to be most wear resistant of silicon nitride materials tested. Was it also the hardest material tested?

T. Buljan

Silicon nitride based composition #3 exhibits the highest wear resistance of all materials examined in this study. While composition #3 is the hardest silicon nitride based material, its hardness is notably lower than that measured on the Al_2O_3 -TiC composites which exhibit similar or slightly lesser resistance to abrasive wear.

Unidentified

Silicon nitride based ceramics have higher fracture toughness but lower hardness than alumina based ceramic tool materials. How do they behave in actual cutting application?

T. Buljan

Silicon nitride based ceramics due to their higher fracture toughness are less prone to catastrophic failure and are capable of higher feed rates than alumina based ceramics.

The wear resistance of selected compositions (for example #3) in machining of gray cast iron was found comparable to that of Al_2O_3-TiC composites.

Unidentified

What is notch resistance of GTE silicon nitride based cutting tools compared to Al_2O_3-TiC?

T. Buljan

Limited number of tests designed to examine notch resistance of Si_3N_4 based tools have been conducted and indicate improved notch resistance of those tools as compared to Al_2O_3 -TiC. Further examination of the influence of cutting geometry, cutting conditions and workpiece materials on notch formation is necessary in order to define this material and answer this question more fully.

Unidentified

How do properties of cold pressed SiAlON compare to hot pressed?

T. Buljan

Hot pressed silicon nitride materials, including SiAlON usually exhibit higher modulus of rupture and hardness than their conventionally sintered counterparts. Fracture toughness of conventionally sintered materials tends to be somewhat higher than that observed on hot pressed materials.

Unidentified

In the tool machining performance tests were edges prepared equally?

T. Buljan

Geometry of all tested tools as well as their K-Land configuration were equal. It is difficult to say whether grinding procedures used to prepare those are also equal since commercial tools have been used in their as-received state.

RELATIONSHIP BETWEEN THE THERMAL, MECHANICAL PROPERTIES AND CUTTING PERFORMANCE OF TiN-TiC CERMET

HIROSHI TANAKA

NGK SPARK PLUG CO., LTD.

ABSTRACT

The present experiment has been carried out to investigate thermal, mechanical properties and cutting performance of TiN-TiC cermet, and considered the relationship between both. Specimens were prepared 10 different compositions, i.e. TiC-(0-53.1)TiN-5Mo_2C-(5-20)Ni. Thermal conductivity, coefficient of thermal expansion, transverse rupture strength, and Young's modulus were measured. Turning and milling tests cutting alumina compact, gray cast iron, carbon steel and chromium molybdenum steel were carried out.

INTRODUCTION

Titanium nitride (TiN) attracts attention as a suitable material for cutting tools due to its excellent thermal conductivity and high thermal shock resistance (1), (2). For recent years, efforts have been directed toward improving the toughness of cermet by adding up to about 20% of TiN to TiC cermet (3), (4), and several such types are now commercially available. It is however considered necessary to add more TiN than 20% to take advantage of the useful features of TiN. We have then made for testing purposes a cermet which contains an increased quantity of TiN, and examined its thermal, mechanical properties and cutting performance for considering the relationship between both.

EXPERIMENTAL PROCEDURE

A specimen of cermet has been made which consists of commercially available TiN, TiC, Mo_2C, and Ni powders combined in a ratio of TiC-(0-53.1)TiN-5Mo_2C-10Ni. An additional specimen containing TiC-TiN-5Mo_2C-(5-20)Ni has been made to observe the effect of the the varying quantities of Ni.

After conventional ball-milling and pressing, compacts were sintered in the range 1420-1550°C (2588-2822°F) for 40 min in vacuum. The specimen took three different shapes, such as 9 mm (0.354 in.) dia. x 1.5 mm (0.059 in.) thickness for thermal conductivity, 3.5 mm (0.138 in.) dia x 13 mm (0.512 in.) long for thermal expansion, and 4 mm (0.157 in.) x 8 mm (0.315 in.) x 25 mm (0.984 in.) for transverse rupture strength.

Thermal constants instrument PLP-1000 made by Sanki Engineering Company was used to measure the thermal conductivity, the measurement having been taken in the laser beam flush method. Thermal expansion coefficient was measured with Rigaku Denki Company's instrument with alumina used as a standard specimen. Young's modulus measurement was taken by using Instron type testing apparatus, with a strain gage applied to the center of the tensile side of the specimen which is subject to transverse rupture strength. Transverse rupture strength was measured according to JIS (5), and each value presented is the mean from twenty pieces.

Tip SNG432 was used for cutting test with cutting edge chamfering of 25° x 0.07 mm (0.003 in.). The cutting test conditions and cutting geometry are shown in Table 1 and 2 respectively.

Table 1. Cutting test conditions

Test No.	Workpiece material	Cutting method	Cutting speed : V, m/min Depth of cut : d, mm Feed : f, mm/rev Cutting time : T, min
A	Alumina compact 100-20 mm dia.	Facing	V : 100-20 d : 1.0 f : 0.31 T : 10 times
B	Gray cast iron JIS FC20	Turning	V : 160 d : 1.5 f : 0.31 T : 3.0
C	Carbon steel JIS S45C	Turning	V : 200 d : 1.0 f : 0.31 T : 3.0
D	Chromium Molybdenum steel JIS SCM440	Turning	V : 150 d : 1.0 f : 0.31
E	Chromium Molybdenum steel JIS SCM440	Milling	V : 177 d : 1.0 f : 0.41 mm/teeth

Table 2 Cutting geometry

Turning test (A $\sim$ D)	Milling test (E)
Back rake angle, -5°	Face milling cutter
Side rake angle, -7°	axial rake, -5°
End relief angle, 5°	radial rake, -6°
Side relief angle, 7°	cutting edge angle, 25°
End cutting edge angle, 15°	nose radius, 0.8 mm
Side cutting edge angle, 15°	
Nose radius, 0.8 mm	

RESULTS AND DISCUSSION

THERMAL, MECHANICAL PROPERTIES

Examining the sintered specimen by means of X-ray diffraction indicates the hard phase (TiN-TiC) takes a single phase between TiN/(TiC + TiN) = 0/100 and 62.5/100. The result presents a good agreement with the lattice constant of the constituent powdered elements of the specimen used. Thermal, mechanical properties and lattice constant are presented in Table 3.

Table 3 Thermal, mechanical properties

TiN / TiC+TiN	Thermal conductivity, W/m.K		Coefficient of thermal expansion, $\times 10^{-6}$/K	Transverse rupture strength, MPa	Young's modulus, GPa	Lattice constant of hard phase, nm
	298 K	1173K	(298 - 1173 K)			
0	13.39	26.78	8.8	900	420	0.4338
0.125	14.64	28.45	8.9	900	420	0.4321
0.25	14.64	28.45	9.1	1130	440	0.4312
0.375	14.22	28.87	9.2	1280	440	0.4300
0.5	16.74	31.38	9.2	1320	450	0.4293
0.625	17.99	33.89	9.2	1300	470	0.4280
Ceramic	-	7.11	8.4	310	440	-
Carbide	-	51.04	7.0	1330	530	-

Fig 1 shows the relation of thermal conductivity as a function of TiN/(TiC + TiN) ratio. In fig 1, specimen X presents data obtained in the present experiment, and specimen Y presents data on TiC-(0-80) TiN hot-pressed body containing no bonding metal (6). Thermal conductivity increases as TiN/(TiC + TiN) ratio exceeds 0.5, and this increase is assumed to occur due to increase in thermal conductivity of the hard phase.

Fig 2 shows thermal conductivity of TiC-5Mo_2C-Ni and (50TiC-50TiN)-5Mo_2C-Ni, respectively, measured between 25°C and 500°C. In both cases, thermal conductivity is minimal with the quantity of Ni equal to 5 to 15% while it increases with increasing Ni. It is assumed, however, that non linearity of thermal conductivity in relation to the quantity of Ni may exist due to the size of the hard phase particles, that is different thickness of the bonding phase or different composition of the bonding phase.

ASSUMPTION OF THERMAL SHOCK RESISTANCE

Thermal shock resistance, R, is obtained from the following equation using thermal conductivity at 900°C, K, coefficient of thermal expansion (25 - 900°C)α , transverse rupture strength, σ m, and Young's modulus, E.

$$R = \frac{K \cdot \sigma m}{\alpha \cdot E} \quad (W/m \cdot K^2)$$

The result is presented in Table 4, showing the thermal shock resistance increases with increasing content of TiN. Values of commercially available tungsten carbide (74WC-11TiC-8TaC-7Co) and ceramic (99.1%Al_2O^3, cold pressed and sintered) are also given in table 4. From table 4, it can be seen that the thermal shock resistance of the cermet used for this test, even if it containes more TiN, is not better than that of tungsten carbide.

Table 4 Thermal shock resistance R

$\frac{TiN}{TiC+TiN}$	0	0.125	0.25	0.375	0.5	0.625	ceramic	carbide
$R(\frac{W}{m \cdot K^2})$	6.52	6.85	8.03	9.13	10.01	10.19	0.60	18.30

CUTTING PERFORMANCE

Fig 3 presents a relation between TiN/(TiC + TiN) in Test A to C and flank wear after a certain cutting time period. When turning alumina compact obtained by Test A, abrasion is assumed to be a predominant factor for the wear mechanism. Due to the lower hardness, i.e., a value of 1.77-2.06 MPa (256-299 psi) of hard particles of TiN, than that of 2.75-3.24 MPa (398-469 psi) for TiC, the parts of TiN rich abrade more severely. The minimal abrasion was observed at TiN/(TiC + TiN) = 0.25, and this is assumed to be due to the fact that crystal particle sizes become the smallest in this composition.

The relation between TiN/(TiC + TiN) and flank wear in Test B in turning gray cast iron takes a similar behavior to that in Test A, and the abrasion is assumed to be a predominant factor.

Turning carbon steel shows the similar value of flank wear with TiN/(TiC + TiN) of 0 to 0.25, but the value of flank wear tends to get smaller as TiN/(TiC + TiN) ratio exceeds 0.375. The results are different from those in Test A and B, and wear are assumed to be caused by other influential factors than the abrasion. Fig 4 presents a relationship between the thermal conductivity at 900°C and flank wear in Test C. It can be assumed from the results that in turning carbon steel, the cause of the wear is very closely related to heat produced in turning.

Fig 5 shows a relation between TiN/(TiC + TiN) ratio in Test D turning chromium molybdenum steel and tool life. In this case, wear occurs more greatly at the interfacial zone, and therefore the life has expressed as $V_B(max)$ = 0.2 mm (0.008 in.) including this groove wear. It is considered that interface between tool and work material is a part which presents the greatest temperature gradient on the surface of a tip in turning. Fig 6 is a relation obtained between thermal shock resistance and tool life in Test D, showing a good correlation between both.

Chromium molybdenum steel blocks having an area of 100 mm (3.937 in.) x 200 mm (7.874 in.) were cut with a face milling cutter with one blade, and tool life was determined by the frequency of impacts until the tip was damaged in Test E. Fig 7 shows tool life in Test E as a function of TiN/(TiC + TiN) ratio, and the impact frequency is the greater as

TiN/(TiC + TiN) is increased. Fig 8 presents a relation between thermal shock resistance and impact frequency, showing a good correlation between both as in Test D.

Observing the tip during the cutting process shows cracks produced in the crater surface when 50 times of impacts are reached, resulting in flucture caused in the cracks. It can be assumed that the cracks are caused by fatigue due to thermal shock, and are further developed to eventual breakage.

CONCLUSION

After examining the thermal, mechanical properties and cutting performance for the cermet containing TiC-(0-53.1)TiN -5Mo_2C-(5-20)Ni, made for the purposes of the present experiment, the following conclusions are reached :

(1) In facing alumina compact and turning cast iron, flank wear talees a minimal value at TiN/(TiC + TiN) = 0.25. It is caused by the hardness and particle size of the hard phase.

(2) In turning carbon steel, flank wear presents substantially the same values in the range of TiN/(TiC + TiN) ratio of 0 to 0.25, and decreases as TiN is further increased. In this case, flank wear and thermal conductivity of the tip show a good correlation.

(3) In milling chromium molybdenum steel, the life to fracture increases with increasing TiN/(TiC + TiN) ratio. Since the result shows a good correlation with the thermal shock resistance, the fracture due to the thermal fatigue is assumed to become the principal cause.

REFERENCES

(1) R. Kieffer, P. Ettmayer and M. Freudhofmeier, "About Nitrides and Carbonitrides and Nitride-Based Cemented Hard Alloys", Modern Developments in Powder Metallurgy, Vol 5, 1971, p201

(2) H. Tanaka, "High Performance Titanium Nitride Alloy", Kinzoku, special number, 1975, p65

(3) H. Suzuki, K. Hayashi, T. Yamamoto and J. Lee, "Effects of Additional Titanium Nitride on the Strength of TiC-Mo_2C-Ni Alloy", J. Japan Soc. of Powder and Powder

Metallurgy, vol 23, No 7, 1976, P224
(4) E. Rudy, S. Worcester and W. Elbington, 8 th Plansee Seminar, 1974
(5) Japanese Industrial Standard, B 4104 "Cemented Carbide Tips", 1978.
(6) H. Tanaka, unpublished data.

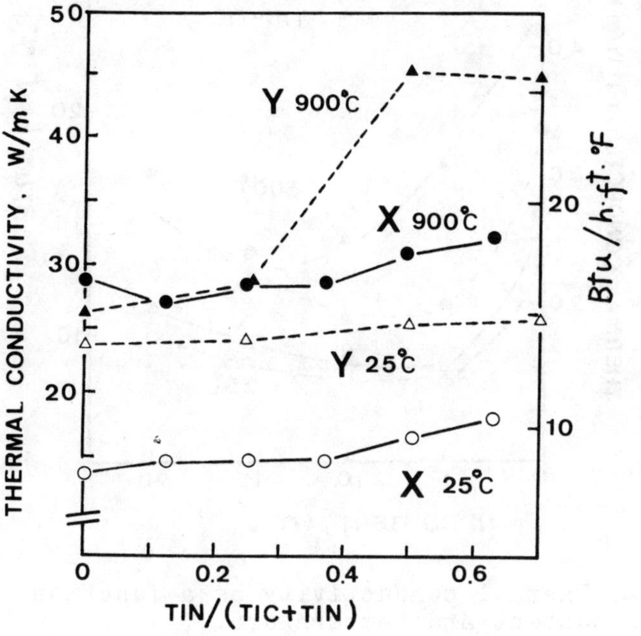

Fig.1. Thermal conductivity as a function of TiN/(TiC+TiN) ratio and temperature.

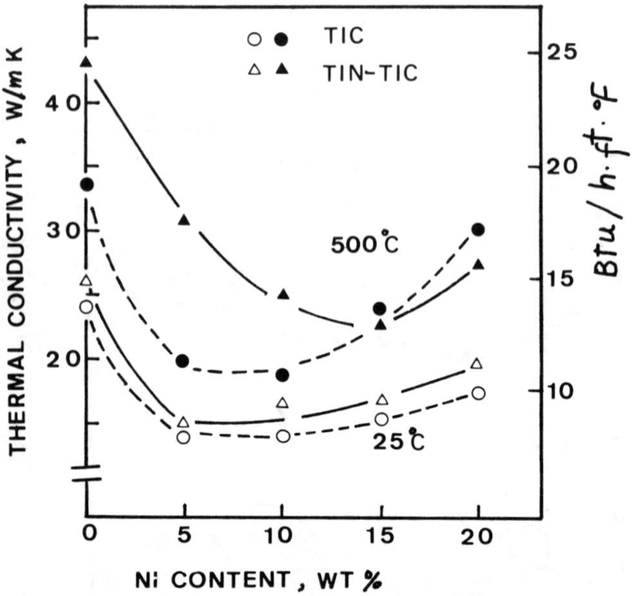

Fig.2. Thermal conductivity as a function of Ni content and temperature.

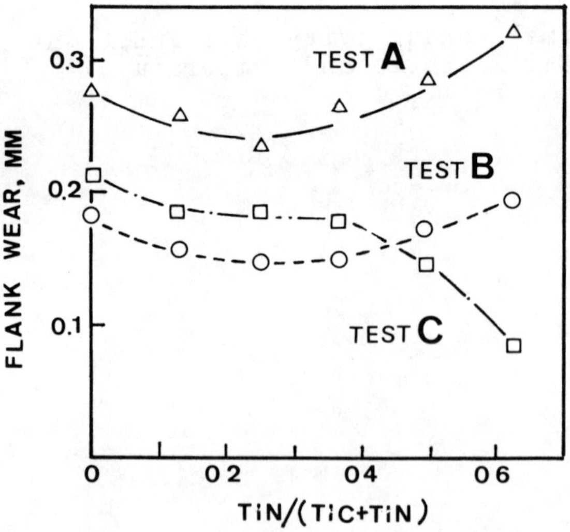

Fig.3. Average flank wear in Test A,B AND C as a function of TiN/(TiC+TiN) ratio.

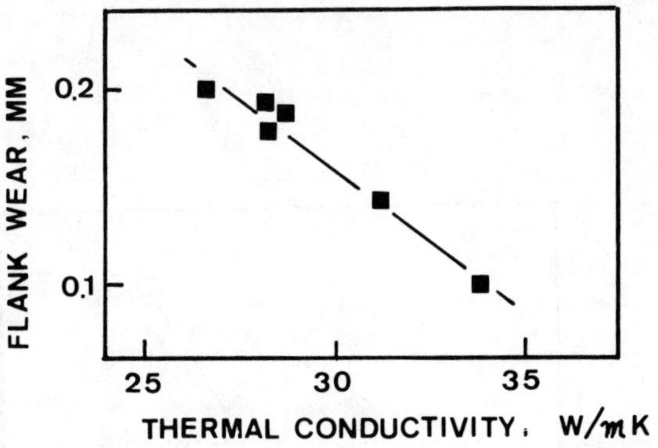

Fig.4. A relationship between the thermal conductivity at 1137K and average flank wear in Test C.

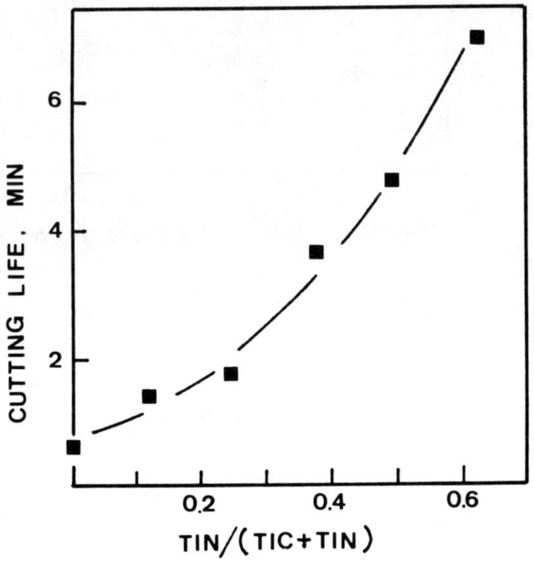

Fig.5. Cutting life as a function of TiN/(TiC+TiN) ratio in Test D.

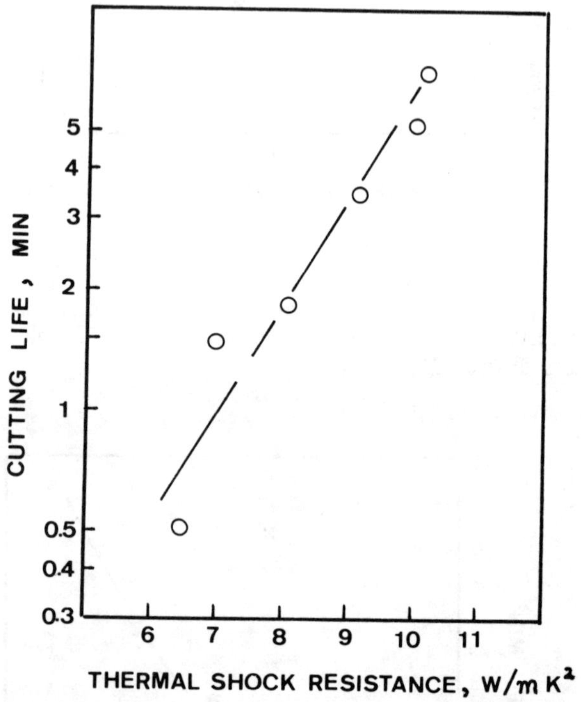

Fig.6. A relationship between the cutting life in Test D and thermal shock resistance.

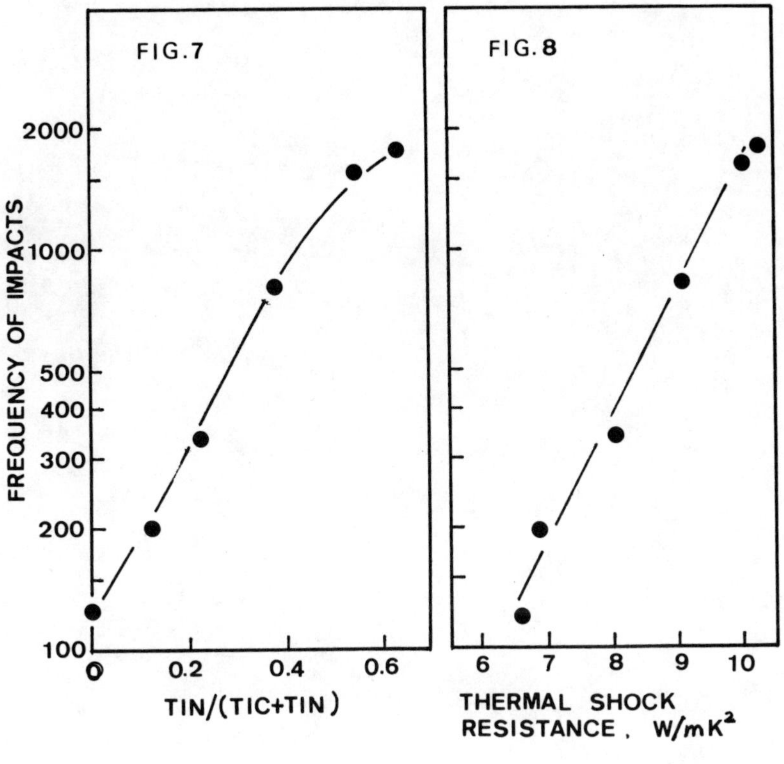

Fig.7. The frequency of impacts in Test E as a function of TiN/(TiC+TiN) ratio.

Fig.8. A relationship between thermal shock resistance and impact frequency.

SESSION VII

APPLICATION OF CUTTING TOOL MATERIALS

SELF GRIP EFFECT OF INDEXABLE CARBIDE BITS PROVIDES HIGHER EFFICIENCY IN PARTING OPERATIONS

R. Wertheim
Iscar Ltd, Nahariya, Israel
Research & Development

This paper describes the development of a new self-grip parting tool for use with both coated and non-coated indexable carbide bits, for machining on a wide variety of materials and on specially difficult applications such as interrupted cuts. Analysis of insertion forces of carbide bit, cutting forces and forces required to remove bit are made theoretically and then proved in performance. Test tools are examined with regard to flow forces, chip shape and ability to machine many materials. Optimization of the self-gripped bit- and tool-geometries facilitates smooth chip flow without damage to machined surfaces and longer tool life.

1. INTRODUCTION

Parting off still remains one of the most problematic of all cutting operations (1,2,3,7). Special systems developed for parting operations, starting with HSS tooling and continuing through brazed carbide tools, and tools with indexable carbide bits, provide only limited solutions to problems which arise during parting. Bearing in mind phenomena created during the parting operation, a number of requisites were selected for a new parting system based on the self-grip phenomena with an indexable carbide bit. The system is derived from the principle of self-gripping with the aid of the friction between the carbide bit and the steel blade. Research and development of this new parting system was carried out in the Laboratory for Metal Cutting at the Technion in Haifa, Israel.

A special effort was made to calculate and measure cutting forces, wear behavior, tool life, surface quality, geometry after machining, chip form and the geometry of the carbide bit itself.

2. CHARACTERISTICS OF PARTING OPERATIONS

The most important characteristics of parting operations in comparison with other cutting operations in turning are shown in Fig. 1, including description of

Fig. 1: Geometry, Forces and Speeds during Two-Dimensional Parting Operations

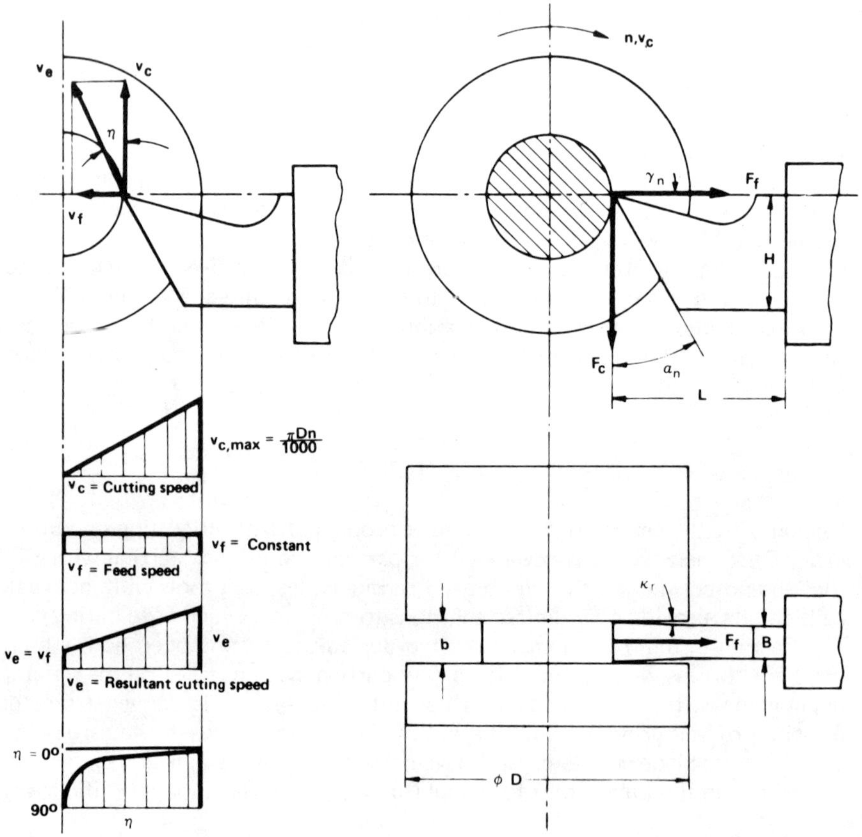

changes in cutting speeds, forces and angle of resulting speeds. The main points are:

a. Chip flow is influenced by the narrow slot and the rake surface of the carbide bit. Chip flow may damage the surface of the workpiece. Therefore, the problem of chip deflection and chip breaking is much more complicated with regard to two-dimensional cutting operations, the narrow slot and the change in cutting speeds during the operation.

b. During parting the cutting speed changes from a maximum $v_{c,max}$ at the outer diameter d_A of the workpiece and decreases to 0° when reaching the center of the workpiece. This phenomenon is only true when a constant n (RPM) is maintained (Fig. 1). Feed speed in the direction of the feed v_f is constant. As a result of the two speeds, the resultant cutting speed v_e decreases from the maximum at the outer diameter to v_f value at the center and the angle between the resulting cutting speed v_e and cutting speed v_c increases from nearly 0° to 90° at the center. The low value of the resulting cutting speed at the center combines with the high radial force (equal to the back force) at the center of the workpiece. This phenomenon means that parting off, especially to the center, can not be considered a normal cutting operation, but more as bending forces. The resultant cutting speed v_e can be defined as a vector of $v_c + v_f$.

c. Parting tool systems are constructed of three main parts: the cutting bit, the element which can be adjusted according to the required diameter which may be called the blade and which holds the bit, and the block which is fastened to the lathe itself. Because of narrow width of cut, temperature changes and changes in cutting forces, a main problem in parting is holding the small cutting bit by the blade itself.

d. The parting system is constructed of a narrow blade with a large overhang in order to be able to cut workpieces with large diameters. These characteristics combined with the forces which act during parting demand a special form, dimensions and material for the cutting blade. The forces acting during parting: cutting force (tangential) F_c, axial force (side force for parting) F_p and radial force (feed force) F_f are described in Fig. 2.

Axial forces are effective only when cutting is done in three dimensional parting operations, meaning that the cutting edge is not parallel to the turning axis (the axis of the workpiece).

The resultant active force in two-dimensional parting operations is:

$$\vec{F}_{a1} = \vec{F}_f + \vec{F}_c \qquad (1)$$

and the resulting direction of the resultant active force is:

$$tg\,\delta_1 = F_c / F_f \qquad (2)$$

The resultant active force in the three-dimensional parting operation is:

$$\vec{F}_{a2} = \vec{F}_p + \vec{F}_c + \vec{F}_f \qquad (3)$$

Fig. 2: Forces Acting During Two- and Three-Dimensional Parting Operations

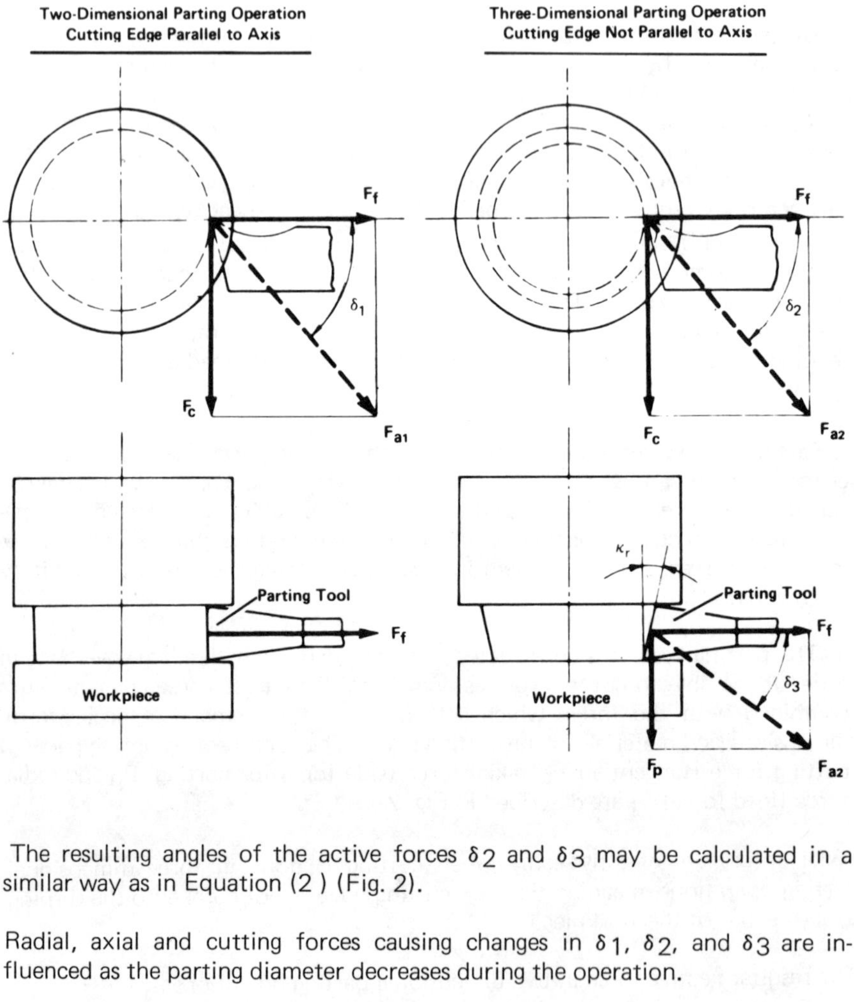

The resulting angles of the active forces δ_2 and δ_3 may be calculated in a similar way as in Equation (2) (Fig. 2).

Radial, axial and cutting forces causing changes in δ_1, δ_2, and δ_3 are influenced as the parting diameter decreases during the operation.

The deflection of the cutting edge in the tangential direction for the two-dimensional cutting operation may be calculated by using the bending moment $M=F_c \cdot L$ and the moment of inertia $I = B \cdot H^2/12$ which will result in:

$$f_c = \frac{F_c \cdot L^3}{3 \cdot E \cdot I} = \frac{4 \cdot F_c}{E \cdot B} \cdot \left(\frac{L}{H}\right)^3 \tag{4}$$

This equation means that for a greater ratio of overhang L to height of blade H, $(L/H)^3$, for a limited deflection f_c, the maximum force F_c is limited. Change of the width of blade B, meaning a change of the normal width of cut, has only a limited influence on the deflection in the tangential direction f_c because the radial and cutting forces are changed in proportion to this variation. Therefore, the influence of the variation in width of cut on the ratio F_c/B is small. Because of bowing, variation of the width of blade B has a large influence on the maximum allowable radial force F_f. In the extreme case where one side of the blade is fixed in the tool block and the other is free to move, the radial force F_f will be:

$$F_f = \frac{\pi^2 \cdot E}{48} \cdot \frac{H \cdot B^3}{L^2} \tag{5}$$

When machining with three-dimensional parting operations, the influence of nominal width of cut and therefore the width of blade is much larger because in addition to bowing, the axial force causes a bending in the same direction. The influence of the bending has a negative result concerning the flatness of the workpiece surface. The deflection in the axial direction as a result of the axial force will be calculated as follows:

$$f_p = \frac{4 \cdot F_p}{E \cdot H} \cdot \frac{L^3}{B^3} \tag{6}$$

An analysis of this equation leads to the conclusion that enlarging the width of blade B causes a proportional increase of the axial force F_p, but at the same time causes a decrease in the deflection in the axial direction f_p. Another possibility to decrease the force F_p and therefore decrease f_p can be the use of a larger tool cutting edge angle κ_r ($\kappa_r = 90^\circ$).

Based on these special characteristics of the parting operation, a new parting system was developed.

3. REQUIREMENTS FOR A NEW OPTIMAL PARTING SYSTEM

For the development of a new parting system, the following criteria were chosen:

a. Unhindered chip flow combined with control of chip form.
b. Minimum number ot parts.
c. Simple and quick method of changing cutting bit with guarantee of maintaining same height of cutting edge.
d. The influence of the forces and temperatures during parting should not limit the feed and the cutting speed.
e. The blade is rigid and at the same time may be easily adjusted to the required cutting diameter.

f. Reliability against breakage of cutting bit and blade at high feeds and speeds.
g. Longer tool life of cutting bit and blade.
h. Undamaged surface quality of workpiece.
i. Ability to machine very hard metals and interrupted cuts which involve deviations in forces and temperatures.
j. Wide range of applications possible including different workpiece materials, and a wide range of diameters and machining conditions.
k. Optimum cost efficiency.

Based on these criteria a number of parting tools were developed in the past. The differences between the tools are in cutting material, clamping and adjusting systems and the range of capabilities.

The first generation of parting tools tested were adjustable HSS blades. Then brazed tools were tested and finally the most modern, those with indexable carbide bits. This last group represent the majority of parting tools in use today because these meet most of the above-mentioned criteria. But it must be pointed out that in most of the solutions using indexable carbide bits, the stability of the clamping system is adversely influenced by forces and temperatures which limit the overhang and machining conditions. The clamping systems include a number of clamp variations, screws, pins and springs using the elasticity of the blade. All the solutions require a larger space for clamping and therefore the minimum width of cut is large.

Upon examination the carbide bits fall into three main types. The first type are brazed carbide bits which may be ground for repeated use. The second type are indexable carbide bits which may also be ground or reground, but which do not allow for either chip deflector or coating, and so their use is limited. The third group are those indexable bits designed for one-time use after which they are discarded. The majority of types in use fall within group three.

Systematic analysis of the criteria demands and the existing parting systems brought about the development of the new parting system based on the self-gripping principle of the carbide bit in the adjustable blade. With this sytem many of the disadvantages of the known systems have been eliminated and nearly all of the criteria and demands are fulfilled.

4. PARTING SYSTEMS BASED ON THE PRINCIPLE OF THE SELF-GRIP

Fig. 3 shows an experimental model of the new parting tool based on the self-grip principle. It is constructed from only three components:

a. Indexable carbide bit with special geometry.
b.Adjustable blade with a special seat with prism on both sides fitted for the carbide bit.

c. Tool block for holding and adjusting the blade with prismatic guides on both sides to insure correct sliding movement and clamping.

Fig. 3: Parting Tool with Indexable Carbide Bit According to the Self-Grip Principle

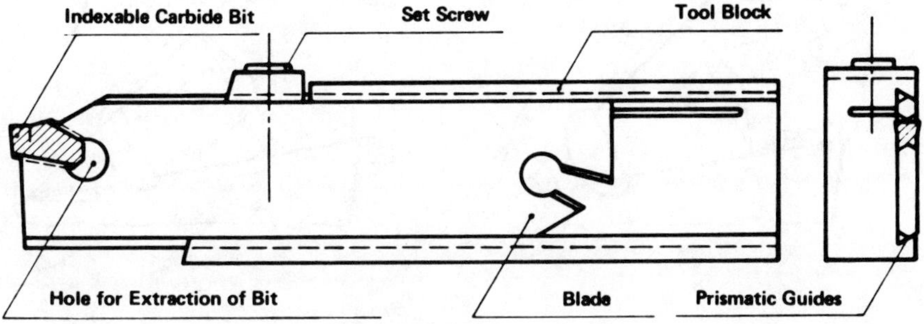

The carbide bit is secured in its seat in accordance with the self-grip phenomena. The blade is made of one piece, requires no clamping element to secure the bit and enables a larger depth of parting operation than all other known tools. The overhang can be adjusted according to the required workpiece diameter. The bit is inserted by hand and once in place may be removed by a slight turn of a simple extractor, which utilizes the open space directly behind the bit itself as shown in Fig. 3.

5. THE PRINCIPLE OF THE SELF-GRIP PHENOMENA

Fig. 4 shows schematically the self-gripping principle during bit insertion and bit removal. In principle the forces acting during parting operations are equal to those described during the insertion operation. Under all machining conditions the frictional forces between the carbide bit and blade seat eliminate any possibility that the carbide bit will fall out from its seat, taking into consideration variations or interruptions of the parting operation itself. On the other hand, the clamping effect of the wedge-shaped carbide bit should remain such as to enable its removal after use in a simple way with low force. As a result of these considerations, the angle of the insert φ and the angle of prisma in insert and blade Θ were optimized so as to give security after insertion of the bit on the one hand and at the same time enabling its easy removal after use. The cutting force F_c and the radial force F_f acting during parting operations improve the clamping of the carbide bit in the seat. As a result it may be concluded that higher cutting speeds, feed rates and forces may be used with this self-gripping system.

Fig. 4: Forces Acting During Bit Insertion and Removal

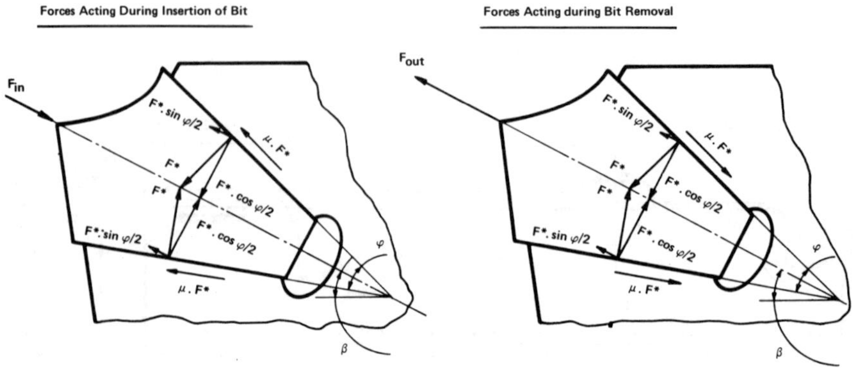

Symbols

F*	: Normal force
F_{in}	: Insertion force
F_{out}	: Exertion force
φ	: Wedge angle of insert
β	: Angle between horizontal plane and insert centerline
μ	: Coefficient of friction

In Fig. 4 the forces acting under the self-gripping principle are described and analyzed. When inserting the carbide bit with the F_{in}, two opposing forces result. The first is a component of the friction forces in axial direction, $2 \cdot \mu \cdot F^* \cdot \cos(\varphi/2)$; the second is the axial component of the normal force F*, both of which act parallel to the centerline of the carbide bit. The resulting equation for the force balance during insertion of the carbide bit will be:

$$F_{in} = 2\mu \cdot F^* \cdot \cos(\varphi/2) + 2F^* \cdot \sin(\varphi/2) \qquad (7)$$

in which φ is the wedge angle of insert and seat, and μ is the coefficient of friction between carbide and steel. The forces acting perpendicular to the center axis of the carbide and the seat, mainly the components $\mu \cdot F^* \cdot \sin(\varphi/2)$ and $F^* \cdot \cos(\varphi/2)$ are equal and therefore cancel each other out.

During parting operations a similar balance of forces acts, but as the radial force F_f is not equal to cutting force F_c, only a limited balance of forces may be made. The two forces depend on the machining conditions of feed, cutting speed, and width of cut. In normal cases the component of the cutting force $F_c . \cos\beta$ is larger than the component of the radial force, $F_f . \sin\beta$. In extreme cases these two are equal, $F_c . \cos\beta = F_f . \sin\beta$ which means that no forces are acting perpendicular to the axis of the seat. In this case the force balance along the bit centerline will be:

$$F_f \cdot \cos\beta + F_c \cdot \sin\beta = 2 \cdot \mu \cdot F^* \cos(\varphi/2) + 2 \cdot F^* \sin(\varphi/2) \quad (8)$$

In normal parting operations because of $F_c.\cos\beta > F_f.\sin\beta$, one can expect that the friction force on the underside of the seat will be larger than on the over-side and the symmetry on which Equations (7) and (8) are based does not exist.

The normal force according to Equation (7) is shown here:

$$F^* = \frac{F_{in}}{2[\mu \cdot \cos(\varphi/2) + \sin(\varphi/2)]} \quad (9)$$

and during parting operations, the normal force of the over- and undersides of the seat will be according to Equation (8):

$$F^* = \frac{F_f \cdot \cos\beta + F_c \cdot \sin\beta}{2[\mu \cdot \cos(\varphi/2) + \sin(\varphi/2)]} \quad (10)$$

After finishing the parting operation, when no forces are acting on the carbide bit, one may calculate the forces needed to remove the bit from the seat according to the right scheme in Fig. 4:

$$F_{out} = 2 \cdot \mu \cdot F^* \cos(\varphi/2) - 2 \cdot F^* \sin(\varphi/2) \quad (11)$$

Combining Equations (9) and (11) using the normal force F^*, the result is the removal force F_{out} as shown here:

$$F_{out} = \frac{F_{in} \cdot [\mu \cdot \cos(\varphi/2) - \sin(\varphi/2)]}{\mu \cdot \cos(\varphi/2) + \sin(\varphi/2)} \quad (12)$$

and using Equation (11) with Equation (10), the F_{out} force as a function of the machining forces F_c and F_f will be:

$$F_{out} = \frac{[F_f \cdot \cos\beta + F_c \cdot \sin\beta] \cdot [\mu \cdot \cos(\varphi/2) - \sin(\varphi/2)]}{\mu \cdot \cos(\varphi/2) + \sin(\varphi/2)} \quad (13)$$

From this Equation (13) one sees that when all geometrical dimensions and parting forces are known or can be calculated or estimated, the forces needed for removing the carbide bit F_{out} can be made known. Normally the forces $F_f.\cos\beta + F_c.\sin\beta > F_{in}$ so that the machining operation itself guarantees a grasp of the bit by the seat. But by using the correct geometry of the seat, a very small F_{in}, achieved when inserting the bit, is enough to insure the grasping of the bit in the seat.

Conditions under which no forces are required to remove the carbide bit from the seat (F_{out} = 0) can be calculated according to Equations (12) or (13) as follows:

$$\mu \cdot \cos(\varphi/2) - \sin(\varphi/2) = 0 \tag{14}$$

Using the known physical rule the coefficient of friction μ is equal to the tangent of the friction angle γ^* resulting in:

$$tg(\varphi/2) = \mu = tg\,\gamma^* \tag{15}$$

The coefficient of friction between carbide and steel depends on both materials and the contact conditions between them. Using different sources, for example [5], this coefficient under lubrication will be μ_{wet}=0.1 to 0.2. Using this value in Equation (16), the correlated angle of the wedge will be 11° to 22.5° (as shown in Fig. 5). The figure shows the limited value which guarantees the

Fig. 5: Range of Wedge Angle for Self-Gripping

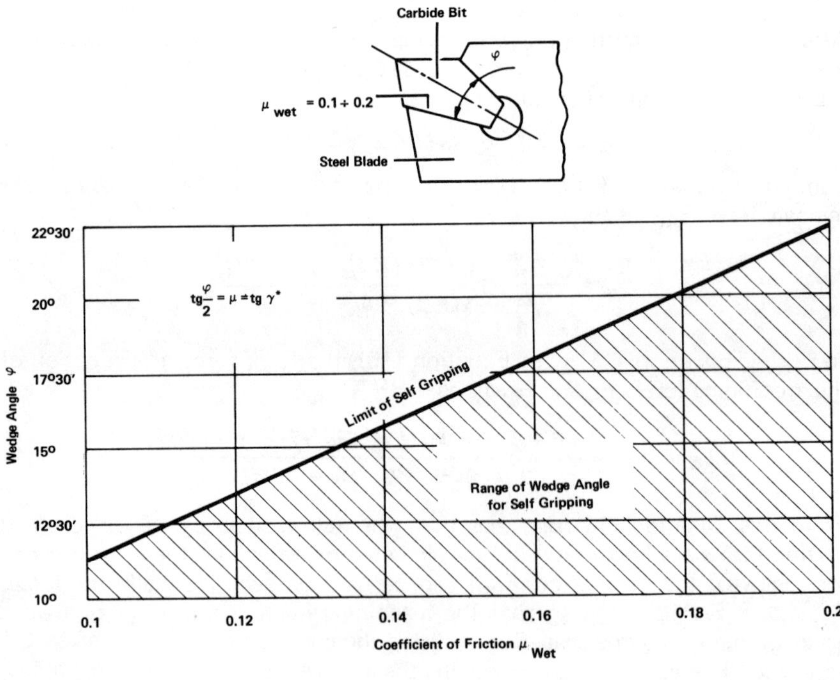

grasping of the carbide insert for any contact conditions. Based on these analyses the optimum value for the wedge angle of the bit was selected to insure a fool-proof grasp on the bit, for any carbide grade and friction conditions and all machining possibilities ($\varphi=10^o$ to 16°).

When selecting the angle between the horizontal plane and seat centerline β, the strength of the blade and the initial reverse movement of the bit in the seat at the start of the parting were taken into account. As a criterion of maximum insertion forces one may calculate with the aid of the differential equation by differentiating according to β in Equation (13) to find:

$$tg\,\beta = F_c/F_f \qquad (16)$$

Using the results which will be described as follows, the angle β should be between 55° and 68°. (See Fig. 4.) But in order to eliminate the influence of the back movement on the perpendicular movement, this angle should be $\beta=0^o$. On the other hand, if $\beta=0^o$ the stability during parting is reduced and problems with the strength of the blade may develop. Taking these variations into account, a compromise value was selected in which β ranges from 15° to 25°.

Comparing the new self-gripping system with the criteria and demands for an ideal parting system, it can be concluded that the principle of the self-gripping system fulfills most of these points, improving on all parting systems known to date. This system gives strong clamping, high stability, minimum number of parts, excellent parting performance, even in difficult machining conditions and interrupted cuts; is simple and easy to index bits; and involves minimum influence of temperatures and forces, all in an economical way.

6. GEOMETRY OF THE CARBIDE BIT

According to the analysis of the self-gripping principle, the geometry of the seat and the bit must be optimized to insure that the carbide bit will remain firmly grasped in the seat and keep a relatively high degree of accuracy of position over all machining conditions. The shape and geometry of the carbide bit with different angles is shown in Fig. 6. The bit has prismatic top and bottom surfaces with the angle Θ. These surfaces align the carbide bit in the center of the seat in the blade and guarantee this position during machining. The two negative prismatic surfaces on the carbide bit, in comparison with the possibility of negative prism in the blade, give a number of advantages. The first is stability. The second is the fact that in the event of damage, for example, because of incorrect set up, the damage will occur in the carbide bit and not in the more expensive blade. The danger of pressure and bending of the blade with negative prism in the bit is smaller than when using positive prism in the bit.

Fig. 6: Geometry of Bit

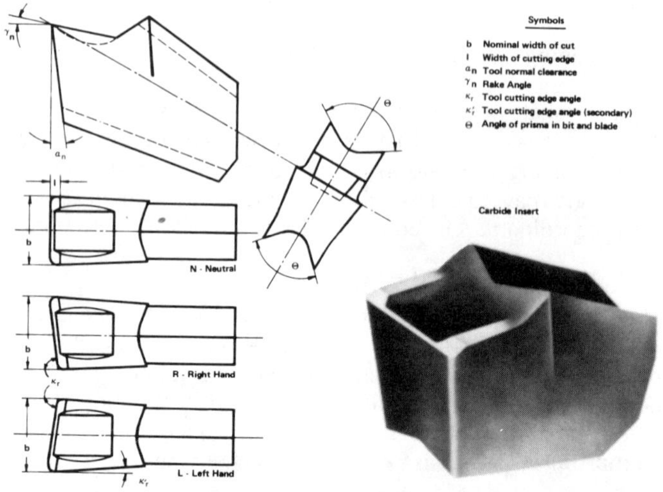

The rake surface of the carbide bit has a special geometry; it is higher at the sides than at the center and the cutting edge is chamfered, with a positive or negative rake angle γ_n according to the machined workpiece material. This chamfer guarantees a higher strength of the cutting edge during heavy machining conditions as well, for example, during interrupted cutting, parting of very hard steels and alloys, and parting on unstable machines [2].

The geometry of the rake surface with the higher walls on both sides and the crater shaped cavity in between results in a narrower chip in comparison with the width of cut and causes the material to spiral and thus be removed mainly into coil like chips. The narrow chip in comparison with the nominal width of cut gives the advantage that the surface of the workpiece is not damaged, combined with ease of chip flow out of the narrow parting slot. (See Fig.7.) The well shaped short, spiral chips give security to the operator and can be disposed of easily.

In addition to using the normal two-dimensional cutting bit N neutral, two other possibilities, the R right-hand and L left-hand, are used in order to get a flat surface with minimum stub. The right and left carbide bits are shown in Fig. 6. This results in a three-dimensional parting operation. The tool cutting

Fig. 7: Typical Chip Shapes

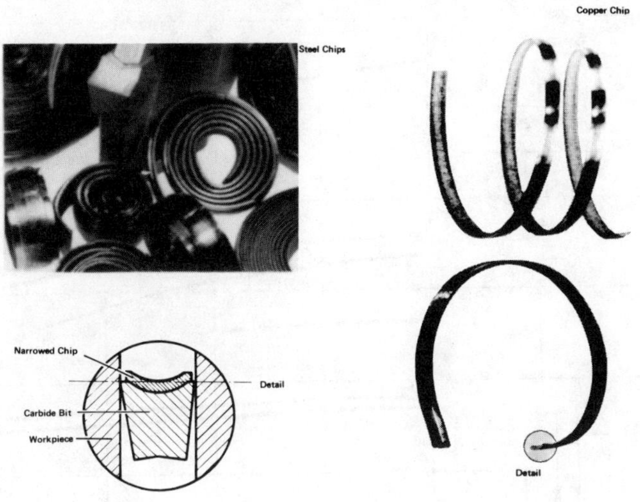

edge angle κ_r of the R right and L left bits ranges between 82° to 87° in accordance with the axial forces and bending of the blade as described above. The selected κ_r angle correlates with the material being machined, the stability of the machine and the machining conditions. The experiments were carried out with ISO P40 and K10 carbide and a TiC/TiN coated carbide bit.

7. FORCES IN PARTING

During parting operations forces act in three main directions: the cutting force F_c tangentially, the axial force F_p axially, and the radial force F_f radially. These are described in Fig. 8. The values of these forces influence the required power of the machine for a given machining condition, the bending of the blade, and the initial reverse movement of the carbide bit in the seat in radial and tangential directions. Therefore, for this system the measurement of these forces is very important.

During the development of this new tool system, the three components F_p, F_c and F_f were measured with the aid of a piezo-electric Dynamometer. Forces for 3 to 6.4mm width of cut were measured. Feeds ranged between 0.05 and 0.25mm/rev, and cutting speeds v_c ranged between 40 to 120m/min at the outer diameter. In all cases the diameter of the workpiece d_A†50mm. Experiments were carried out with a constant RPM so that the cutting speed decreases as moving toward the center of the workpiece. The cutting speed is determined by the outer diameter.

Fig. 8: Influence of Cutting Speed on Radial and Tangential Forces

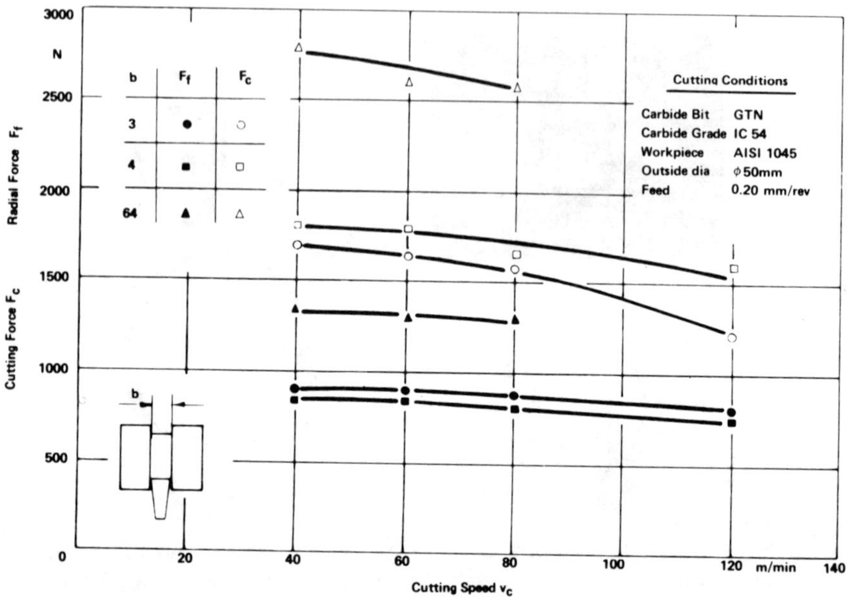

As a first step experiments with neutral N carbide bits were carried out. The axial force F_p acting in this case was maximum 3% of the value of the F_c and F_f forces. Therefore, it can be ignored for machining with the neutral bit. With the aid of measured forces, it was possible to analyse the influence of the different machining parameters and to optimize the geometry of carbide bit and blade.

The influence of the cutting speed on the tangential cutting force F_c and the radial force F_f is given in Fig. 8. From the diagram it is clear that with the higher cutting speed, the cutting forces and the radial forces were reduced. The maximum of the cutting speed is selected in accordance with the carbide grade ISO P40. A higher cutting speed will cause higher wear for this carbide grade and therefore higher forces.

The decrease of the cutting force as shown in the figure is explained by the decrease of the shearing force in the shearing plane because of the higher temperature in the contact area between tool, chip and workpiece [5]. Also the decrease in the radial force F_f may be explained with the increase in the temperature in the contact area. This phenomenon was observed in all widths of cut (3,4,5,6.4mm). A wider cut, of course, increases the measured forces.

Fig. 9: Influence of Feed on Radial and Tangential Forces

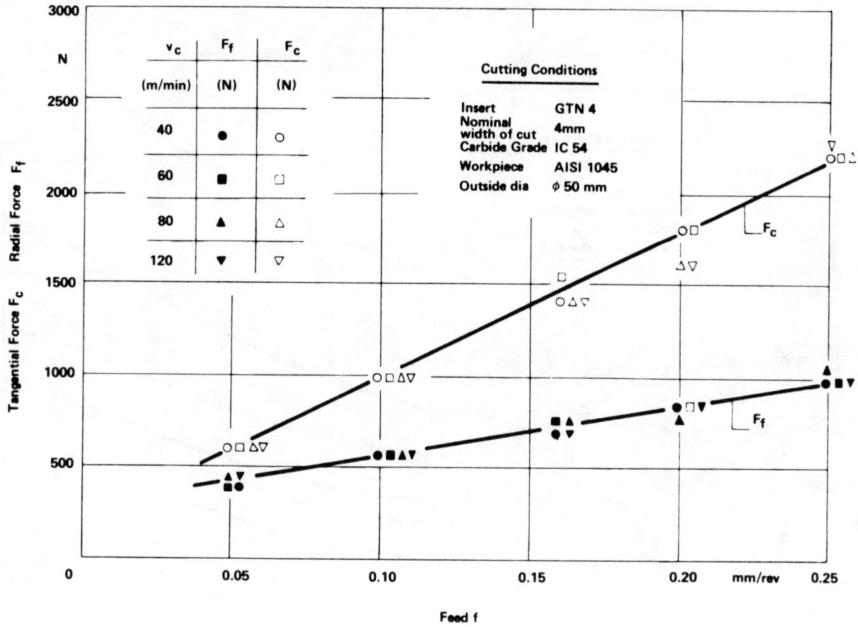

The influence of the feed is illustrated in Fig. 9. At all cutting speeds an increase in cutting forces can be observed with the increase of the feed. The increase of the cutting force F_c is twice as high as the increase of the radial force F_p. This relation is explained by the higher cutting surface area when using higher feeds in comparison with relatively little change of the cutting edge length. In comparison with this the radial pressure is only influenced in a small way by the feed, and therefore the increase is much smaller. The increase in the cutting force F_c depends not only on the radial pressure, but also on the shearing force. The shearing force is not influenced as much as one would expect from the increases in feed. Therefore, the doubling of the feed does not result in a doubling of the cutting force; and parting operations with higher cutting speeds and low feed can reduce the power requirements of the machine.

In Fig. 10 the influence of the width of cut on the radial and cutting force F_c is described. As the width of cut increases both forces also increase. The increase is not directly linear, a phenomenon which can be explained by the lower specific cutting force κ_r. With a higher feed the specific cutting force is smaller, and therefore the increase in the forces and required power do not increase linearly.

Fig. 10: Tangential and Radial Cutting Forces as a Function of Nominal Width of Cut

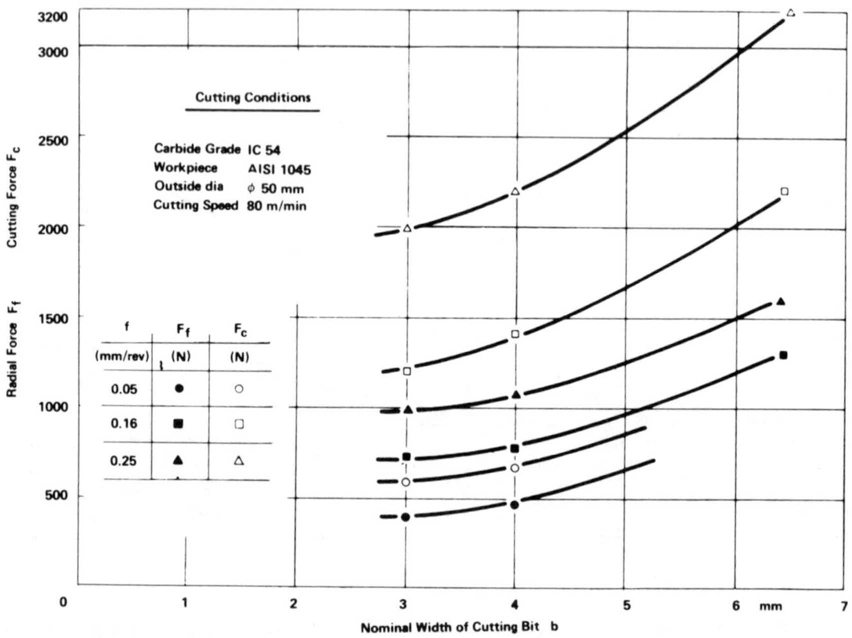

The angle of the active force δ_1, which describes the angle between the active force and the radial force as a function of the machining condition may be analysed with the aid of Fig. 11. A knowledge of this angle is needed in order to optimize the geometry of the tool as already mentioned above. After selecting the right angle β in accordance with δ_1, the initial movement in the tangential and radial directions can be calculated.

From the diagram it may be concluded that δ_1 is in the range of 55° to 68° for the given machining conditions. The increase of feed which caused a large increase of F_c in comparison with F_f results in an increase in this angle. The increase of the cutting speed in opposition to the increase of the cutting speed decreases the force F_c in correlation to F_f and therefore results in the decrease of the angle.

Fig. 11: Angle of Resultant Active Force during Two-Dimensional Parting as a Function of Feed and Carbide Bit Dimensions

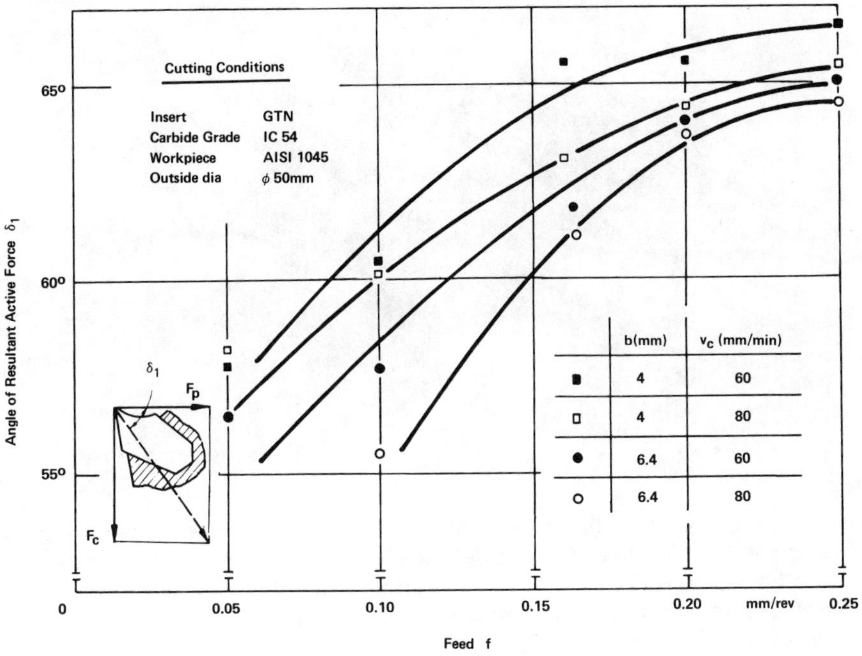

8. TYPES OF SELF-GRIP TOOLHOLDERS

As a result of the theoretical and experimental investigations a number of different parting tools were developed. The three types are shown in Fig. 12.

The first type consists of only two elements: the shaft, a steel body with one gripping seat for the carbide bit. This tool is either left or right, is reasonable in price, its range of application is limited to small diameters. It is the most simple of existing parting tools, and the time required to replace a carbide bit is very short.

The other two types consist of a tool block, blade and the carbide bit. The holder blade is rigid and adjustable to extension required. There are no loose or flexible parts: the insert gripping seat is integral to the body of the blade. All blades feature top and bottom prisms which allow for use as either a right- or left-hand parting tool. The blades are designed with a seat at each end, and this offers double tool life, An extra cavity is featured at the bottom end of the seat. Its main function is to allow space behind the bit for the extractor tool.

Fig. 12: The Three Types of Self-Grip Toolholders

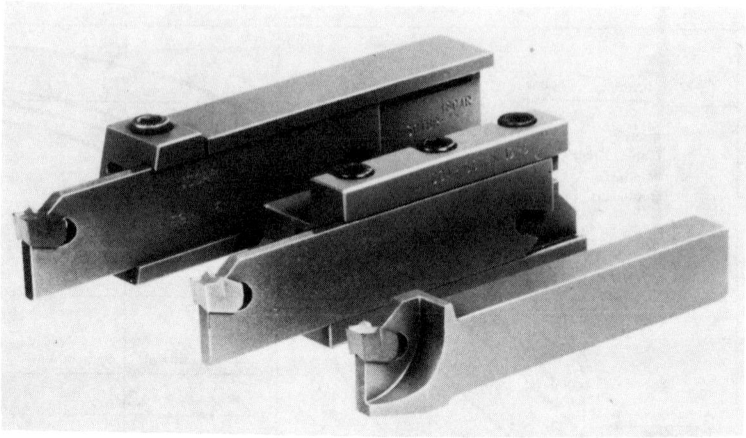

The tool block which holds the blade is a single integral steel body having no loose parts. A dovetail on each side matching the size of the blade permits it to slide freely. Bolts provide for a secure grip on the blade. The block is made of steel specially hardened and tempered to provide optimum flexibility of fastening.

Two basic designs of tool block are available: One type is a compact block, its total width fitting into the lathe support holder. Three bolts serve to fasten the blade in the block and the block in the lathe. Loosening of the first two bolts will allow the blade to slide to a new location with no risk of change in alignment. This block is neutral which means that it may be used for either right or left.

The second type block is clamped to the machine, the blade again held in place by only one bolt which in no way interferes with the clamping action.

Range and Application

The new Self-Grip tools have succeeded in efficiently parting practically all machinable materials. At various laboratories and enterprises, complete process systems have been revised to profit from the extended horizons made possible by these tools. Depth of cut and feed rates reached are far beyond those known before with regard to other parting tool systems..

Carbide Grades

Different carbide grades were tested and for the commercial use of this tool, three different carbide grades are available, as described below:

IC20 (ISO K20) grade: A high wear-resistant grade (HRa 92.6 and TRS 2100 N/mm^2) for machining of cast iron, non-ferric metals, abrasive and high temperature resistant materials.

IC54 (ISO P40) grade: A shock resistant, universal steel machining grade. Steels, alloy castings, forged, stainless cast or austenitic, other-than-round workpieces (interrupted cuts) and other tough metals are all successfully cut by this grade. IC54 is considered as being in the low-wear resistance range; but in parting operations wear is of minor importance as compared with the risk of rupture. Therefore, this grade proved itself as the most suitable for the largest range of steel parting.

IC656 (ISO P20-P40) grade: This TiC/TiN coated grade is shock resistant and most suitable for variable cutting speeds on steels; carbon, tool, alloy, castings; malleable cast iron; bearing steels; austenitic and martensitic stainless steels. It is recommended when grades IC20 and IC54 are not suitable for the job, as it has a dull cutting edge. This grade has a hardness 90.5 HRa and a TRS of 2600 N/mm^2.

Carbide bits are available with cutting widths of 2.2, 3, 4, 5, 6.35 and 9.5mm and holder blades for each size are furnished as standard. A range of tool blocks with various sizes, for each width of cut, are also standard. Full diameters up to 160mm have been successfully parted with the 6.35mm standard bit. Full steel bars of 100mm diameter have been parted with the 2.2mm bit, using some elementary extra care during testing.

Recommended speeds and feeds are given in Table 1 below: As a general rule, 80 per cent of the normal turning cutting speed and twice the feed previously used with other parting tools, are considered a sound recommendation. Chip form control mechanisms perform better at higher feeds. When parting with right- or left-hand bits, the feeds and depths of cut should be in the lower range.

For all machining operations the use of coolant improves the surface quality, reduces the wear and cutting forces. Flushing coolant, either oil or water base, should be applied at parting location and will improve cutting performance on all the metals mentioned in Table 1, except for brass which is normally machined without a coolant. Especially when machining alloy steels like SAE 8620 or steel with hardness of more than 250HB, coolant should be used which contains chloride and sulphur additives which cause a lower wear coefficient and less wear.

Table 1: Recommended Speeds and Feeds

Material Machined	Brinell Hardness	Recommended Carbide Grade	Cutting Speed at outside dia m/min	Width of cut b (mm) 2.2	3;4	5;6.35
				Feed range: Dia mm/rev		
Steels, Alloy steels Martensitic & Ferritic stainless	Up to 200	IC54	70 ÷ 110	0.07 ÷ 0.20	0.12 ÷ 0.35	0.12 ÷ 0.45
	250		70 ÷ 90	0.07 ÷ 0.18	0.12 ÷ 0.32	0.12 ÷ 0.42
	325		55 ÷ 80	0.07 ÷ 0.15	0.10 ÷ 0.30	0.10 ÷ 0.40
	325 over		45 ÷ 60	0.05 ÷ 0.13	0.10 ÷ 0.25	0.10 ÷ 0.35
steels	325 over	IC656	30 ÷ 100	--------	0.15 ÷ 0.25	0.15 ÷ 0.35
Austenitic Stainless Steel	Up to 300	IC20	85 ÷ 110	0.05 ÷ 0.15	0.10 ÷ 0.25	0.10 ÷ 0.30
Cast Iron	Up to 200	IC20	70 ÷ 95	0.12 ÷ 0.30	0.20 ÷ 0.60	0.20 ÷ 0.60
	300		45 ÷ 65	0.12 ÷ 0.25	0.20 ÷ 0.50	0.20 ÷ 0.50
Bronze	Unlimited	IC20	130	0.07 ÷ 0.25	0.10 ÷ 0.40	0.10 ÷ 0.45
Brass	Unlimited	IC20	250	0.07 ÷ 0.40	0.10 ÷ 0.70	0.12 ÷ 0.75
Aluminium	Unlimited	IC20	300	0.07 ÷ 0.45	0.10 ÷ 0.50	0.15 ÷ 0.60

NOMENCLATURE

b = Nominal width of cut
B = Width of blade
d_A = Outer diameter
D = Workpiece diameter
E = Modulus of elasticity
f = Feed per revolution
f_p = Deflection in axial direction
f_c = Deflection in tangential direction
F^* = Normal force
F_{a1} = Resultant active force in two-dimensional parting
F_{a2} = Resultant active force in three-dimensional parting
F_c = Cutting force (tangential force)
F_f = Radial force (back force equals feed force in parting)
F_p = Axial force (side force for parting operations)
F_{in} = Insertion force
F_{out} = Exertion force (removal)
H = Height of blade
I = Moment of inertia
K_s = Specific cutting force
l = Width of cutting edge
L = Overhang
M = Bending moment
n = Revolution per minute
v_c = Cutting speed

v_e = Resultant cutting speed
v_f = Feed speed
φ = Wedge angle of insert
α = Clearance angle
α_n = Tool - normal clearance
β = Angle between horizontal plane and insert centerline
γ_n = Rake angle
γ^* = Friction angle
Θ = Angle of prisma in bit and blade
δ_1 = Angle of resultant active force in normal parting operations
δ_2 = Angle of resultant active force in non-parallel parting operations
δ_3 = " " " " " " "
η = Angle between resulting cutting speed and cutting speed
κ_r = Tool cutting edge angle
κ'_r = Tool cutting edge angle (secondary)
μ = Coefficient of friction

REFERENCES

1 Anonym. 'Von Futter-zu Stangenarbeit durch HM-Abstechwerkzeug', moderne fertigung, Juli/August 1975, s. 30-32.

2 Tippnis, V.A. and Joseph, R.A.'Einstechbearbeitung auf Mehrspindel Drehautomaten', Werkstatt und Betrieb 105/1972, 5, s. 367-377.

3 Dueniss, Neumann, Schwarz. 'Fertigungstechnik-Trennen', 'VEB Verlag', Technik, Berlin, 1974.

4 Schleiffer, C.J. 'Berechnung der aufzubringenden Spannkraft bei Spannzangen', Technische Rundschau 35(1973), 1, s.17-18.

5 Marks. Standard Handbook for Mechanical Engineers, McGraw-Hill, 1967 Edition, USA.

6 Langhammer, K. Die Zerspankraftkomponenten als Kenngroessen zur Verschleissbestimmung an Hartmetall-Drehwerkzeugen. Dissertation, T.H. Aachen, 1972.

7 Malle, K. 'Werkzeuge und Spannzeuge', Wt-Z ind Fertigung 68 (1978), Nr. 12, s.776-781.

8 Wertheim, R., Pano, J., Stambler, A. 'Abstechwerkzeuge mit selbstklemmenden, auswechselbaren Hartmetall-Schneidplatten', Werkstatt und Betrieb 111(1978) 5, s. 297-303.

UNIQUE CUTTING ACTIONS AND PERFORMANCES OF CARBIDE BALL NOSE END MILL WITH A SPIRAL EDGE

Otohiko Kasukawa
Manager of Engineering Dept.
Dijet Industrial Co., Ltd.

INTRODUCTION

Under the title of the same new tool, Mr. Hosoi (the inventer) and Dr. Hoshi have presented several papers at SME, CIRP, and ICPE Conference since 1977. Dijet Industrial Co., Ltd. (one of the four big manufactuers of cemented carbides & tools in Japan, and located near from Hosoi Machine Works Co., Ltd.) has been developing this carbide tool from the beginning as the manufacturer of this tool.

Cutting actions of this new tool (U.S. Patent No. 4132493 Jan. 2, 1979) were studied scientifically in the said papers, particualrly in "Annals of CIRP Vol. 25/1/1977, p.49-53" and "SME Technical Paper MR78-341". The author feels the need to explain these unique actions more popularly for practical use in the machining field. Therefore, in addition to these studies, the author presents more popular and intuitional analysis about the unique cutting actions of this new tool by schematic illustrations.

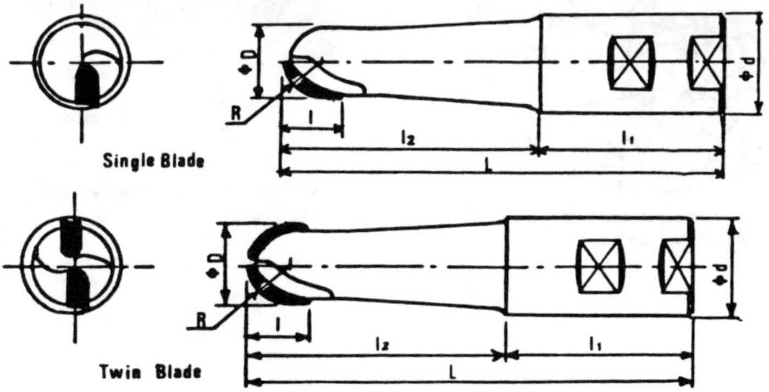

Fig. 1 Carbide Ball Nose End Mill with A Spiral Edge

ANALYTICAL OBSERVATION OF CUTTING ACTION

Fig. 1 illustrates this tool which comes out with single blade and twin blades (or single with auxiliary blade). Fig. 2 shows a bird's-eye view of the ball nose part with a single blade. Fig. 3 is a schematic comparison of nose cutting edge (single blade type) viewed axially from the bottom end. In this figure, the gray colored part indicates the carbide blade of this new end mill and spotted part indicates the carbide blade of the traditional ball nose end mill, and line A indicates the cutting edge line of this new end mill and line B shows that of the traditional ball nose end mill.

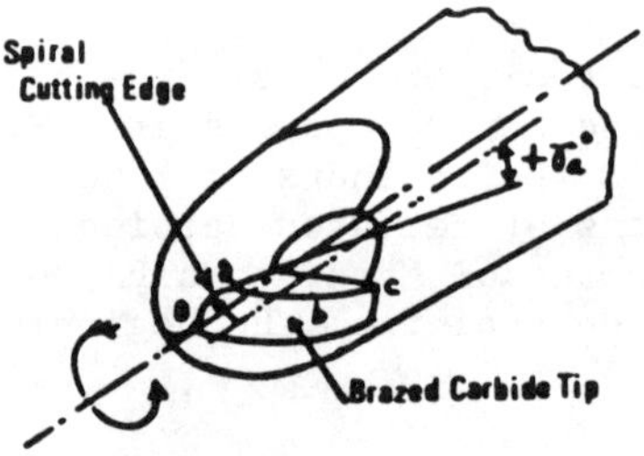

Fig. 2 Bird's-eye View of Ball Nose

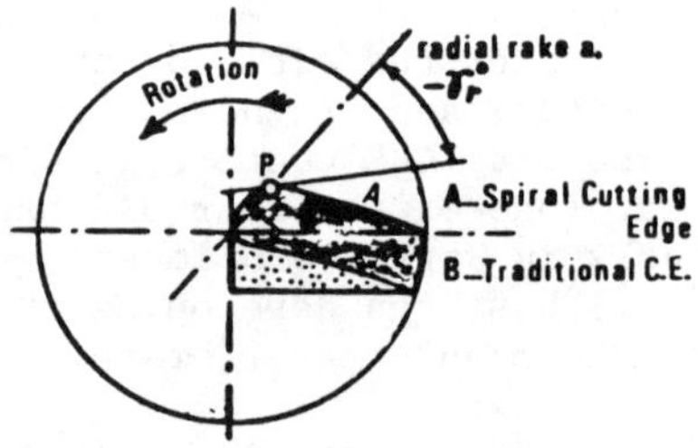

Fig. 3 Comparison of Nose Cutting Edge

Fig. 4 Three-dimensional Image of Spiral Cutting Edge Line

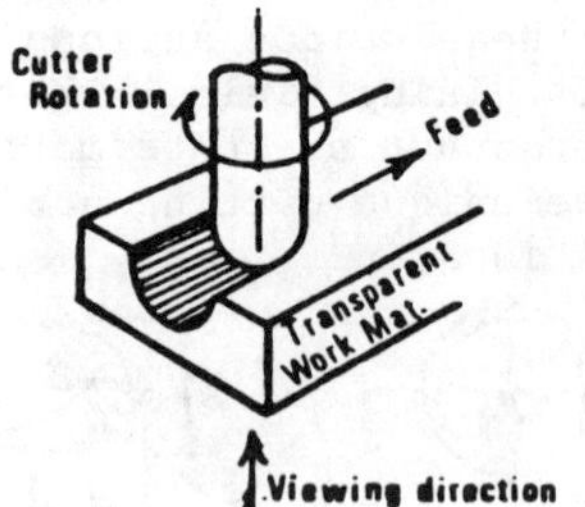

Fig. 5 Cutting Circumstance Supposed for Fig. 6

The cutting edge line of this new end mill seems like the central part of spiral line, while that of the traditional ball nose end mill is straight line. The unique cutting edge curve may be imagined three-dimensionally by the thick line as shown in Fig. 4.

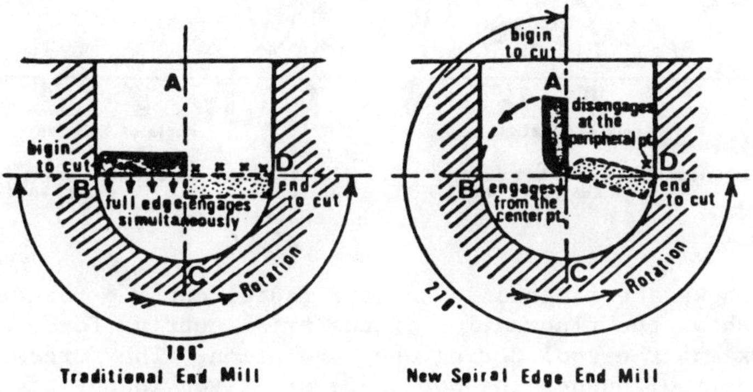

Fig. 6 Comparison of Cutting Action

For the purpose of intuitional understanding, just imagine a cutting circumstance as shown in Fig. 5, and suppose the work material is transparent like acryl plastics. And also suppose you are looking at the cutting edge actions from the bottom end of the end mill axially. Then, as shown in Fig. 6 Left, the traditional end mill begins to cut actually at rotative position B and ends to cut at position D along entire length of cutting edge simultaneously. Angle of rotation between B and D is 180° (= 1/2 revolution) in this case. In case of new end mill (Fig. 6 Right), the cutting edge begins to cut actually at position A and ends to cut at position D, and begins to engage with the work material at A from the center point of the cutting edge which is about perpendicular to feed direction partially at this point, and disengages at D at the peripheral point of the cutting edge. Angle of rotation between A and D is 270° (= 3/4 revolution) in this case, and full length of the cutting edge engages with the work material between B and C (= 90° or 1/4 revolution) only. Above-stated intuitional analysis is proved by Dr. Hoshi and Mr. Hosoi in their studies on cutting force analysis, an example of which is shown in Fig. 7 and Fig. 8.

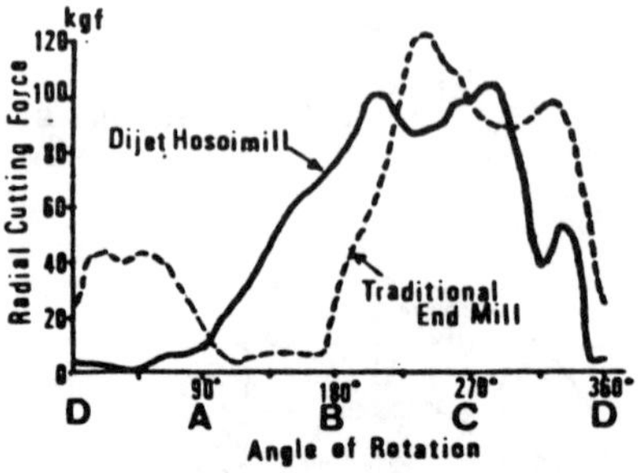

Fig. 7 Comparison of Radial Cutting Force (Hosei and Hoshi)

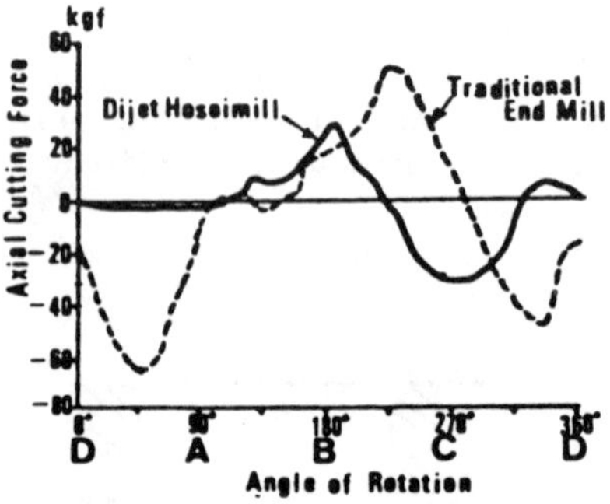

Fig. 8 Comparison of Axial Cutting Force (Hosei and Hoshi)

Fig. 7 shows the fluctuation of the radial cutting force and Fig. 8 shows the fluctuation of the axial cutting force between the work and the tool during one revolution. The forces with which the tool pushes the workpiece are given in positive values, while the force pulling the workpiece are given in negative values in Fig. 8. These figures indicate that the cutting force of the traditional end mill rises sharply during shorter cutting period and its maximum value is higher than that of the new end mill. And in case of the traditional end mill, fairly high cutting force is indicated during non-cutting period (D - A - B) by the influence of built-up edges. While in case of new end mill, the cutting force increases and decreases more slowly and the influence of built-up edges is very little, and the positive axial rake angle of the new end mill causes negative axial cutting force at about the position C. The cutting conditions in this figure are;

Tool	Dijet Hosoimill ØD: 20 mm (single blade)
Work Material . .	S45C (equiv. to SAE1045) Brinell Hardness 176
Cutting Speed . .	1,500 rpm (94 m/min at peripheral)
Feed	500 mm/min (0.33 mm/tooth)
Depth of Cut . . .	4 mm
Width of Cut . . .	15 mm down cut
Cooling	Compressed Air Jet

Through these cutting action analyses, we can understand the reason why the mechanical and thermal shocks for the new spiral edge ball nose end mill are lower and chipping and cracking troubles of the carbide cutting edge can be avoided.

Therefore, to start cutting from the center cutting edge is very effective and important for this new type end mill, and it is perferable to perform plunge milling at first cutting in, and then to feed to the radial direction.

As shown in Fig. 3, the radial rake angle at any point P on the cutting edge of this new end mill is negative along full length of the edge except for the center point, and the axial rake angle is positive as shown in Fig. 2. In case of face milling cutters, it is well known that the combination of negative radial rake and positive axial rake results in clearing chip flow by the resultant positive inclination angle as shown in Fig. 9. Therefore, chips generated by this new ball nose end mill are ejected easily by the centrifugal force caused by the high-speed revolution, which is a peculiarity of carbide tools, and by the said well known action of cutting edge geometry.

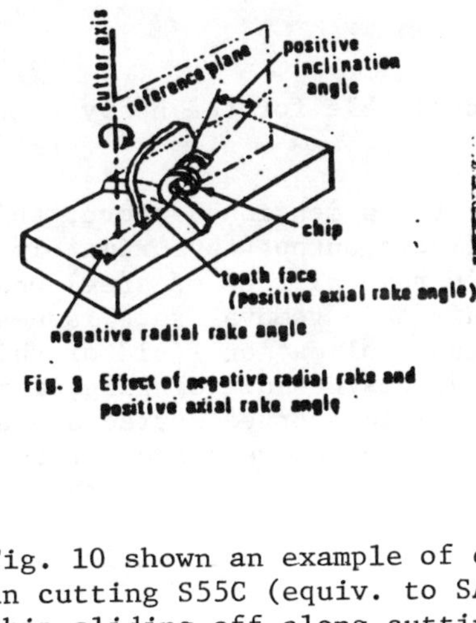

Fig. 9 Effect of negative radial rake and positive axial rake angle

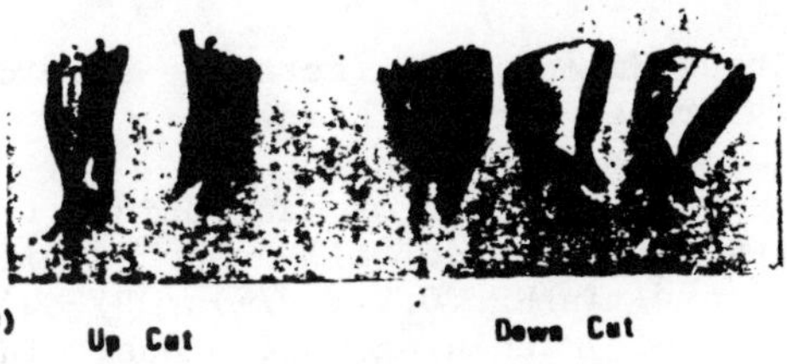

Fig. 10 Chips of New Spiral Edge End Mill (medium carbon steel)

Fig. 10 shown an example of chips generated by this new end mill in cutting S55C (equiv. to SAE1055) steel. This behavior of a chip sliding off along cutting edge line restrains the growth of pressure welded built-up edges and prevents from chipping damages.

In addition to these special features, this new type carbide ball nose end mill has very strong and rigid design as shown in Fig. 11. The direction of cutting force of this new tool is dispersed as shown in this figure, and the cutting edge of central part is supported rigidly for the direction of cutting force by the thicker dimension of the carbide blade. While the direction of cutting force of traditional ball nose end mills is concentrated to one direction along the full length of cutting edge as shown in this left figure. Moreover, as a result of convex curved cutting edge, the steel back metal to support the carbide blade is thicker and stronger than in case of the traditional end mill as shown in this right figure.

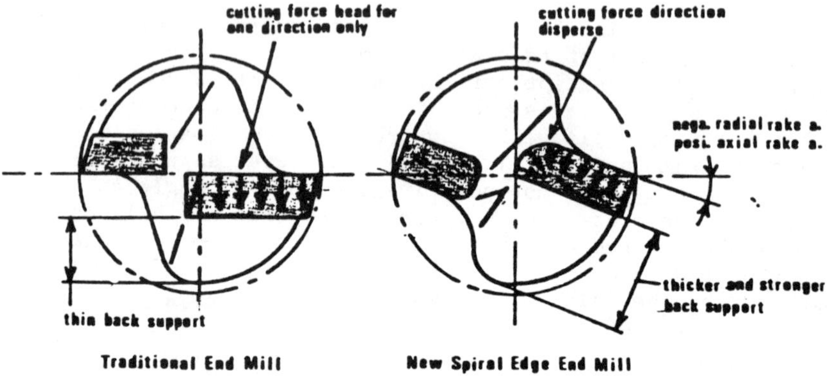

Fig. 11 Strong and Rigid Design

These dimensional facts are also desirable for the heavy cutting by brazed carbide blade.

As a result of many unique features mentioned above, this new type carbide ball nose end mill can output approximately 15 cm^3/min/HP (0.9 in^3/min/HP) chip removal rate of steel under suitable machining conditions. This chip removal rate is nearly same as face milling cutters, so the application field of this new tool will certainly be expanded, besides die sinking, for machining of general machine parts having curved surfaces and for machining the machanical structure parts required for the weight reduction.

The practical examples using this new carbide ball nose end mills are as follows:

PRACTICAL EXAMPLES

Example 1. Forging die for knuckle

Work material: Die steel (JIS SKT6, Fe-.75C-.80Mn-2.75Ni-.95Cr-.40Mo)
Scleroscope hardness 62
Shape: Fig. 12 and Fig. 13
Machine tool: HC450 tracer-controlled milling machine-(Hosoi Machine Works Co., Ltd.)

Cutting cond. / Phase	Mode	Tool size	Spindle rpm	Feed mm/min	Pick feed mm	Mach. time min.
1 Rough machining	X-Y contouring	D16mm, twin blades	1,500	450	2.5	25
2 Semi-finishing	X-Y contouring	D10mm, twin blades	1,800	350	1.0	15
3 Finishing	X-Y contouring	D6mm, twin blades	1,800	1,000	0.5	24
4 Finishing	X-Y contouring	D4.5mm, twin blades	1,800	1,000	0.4	60
5 Final finishing	X-Z surface profiling	D4.5mm, twin blades	1,800	Y:1,000 Z:1,500	0.3	90
					Total	214 min

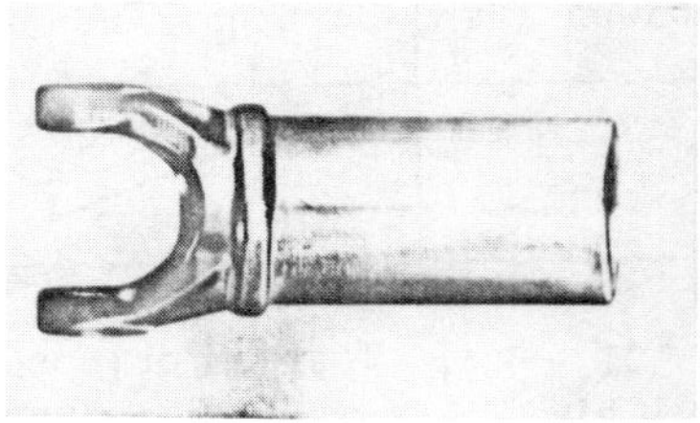

Fig. 12

Fig. 13
close up

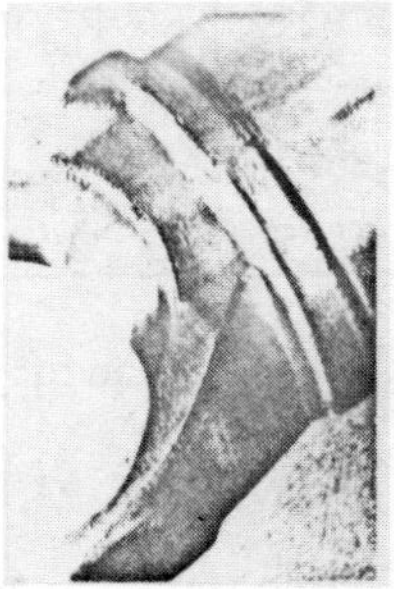

(Source: Mr. T. Hosoi)

Example 2. Forging die for crank shaft

Work material: Die steel (JIS SKT4 equivalent to AISI 6F2)
Shape: Fig. 14 roughly
Machine tool: "MH-6 TC" OKK-HOSOI milling machine (Osaka Kiko Co., Ltd.)

Phase \ Cutting cond.	Mode	Tool size	Spindle rpm	Feed mm/min	Pick feed mm	Mach. time min.
1 Rough machining out of plain	X-Y contouring	D20mm, twin blades	1,650	350	5	150
2 Rough machining except for bottom surface	X-Y contouring	D20mm, twin blades	1,650	350	2	60
3 Rough machining of bottom surface	Y-Z surface profiling	D20mm, twin blades	1,650	Y:300 Z:600	2	60
4 Semi-finishing except for bottom surface	X-Y contouring	D12mm, twin blades	2,000	400	.8	210
5 Finishing of bottom surface	Y-Z surface	D12mm, twin blades	2,000	Y:400 Z:800	.8	120
					Total	600 min

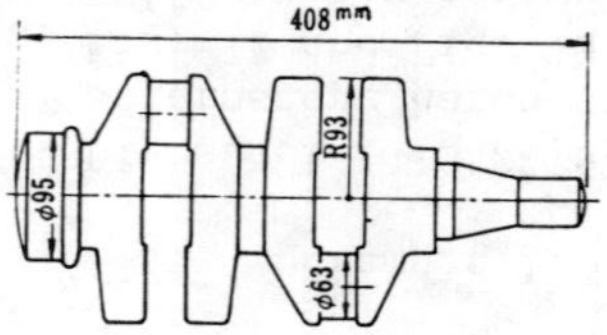

Fig.14
forged shaft

(Source: Mr. T. Kotani
Osaka Kiko Co., Ltd.)

Example 3. Plastic mold die for telephone receiver

Work material: Medium carbon steel (JIS S55C equivalent to AISI 1055)
Scleroscope hardness 30
Shape: Fig. 15 and Fig. 16
Machine tool: HC450 Tracer-controlled milling machine (Hosoi Machine Work Co., Ltd.)

Cutting cond. / Phase	Mode	Tool size	Spindle rpm	Feed mm/min	Pick feed mm	Mach. time min.
1 Rough machining out of plain	X-Y contouring	D25mm, twin blades special length	1,500	1,000	5	50
2 Rough machinine	X-Y contouring	D20mm, twin blades special length	1,800	1,500	1	50
3 Semi-finishing	X-Y contouring	D8mm, twin blades special length	2,000	1,500	.6	90
4 Finishing	X-Y contouring	D4mm, twin blades solid carbide	2,000	1,000	.3	130
5 Finishing	X-Z surface profiling	D4mm, twin blades solid carbide	2,000	X:1,000 Z:1,200	.3	160
					Total	480 min

Fig.15

Fig.16
close up

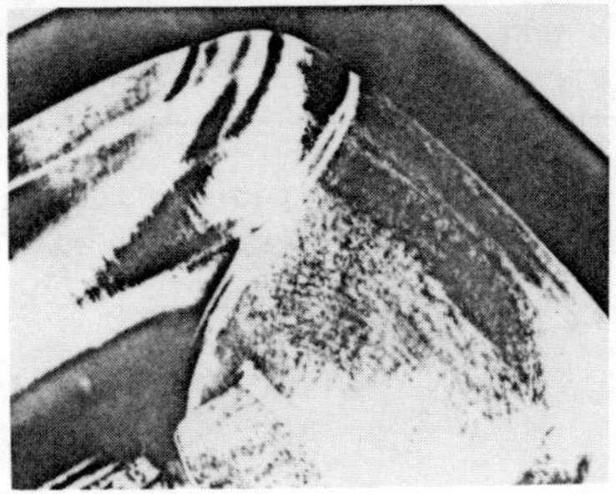

(Source: Mr. T. Hosoi)

Example 4. Glass mold die for liquor bottle

Work material: Gray cast iron (JIS FC35, min. tensile strength 32 kgf/mm)
Shape: Die for 2/3 quart square bottles with surface of coarse net pattern
Machine tool: "MH-8 TC" OKK-HOSOI milling machine (Osaka Kiko Co., Ltd.)

Phase \ Cutting cond.	Mode	Tool size	Spindle rpm	Feed mm/min	Pick feed mm	Mach. time min.
1 Rough machining	X-Y contouring	D16mm, twin blades	1,800	1,000	4	12
2 Rough machining	Y-Z surface profiling	D16mm, twin blades	1,800	Y: 750 Z:1,500	6	7
3 Finishing	X-Y contouring	D8mm, twin blades	3,360	1,500	.2	130
4 Finishing	Y-Z surface profiling	D8mm, twin blades	3,360	Y: 750 Z:1,500	.2	70
					Total	219 min

(Source: Mr. T. Kotani, Osaka Kiko Co., Ltd.)

ACKNOWLEDGEMENTS

The author expresses his sincere appreciation and acknowledgement to Mr. Toshiaki Hosoi, Dr. Tetsutaro Hoshi, and Mr. Teruo Kotani for their studies and favors.

REFERENCES

(1) T. Hosoi and T. Hoshi, "Cutting Actions of Ball End Mill with a Spiral Edge," Annals of CIRP, vol. 25, No. 1, 1977, p. 49-53.

(2) T. Hosoi, R. Hosoi and T. Hoshi, "Cutting Performance of Carbide Tipped Ball End Mill with a Spiral Edge", Proceedings of International Conf. on Prod. Engg., New Delhi 1977, vol. 1, p. v-190-196.

(3) T. Hosoi and T. Hoshi, "High Speed Tracer-controlled Milling - A new Technology of Diemaking -", SME paper MR78-341, 1978, p. 1-15.

(4) V.A.Tipnis, "High Rate Copy Milling Carves New Niche", Modern Machine Shop, Feb. 1979, p. 105-117.

(5) T. Hosoi, T. Hoshi and V. Tipnis, "Comparative Evaluation of New Technologies for Copy Milling and Die Sinking", SME paper MR79-344, 1979, p. 1-28

HIGH RATE DRILLING WITH A NEW TYPE OF CARBIDE-TIPPED TWIST DRILL

Toshiaki Hosoi (Hosoi Machine Works Co., Ltd.)
Ryosuke Hosoi (The Pennsylvania State University)
Kozo Asano (Hosoi Machine Works Co., Ltd.)
Osamu Horiuchi (Toyohashi University of Technology)
Yoshio Hasegawa (Osaka University)

INTRODUCTION

Hole making is one of the most frequently and widely performed operation among the metal cutting processes. The conventional HSS (high-speed steel) twist drills are commonly used for making new holes at the present time. This operation, however, is becoming a serious bottle neck in achieving a higher production, because of limited possibilities of increasing the feed rate. The application of carbide tips to twist drills has been keenly needed, but it has not yet succeeded because the cutting edges are damaged at the chisel especially in the case of drilling steel workpieces.

Described in this paper is a new type of carbide-tipped twist drill developed successfully for drilling steel work-pieces at considerably higher feed than that possible with the conventional HSS twist drills.

Design principle, cutting mechanism and capability of this new carbide drill will be discussed with reference to a conventional HSS drill.

GEOMETRIES OF THE CARBIDE DRILL

Cutting tests described in this study employed carbide drills of a specific size listed in Table 1 along with a reference HSS drill. Two pieces of ISO grade M carbide are brazed as seen in Fig. 1 to the end of the drill shank of the nickel-chrome-molybdenum steel. It appears that the cutting

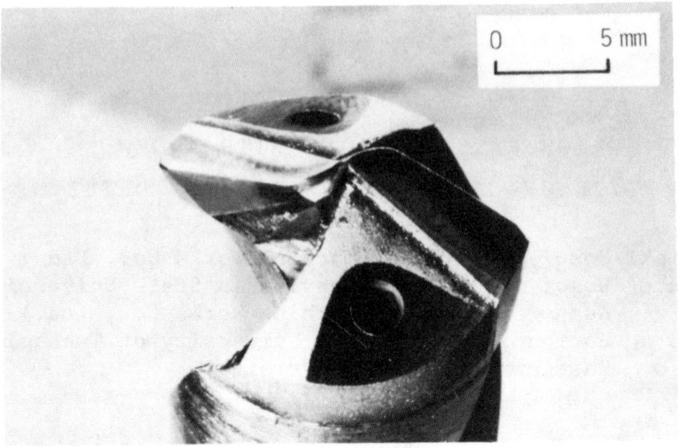

Fig. 1. Cutting edges of a carbide drill, 20 mm (0.79 in) diameter

edge of the carbide drill is ground to a special shape, and it has no chisel but the cutting edges extend up to the center of the drill. The cutting edges near the center form connected circular arcs and the outer parts are linear. This is the spiral edge design known from Hosoi's previous experience with the carbide-tipped ball-nosed end mill to prevent edge chipping near the tool center where the cutting speed is low and to ensure a long tool life (1). The rake angle of the linear cutting edge is 20 deg and that of the circular part is nearly 0 deg. The cutting edges are provided with a narrow negative land. The negative land strengthens the cutting edge and produces stable built-up edge which is effective in chip formation at reduced cutting energy. Width of the negative land should be adjusted depending on the work

Table 1. Drill size used in the study

item	carbide drill	HSS drill
diameter	20 mm (0.79 in)	20 mm (0.79 in)
overall length	155 mm (6.10 in)	255 mm (10.04 in)
flute length	100 mm (3.94 in)	155 mm (6.10 in)
point angle	140 deg	118 deg
helix angle	20 deg	30 deg
web thickness	6 mm (0.24 in)	2.5 mm (0.10 in)

material: for example, 0.4 mm (0.016 in) for low carbon steel and 0.2 mm (0.008 in) for medium to high carbon steels.

The flute of the carbide drill is not smoothly connected to the rake face, but a specially designed chip pocket is provided in between. The produced chips are securely curled, torn and broken as they proceed into the flute through the chip pocket.

There are two different designs of the margin, i.e., single and double margin-types. In the final grinding process of the drill edge, the peripheral surface of the brazed carbide tips are first ground into a circular cylinder concentric with the drill axis. Then the trailing part of the cylindrical surface is recessed, leaving a narrow concentric land at the leading side. This is the single margin-type. The double margin, on the other hand, is formed by flatly grinding the top of the cylindrical surface so as to divide it into the leading and trailing parts. The plateau between the two cylindrical surfaces or margins provides in practice a clearance where the cutting fluid penetrates. The double margin type is intented for drilling holes to a closer accuracy.

Fluid holes are provided through the drill. Cutting fluid is supplied through the holes by a quarter kilowatt fluid pump. Although the supply quantity is only a few liters (or quarts) per minute, the fluid supply is indispensable to keep a successfull drilling especially on steel workpieces.

Tests have been performed on a prototype of a new CNC drilling machine being developed for machining with the carbide drill. The spindle of the machine is driven up to 2500 rpm by a variable speed hydraulic motor powered by a 15 HP hydraulic pump.

CUTTING MECHANISMS OF THE CARBIDE DRILL

In order to investigate drilling mechanisms, specifically the formation, breaking off and transfer of the chip, quick-stopping of the drilling process was performed by an apparatus as illustrated in Fig. 2. A workpiece (50 mm or 1.97 in diameter, 50 mm or 1.97 in height) is tightly held between clampers by the clamping force introduced on stretching the pull rod and the cast iron bolt by the hydraulic cylinder. After the drilling is started and when the drill has penetrated some 20 mm, the hydraulic pressure is suddenly raised

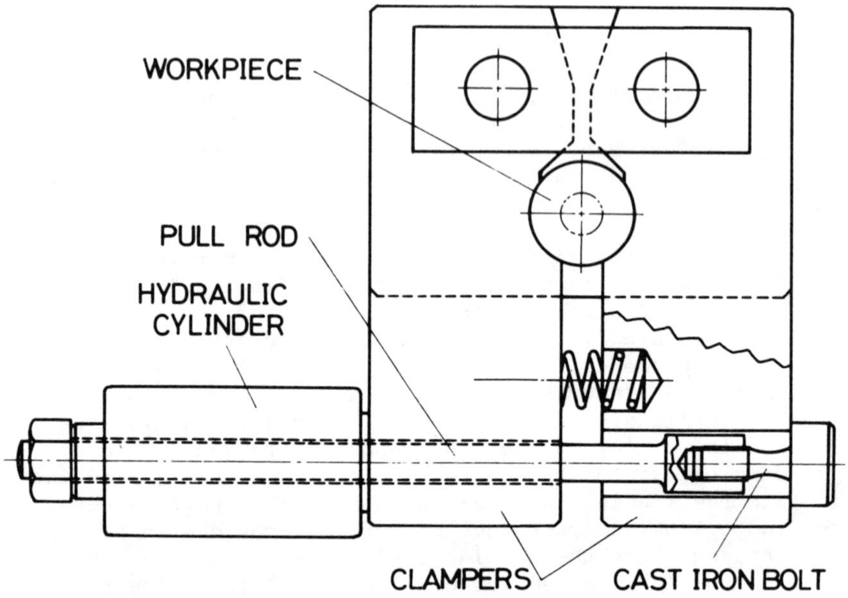

Fig. 2. Quik-stop apparatus.

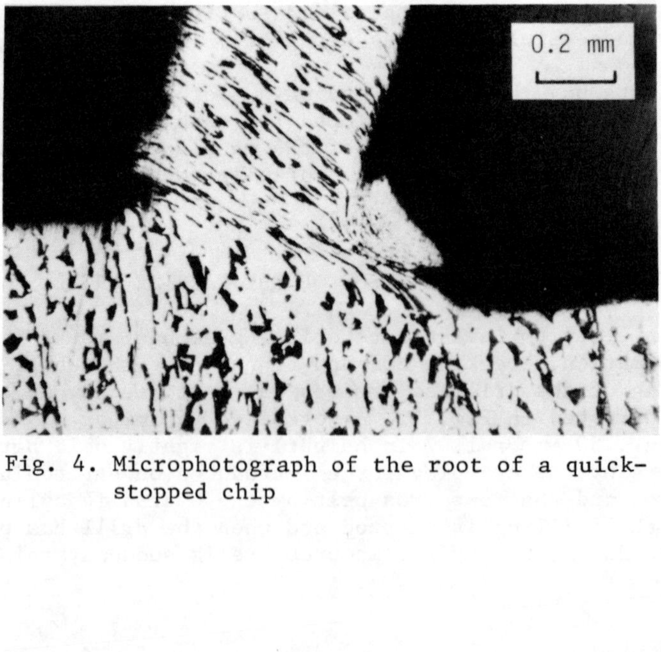

Fig. 4. Microphotograph of the root of a quick-stopped chip

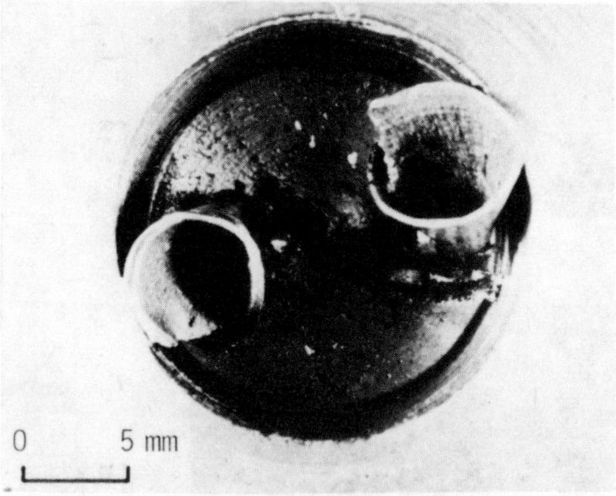

Fig. 3. Quick-stopped specimen.

so that the cast iron bolt fractures by tension. At this moment the workpiece is released and starts to rotate with the drill. Since the acceleration time is very short (0.4 msec in time or 2.3 arc deg of drill rotation at the spindle speed of 1000 rpm) the drilling process is eventually quick-stopped. A specimen as shown in Fig. 3 is obtained. It is cut to expose a cross-section through the root of the chip. Microphotograph of the cross-section, Fig. 4, clearly shows that the chip formation is carried out with the built-up edge being formed on the negative land of cutting edge. This is an application of Silver White Chip (SWC) cutting that positively utilizes the built-up edge formed on the negative land as a part of the true cutting edge (2).

Photographs of Fig. 5 show quick-stopped chips after they have been uncovered by removing the work metal around the drilled hole. The sequence (A) to (D) illustrates the curling and the fracture of the chip immediately after it has been produced by the cutting edge. Fig. 5(A) shows the first step, where the chip is growing in a sector form. Since the outer portion of the chip grows proportionally faster than the inner portion, the chip curls into the chip pocket provided between the rake face and the drill flute and forms a circular cone as seen in Fig. 5(B). As the conical chip is pushed up and enters the flute, due to the velocity difference between the outer and inner portions of the growing chip, it is bent toward outer side of the drill. Tensile stress is induced on the inner side of the growing chip where the chip begins to tear as seen in Fig. 5(C). The tearing crack has

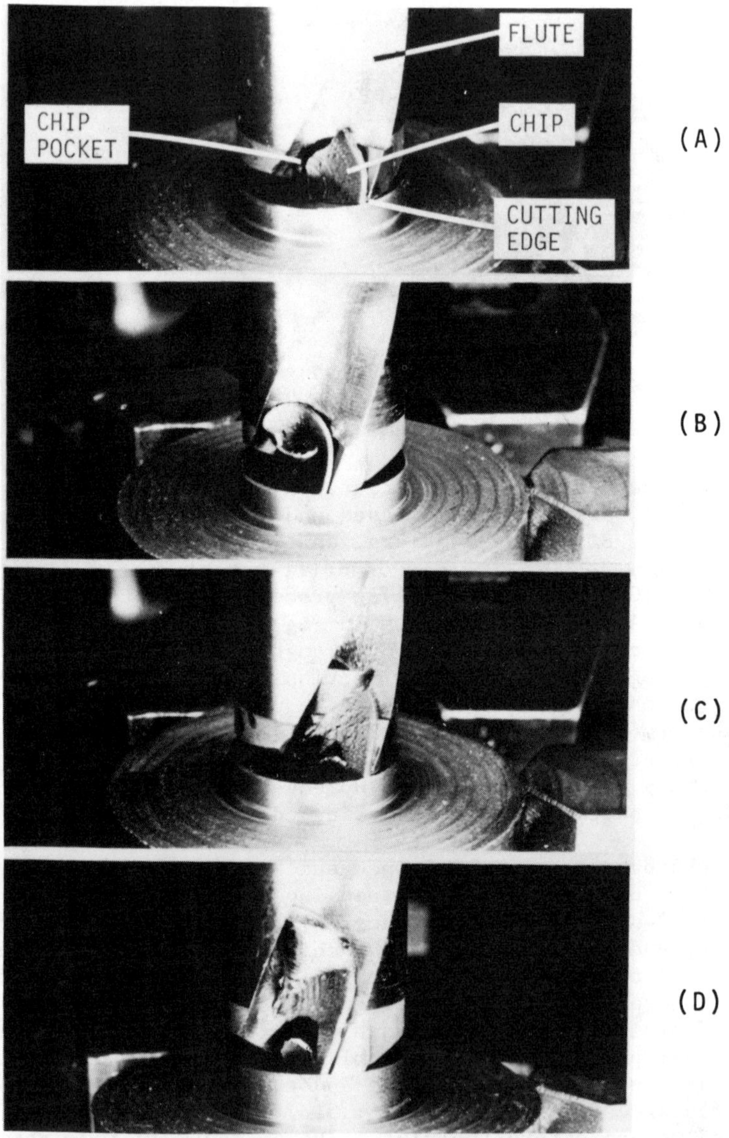

Fig. 5. A sequence of photographs showing the chip flow and breaking. The quick-stopped chip has been uncovered by removing work metal around the drilled hole.

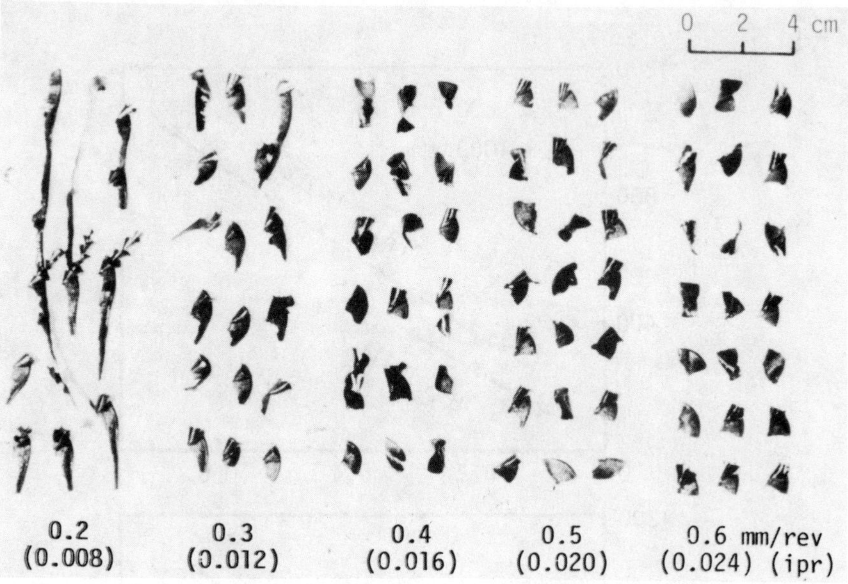

Fig. 6. Chips obtained under various feed per revolution for 0.45%C plain steel. Spindle speed, 800 rpm.

an initial tendency to run parallel to the cutting edge but afterwards it proceeds toward and normal to the cutting edge as seen in Fig. 5(D), an instance just before a break away of the chip. In this case the tearing crack runs toward the outer end of the cutting edge so as to form a broken chip with a tail.

Fig. 6 shows chips obtained under various feed per revolution. It is noted that the chips become shorter with an increased feed per revolution. It is clearly seen that a higher feed is favorable for the chip breaking. Pieces of chip are lined up in the flute and pushed up by the chip growing at the root. The chips are thus continually transfered to the outside without jamming in the flute. A high feed up to 0.6 mm/rev (0.024 ipr) is possible to take without need for stepping the feed of the drill.

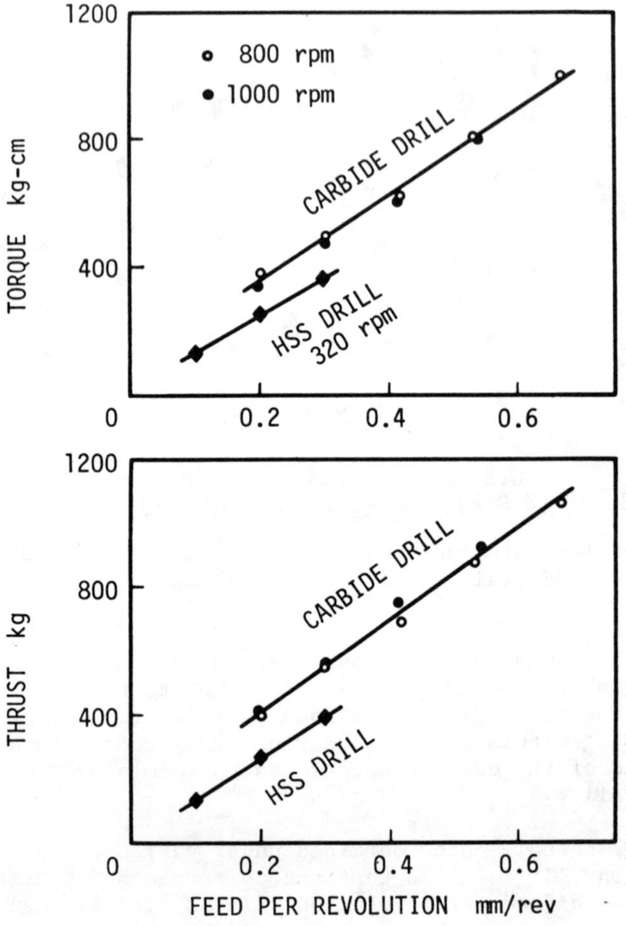

Fig. 7. Drilling forces vs. feed per revolution for a single margin-type carbide drill and the reference HSS drill. Work: Low carbon steel commonly used for weldments, SS41. Drill diameter 20 mm (0.79 in), depth of hole 60 mm (2.36 in).

DRILLING CAPABILITY

In order to investigate capability of the carbide drill, cutting forces and hole accuracy were measured under various drilling conditions. A low carbon steel commonly used for weldments SS41 (167 Bhn) and a 0.45%C plain steel S45C (227 Bhn) were drilled in the test. The negative land width of 0.4 mm (0.016 in) was employed for the low carbon steel and 0.2 mm (0.008 in) for the 0.45%C plain steel. Sulpho-chlorinated water-miscible oil was supplied in drilling experiments by both carbide and conventional HSS drills at a rate of 1.1 l/min (1.2 qt/min).

Fig. 7 shows the results of the drilling force measurement. Each point represents an average of three tests performed at each test condition. It can be clearly seen from the graphs that the drilling forces increase linearly with the feed per revolution. The drilling forces of carbide drills are apparently greater than those of the HSS drill. In Fig. 7 the results for the 0.45%C plain steel S45C are not shown, but they were nearly equal with those for the low carbon steel SS41. Also differences in the drilling forces were very small between the single margin- and double margin -types of the carbide drill.

It is a characteristic benefit of the carbide drills that, as the chips are ejected without a tendency to jam in the flute, the drilling forces stay constant during cut irrespective of the depth being drilled, while forces commonly increase as the HSS drill penetrates deeper.

Fig. 8 shows the average and the range of the hole oversize measured for holes obtained at each test condition. Hole oversize was measured at the depth of 10, 30, 50 mm (0.39, 1.18, 1.96 in) for each hole. As shown in the figure, hole oversize for carbide drills is considerably smaller than that for the HSS drill. Hole oversize for a double margin-type carbide drill is clearly smaller than that of a single margin-type. In case of a HSS drill, hole oversize for the low carbon steel SS41 is greater than that for the 0.45%C plain steel S45C. With the carbide drill, however, the differences of hole oversize between the two work materials are small.

The roughness of the inner surface of the holes made by the carbide drills at a typical drilling condition of 1000 rpm and 0.4 mm/rev (0.016 ipr) is shown in Table 2. It is

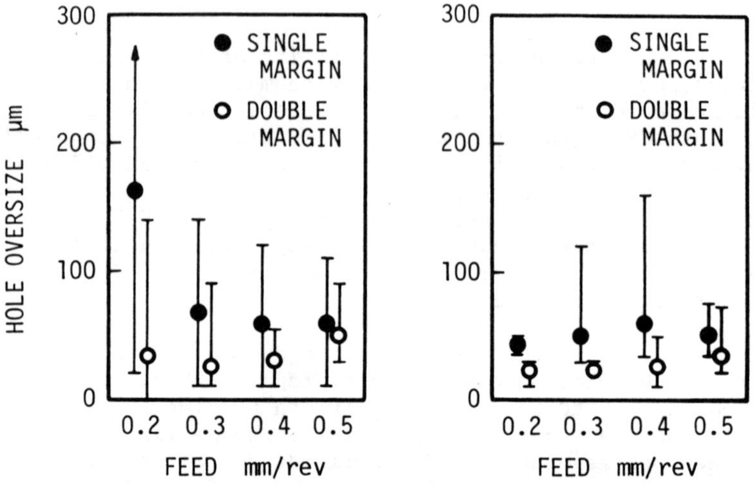

(A) CARBIDE DRILL, SS41, 1000 rpm.

(B) CARBIDE DRILL, S45C, 1000 rpm.

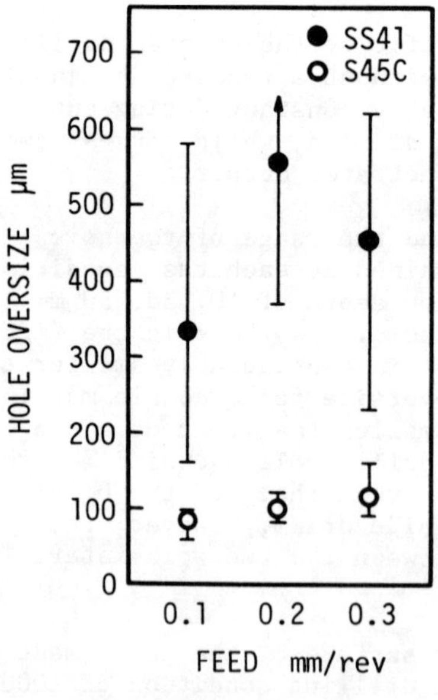

(C) HSS DRILL, 320 rpm.

Fig. 8. Hole oversize vs. feed per revolution for two types of the carbide drill and the reference HSS drill. SS41 denotes low carbon steel commonly used for weldments. S45C denotes 0.45%C plain steel. Drill diameter 20 mm (0.79 in), depth of hole 60 mm (2.36 in).

Table 2. Surface roughness of the hole made by single and double margin-type carbide drills.

Work material	Single margin	Double margin
Low carbon steel SS41	7-12 μmRa	2-8 μmRa
0.45%C plain steel S45C	2-9 μmRa	1-3 μmRa

noted that the double margin-type is superior in making holes to closer accuracy than the single margin-type.

CONCLUSIONS

The new carbide-tipped twist drill is characterized by special shapes designed to the cutting edge, the chip pocket and the flute. It has been ascertained experimentally that the special edge shape makes the tool free from edge chipping which occurs otherwise in steel cutting. Also the chip pocket and the flute design guarantees periodic chip fracture and smooth transport of the chip along the flute.

Recommended cutting conditions for practical use of the tested 20 mm (0.79 in) diameter carbide drill that guarantee the tool life of several hundred to over thousand holes are:

work materials --------	low and medium carbon plain steels (170 to 230 Bhn)
cutting speed ---------	800 to 1200 rpm (50 to 75 m/min or 165 to 250 fpm)
feed per revolution ---	0.25 to 0.4 mm/rev (0.010 to 0.016 ipr)
feed speed ------------	200 to 480 mm/min (8 to 19 ipm)
depth of hole ---------	up to 100 mm (3.94 in)
cutting fluid ---------	low pressure supply of sulpho-chlorinated water-miscible oil at over 1 l/min (1.1 qt/min).

Feed speed indicated in the above is five to ten times faster than that practiced with conventional HSS drill and is comparable to that of Ejector Drills which is only possible to use on dedicated machines equipped with high pressure fluid supply.

Drilling forces for the carbide drill have been experimentally known to be greater than that of conventional HSS twist drills.

The carbide drill gives fairly closer hole accuracy than the HSS drill.

ACKNOWLEDGEMENTS

The authors wish to express their gratitude to Dijet Industrial Co., Ltd. for manufacturing and providing tools tested in this study. They are also grateful to Dr. Tetsutaro Hoshi, Proffessor of Toyohashi University of Technology, for his useful advice and encouragement in this study.

REFERENCES

(1) T. Hosoi and T. Hoshi, "Cutting Action of Ball End Mill With a Spiral Edge", Annals of the CIRP, 25-1 (1977) P. 49.

(2) T. Hoshi, "Recent Development in Silver White Chip (SWC) Cutting", a paper presented to the ASM/SME International Conference on Cutting Tool Materials, September 1980, Cincinnati, Ohio.

DISCUSSION

Mike Schmenk (Cincinnati Milacron)

W.A. Haggerty of my company has done extensive research on both spiral point drills and coolant-fed drills. Based on his results comparable stock removal rates and tool performance could likely be obtained using a properly ground H.S.S. tool without the use of carbide. Why were your cutting speeds with carbide drills limited to 50-75 meters/minute?

Dr. Horiuchi

Although that speed might seem moderate it is the recommendation for industrial use where 1000 holes per sharpening are required. Further the accuracy of the hole is greater with the carbide drill.

RECENT DEVELOPMENT IN SILVER WHITE CHIP (SWC) CUTTING

Tetsutaro Hoshi
Toyohashi University of Technology

INTRODUCTION

Silver White Chip (SWC) cutting refers to the use of cutting tools of tungsten carbide or harder tool materials pointed to a specially designed double-rake edge with an intention of positively utilizing the built-up edge formed on its primary rake face (or the negative land) as the true cutting edge.

The technology was invented in 1955 by Koichi Hoshi (1) and so named after the brilliant metal white appearance of the chip produced in machining plain steels. The process has been finding high potentials in the following machining situations:

1) High feed roughing of plain and alloy steels.
2) Soft but tough work materials such as low carbon steel weldments and austenetic stainless steels.
3) Machining hardened steel parts.
4) Parts having configurations sensitive to work distortions.

The paper reviews backgrounds of SWC cutting, basic principle of edge design, applications in turning, recent achievements in indexable tooling and rail planing.

BACKGROUNDS

Experiences show that a cutting tool with a small (or large negative) rake angle is apt to form a large size built-up edge in front of the cutting edge, while that with a high positive rake angle is often free from the built-up edge formation. The idea of combining the two rake angles into a double-rake edge design so as to stabilize the built-up edge dates back to the 1920's. The high speed steel was the best performing tool material available at that time. The tools allowed only low cutting speeds where the built-up edge formation was the primary obstacle against the smooth surface finish.

Klopstock's tool (2,3) employed a 1mm (0.04 in) wide primary rake (land) with γ_{n1} = +15deg rake angle combined with a γ_{n2} = 60deg secondary rake. HSS tools with this sort of design (4,5) were known to reduce cutting force, extend tool life, and control most fragmental residues of the separated built-up edge to flow out with the chip and not to remain on the finished surface. In recent years, carbide tools disigned by the similar concept but with a slightly negative primary rake angle (γ_{n1} = 0 to - 15deg) are known as the controlled contact cutting tools. Cutting mechanism and performance of this tool design have been reported by many authors (6-10).

SWC edge is designed with a greater negative primary rake angle γ_{n1} = -30deg as illustrated in Fig.1a, and a positive secondary rake angle selected in the range of γ_{n2} = 15 to 30deg depending on the machining situation. A chip former BC is provided in the case of turning tool in order to control the size of chip. When viewed in a cross section, the narrow

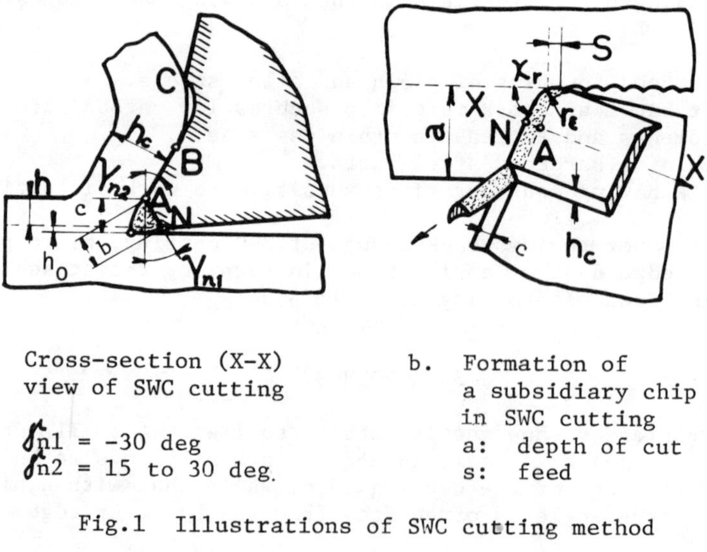

a. Cross-section (X-X) view of SWC cutting

γ_{n1} = -30 deg
γ_{n2} = 15 to 30 deg.

b. Formation of a subsidiary chip in SWC cutting
a: depth of cut
s: feed

Fig.1 Illustrations of SWC cutting method

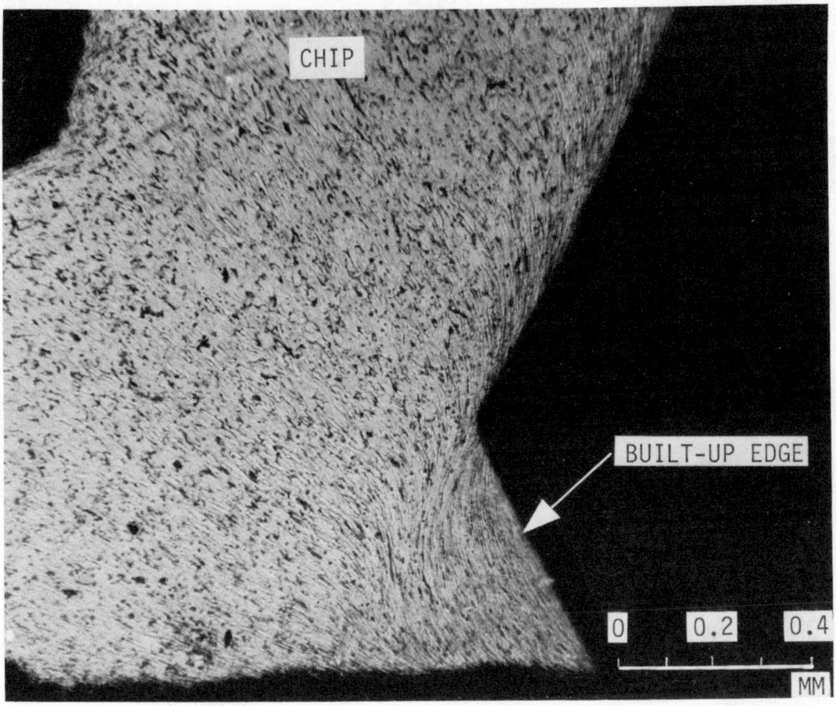

Fig. 2 Cross-section of the end of the planer chip shown in Fig. 3.

negative land (primary rake face) having -30deg rake angle is capped by the built-up edge. The blunt boundary of the built -up edge smoothly blends into the secondary rake face having a high positive rake angle as seen in a photograph of Fig.2. This high positive rake angle reduces the cutting energy by at least 20%.

Different from the built-up edge formed on conventional tools that continually fluctuates in size and shape, the SWC edge controls the built-up edge to retain a predetermined size and shape in the cross-section. But it grows laterally so as to flow out of the cutting area in the direction along the negative land and forms a thin wire-like subsidiary chip as illustrated in Fig.1b, besides the principal body of the chip of the ordinary sence (see Fig.3). The subsidiary chip is a unique characteristic of the SWC cutting. This should

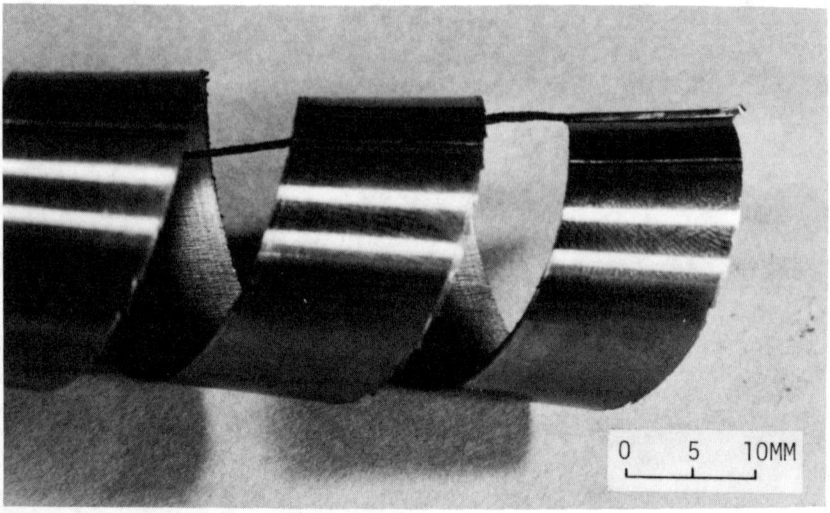

Fig.3. End part of a chip produced in SWC orthogonal cutting using a planer.
Cutting speed 50m/min (144 fpm), width of cut 13mm(0.51in), depth of cut 0.8mm(0.031in), tool material M40 carbide, land width b=0.6mm(0.024 in), work material low carbon steel (C 0.3%).

occur by selecting a sufficiently large feed. The necessary condition is, referring to Fig.1a, that the uncut chip thickness h is greater than the chamfer width c of the negative land.

Substantial portion of the cutting energy goes into the heat carried by the subsidiary chip, while the amount of heat transmitted to the workpiece is greatly reduced. Also the principal chip does not colour by burning but stays brilliant metal white in the case of machining plain steels: the reason for naming the technology "Silver White Chip cutting".

TURNING APPLICATIONS

Since it was proposed in 1955, SWC technology found fair acceptance in domestic industries. Applications were mostly in rough turning, where the following features were benefitial:

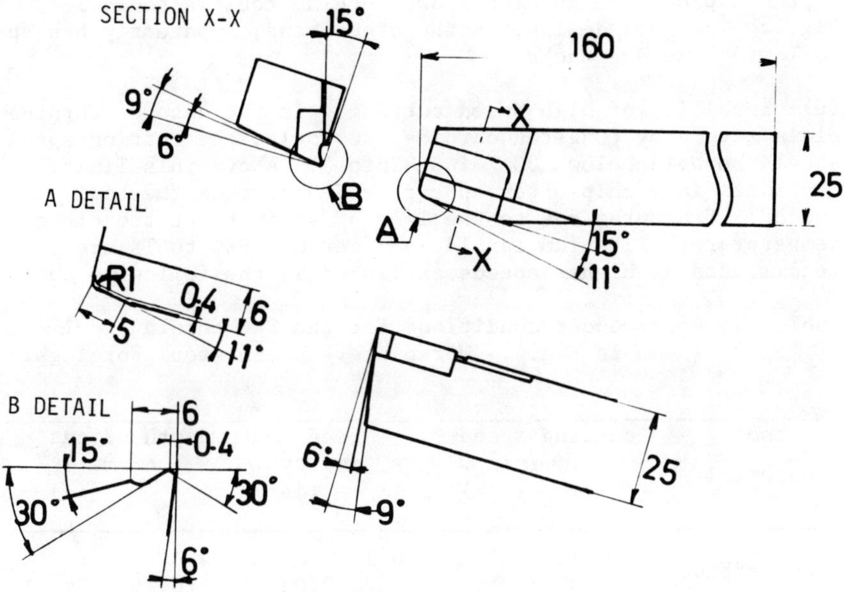

Fig.4. Typical design of SWC turning tool.
Dimensions are shown in mm.

(a) The narrow negative land (primary rake) provided sufficient shock resistance to the cutting edge, so that the tungsten carbide, a new but still fragile tool material at the time, could be successfully used for rough turning of steel parts. Tool life was longer than conventional tools in most cases. The SWC tools could take even interrupted cuts on forged, cast, or welded steel parts.

(b) Saving in horsepower due to high-rake cutting made it possible to utilize the under-rated lathes with insufficient rigidity available at the time for high-rate rough machining.

(c) Controlled formation of the built-up edge on the negative land was so stable that the finished surface showed no sign of the built-up edge residue.

(d) Workpiece was not affected by the heat generated by cutting so that the cutting fluid was not neccessary in most cases.

A typical design of the SWC turning tool is shown in Fig.4. The tool design, on the other hand, apparently has the following two drawbacks:

(e) Inability of high speed cutting. In the case of turning plain steels by tungsten carbide SWC tools, the cutting speed should be held below 100m/min (328fpm). Above this limit, the subsidiary chip often appears red hot, then the tool suddenly fractures due to the loss of strength at the elevated temperature. Titanium carbide and ceramic SWC tools are recommended at higher speeds as listed in the following table:

Table 1. Recommended conditions for the SWC turning tool shown in Fig.4. Workpiece: plain steel forgings.

tool material	cutting speed v, m/min (fpm)	feed s, mm/rev (ipr)	depth of cut a, mm (in)
P 20 carbides	60-100 (197-328)	0.6-1 (0.024-0.039)	4-6 (0.16-0.24)
Titanium carbides	100-150 (328-492)	0.6-1 (0.024-0.039)	3-4 (0.12-0.16)
Ceramics	120-180 (394-591)	0.6-1 (0.024-0.039)	3-4 (0.12-0.16)

(f) Another drawback of the SWC tool is the difficulty of tool grinding because of its intricate shape.

SWC turning seems to have become gradually less popular in recent years, as high-speed high-power lathes replaced old ones and new tool materials with improved shock and wear resistance were marketed. Another reason is that it is now more difficult for most machine shops to secure skilled technicians for the tool grinding job. Although turning applications decline, SWC cutting is finding values in other processes.

INDEXABLE-TYPE PLANING TOOLS

Indexable design eliminates tool grinding in the user's shop. Factory ground SWC carbide inserts are recently available principally for planing works, see Fig.5.

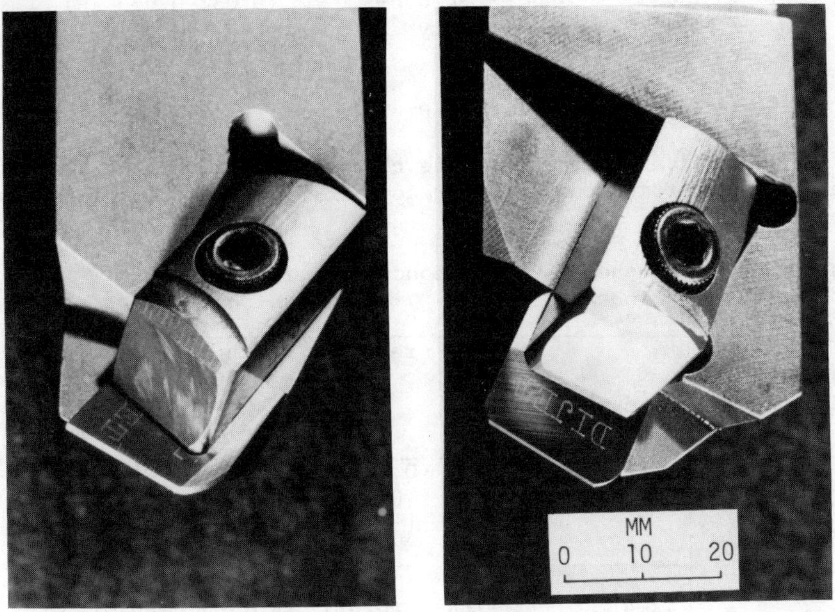

Type-U multi-purpose Type-S heavy duty

Fig.5 Indexable-type SWC planing tools

Planing is suited for SWC application for it is by nature a low cutting speed process, and cooling alternates with cutting for every table stroke, such that the temperature of the tool insert tends to be lower than in turning. Following are SWC cutting data commonly practiced in planing:

Low Carbon Steel and Weldments

Due to the high adhesion of low carbon steel chip to the tool rake face, carbide tools with conventional edge design fail in rough planing. SWC tools produce chips integral with the built-up edge (subsidiary chip) that separates by itself from the tool edge at every end of cutting.

Recommended conditions are:

cutting speed ---- 50-100 m/min (164-328 fpm)
feed per stroke -- 0.8-2 mm (0.031-0.079 in)
depth of cut ----- 6 mm (0.24 in) with type-U tool

width of negative land ----b=0.8-1.0 mm (0.031-0.039 in)
cutting fluid ------------ dry

Resurfacing Die Blocks

Hardened and unhardened die blocks are reworked by planing at the cutting conditions recommended in the following table:

Table 2. Recommended cutting conditions for planing die blocks

kind of die blocks	hardness	cutting speed m/min (fpm)	feed per stroke mm (in)	depth of cut mm (in)	land width b, mm (in)
medium size press forge dies	up to 53 R_c	25-35 (82-115)	0.2-0.5 (0.008-0.002)	3 (0.12)	0.2-0.3 (0.008-0.012)
	42-47 R_c	40-50 (131-164)	0.5 (0.02)	4 (0.16)	0.3-0.4 (0.012-0.016)
large size drop forge dies	32-36 R_c	40-50 (131-164)	0.5-0.8 (0.02-0.03)	6 (0.24)	0.4 (0.016)
0.5% C plain steel	250 Bhn	50-100 (164-328)	0.8-1.8 (0.03-0.07)	6-7 (0.24-0.28)	0.4 (0.016)

note: Use type-S (heavy duty) tools for works listed in the table. Cutting fluid is not necessary.

RAIL PLANING (A CASE STUDY)

Top and sides of inverted T-beams are machined by a planer to manufacture elevator rails. In an elevator manufacturer, conventional process is to machine eight rails at a time using carbide tools and insoluble cutting oil.

Fig.6 SWC planing top of elevator rails (phase 1).

The machine tool used in the process is a double-column-type planer with a 6m (236in) long x 2m (79in) wide table powered by a 45 kw dc motor, 80m/min (262fpm) maximum table speed.

A new set of tooling has been developed that accommodates SWC cutting tools in order to increase the production rate. As seen in Fig.6, three rails are machined at a time by the SWC tooling without cutting oil. The work proceeds in a sequence as illustrated in Fig.7, and cutting conditions listed in Table 3. In spite of the poor machinability of the low carbon steel workpiece, SWC tools take several folds chip loads as compared to conventional tools.

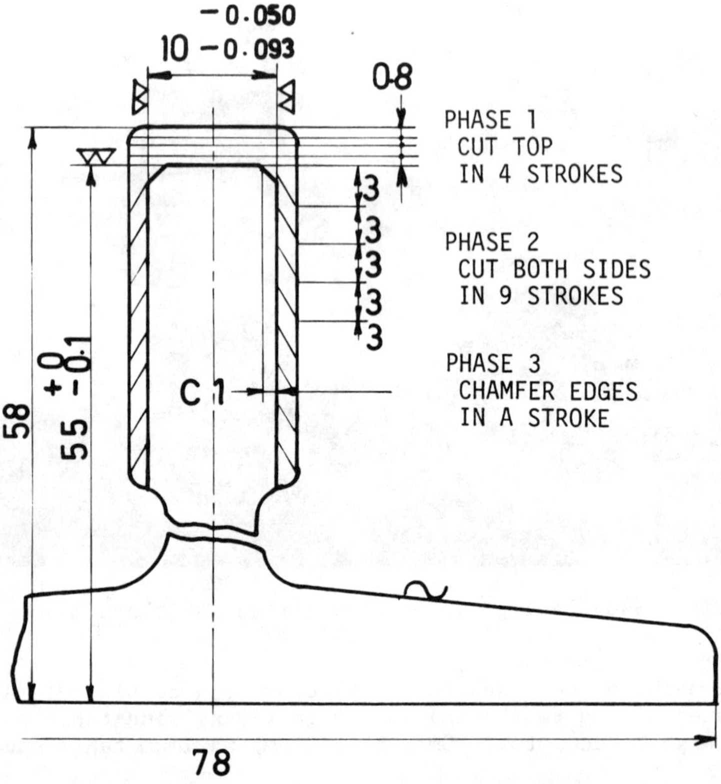

Fig.7 Sequence and cut distribution of SWC rail planing. Length of rails 5,010mm (197in). Dimensions are shown in mm.

Table 3. Cutting conditions for conventional and SWC rail planing. Phase 2, sides cutting.

item	conventional tooling	SWC tooling
cutting speed	40m/min (131 fpm)	50 m/min (164 fpm)
depth of cut	1.5mm (0.059in) parallel to cutting edge	1.5mm (0.059in) normal to cutting edge
feed	0.35mm(0.014in) / stroke normal to cutting edge	3mm(0.12in) / stroke parallel to cutting edge

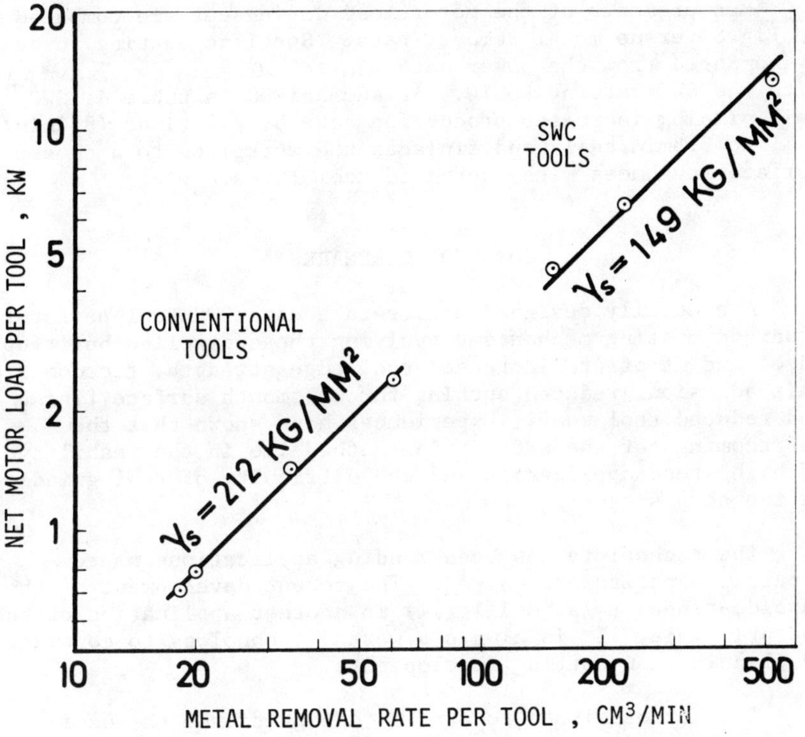

Fig.8 Comparison of SWC and conventional tools with respect to metal removal rates and power requirements in the rail plaining. Idle running power has been subtracted from the motor load shown in the ordinate. Mechanical efficiency $\eta = 0.87$ is employed in computing the specific cutting force γ_s.

Table.4 Performance of conventional and SWC rail planing.

item	conventional tooling	SWC tooling
number of rails machined at a time	8	3
cycle time	65.5min	10.0min
production time per pc.	8.2min/rail	3.3min/rail
surface roughness	18.5μm H_{max}	2.5μm H_{max}
tool life	10 rails	30 rails

Measurements of the motor load during cut are compared in Fig.8 versus metal removal rate. Specific cutting force k_s computed from the power data shows 30% energy saving with the SWC cutting tools. As summarized in table 4, SWC rail planing increases production rate by 2.5 times (8.2min/rail to 3.3min/rail) and finishes the workpiece to a closer surface roughness with increased tool lives.

CONCLUDING REMARKS

A specially designed double-rake edge SWC tool performs a unique cutting mechanism involving the controlled built-up edge, and it offers increased tool edge strength, freedom from chip adhesion, reduced cutting force, smooth surface finish and reduced tool wear. Experiences have shown that the shortcomings of the SWC cutting method lie in the inability of high speed application and the difficulty of tool grinding at the shop site.

The technoloty has been finding applications where apparent benefits are gained. The recent development of the carbide-tipped twist drill (11) is another application of the SWC principle. It is always a worthy technology to consider in planning new cutting developments.

The author wishes to express his gratitude to Hosoi Machine Works Co., Ltd, Dijet Industrial Co., Ltd, and Fujitec Co., Ltd. for generous co-operations. He is also grateful to his father Dr. Koichi Hoshi, Honorary Professor of Hokkaido University, and inventor of the SWC technology, for his stimulations and encouragement in writing this paper.

REFERENCES

(1) K. Hoshi and T. Hoshi, "Metal Cutting Techonology", Kogyo-Chosa-Kai, 1969, (in Japanese).

(2) H. Klopstock, "Recent Investigations in turning and planing and a new form of cutting tool", Paper presented at the Spring Meeting of ASME, 1925.

(3) H. Klopstock, "Berechte des Versuchsfeldes für Werkzeugmaschinen an der Tech.Hochschule Berlin", Vol.8, 1926.

(4) Ripper, "Von Forschung und Forshungsergebnisse zur Schnitt-Theorie (F. Schwerd), V.D.I., Vol.76, No.52, 1932, p.1263.

(5) K. Hoshi, "On the Built-up Edge and Counter-Plot for it", Transactions of The Japan Society of Mechanical Engineering, Vol.5, No.18, 1938, p. 205,(in Japanese).
(6) H. Takeyama and E. Usui, "The effect of tool-chip contact area in metal cutting", Trans.ASME, Vol.80, 1958, p.1089.
(7) B.T.Chao and K.J.Trigger, "Controlled contact cutting tools", Trans.ASME, Vol.81, No.5, 1959, p.139.
(8) E. Usui and M. C. Shaw, "Free machining steel IV, tools with reduced contact length", Trans.ASME, Vol.84, No.1, 1962, p.89.
(9) K. Hoshi and E. Usui, "Wear Characteristics of Carbide Tools with Artificially Controlled Tool-Chip Contact Length", Proc.3rd Int.MTDR, 1962, p.121.
(10) E. Usui, K. Kikuchi and K. Hoshi, "The Theory of Plasticity Applied to Machining With Cut-Away Tools", ASME Paper No.63-Prod.5, 1963, p.1.
(11) T. Hosoi and others, "High-Rate Drilling with a New Type of Carbide-Tipped Twist Drill", a paper presented to the ASM/SME International Conference on Cutting Tool Materials, September 1980, Cincinnati, Ohio.

[Discussion is presented on the next page.]

DISCUSSION

Dr. Ranga Komanduri (General Electric - CRD)

It is known that b.u.e. occurrs in certain materials over a limited speed range. Your speed range appears to be above this. Do you really think a b.u.e. is formed with this high negative primary rake angle? How does your geometry differ from a chamfer of "K" land tool? Would this be applicable to milling?

Professor Hoshi

We think it is b.u.e. since this negative rake tends to promote its formation but there is a possibility that it is extruded nickel.

Tom Hale (General Electric - Carboloy)

What type of flank wear occurs with SWC machining and is the rate of flank wear high?

Professor Hoshi

Flank wear is very uniform and similar to conventional carbide tools. In most cases the rate of wear is lower for similar cutting conditions.

PRODUCTION APPLICATIONS OF HIGH-RATE HOLEMAKING WITH CARBIDE-TIPPED TWIST DRILLS

Mamoru DAIMON* Hideyuki YAMAMOTO**
Hiroshi OZAKI****, and Masataka KASHIMOTO***

1. INTRODUCTION

The drilling process is one of the oldest and most widely used machining processes. The drilling process time is over 10% of all machining time.

Importance of the drilling technology development is ever increasing to achieve higher productivity. With the advent of Hosoi's carbide-tipped drill specifficaly desighned for steel workpieces, high-speed drilling technology has been recently made available.

This new type of carbide-tipped drills offers high-rate holemaking with superb finish accuracy and prolonged tool life as compared with conventional carbide-tipped twist drills.

*	Senior Researcher,	Manufacturing Engineering Research Center, Komatsu Ltd.
**	Researcher,	Manufacturing Engineering Research Center, Komatsu Ltd.
***	Gerneral Manager,	Manufacturing Engineering Research Center, Komatsu Ltd.
****	Tool Engineer,	Productive Engineering Dep. Osaka Plant, Komatsu Ltd.

Potential impacts on part manufacturing are demonstrated in this paper by presenting two examples of production applications.

2. APPLICATION OF HIGH FEED CARBIDE DRILLS

2-1 Case Study 1, Application for the welded structure parts

The welded structure parts have many hole-making processes. The parts described here is one of the welded structures made up of two kinds of materials different in hardness, one is High-tension Steel and the other is Mild Steel. In order to apply carbide drills for this welded structure parts, a dedicated machine was produced.

Fig.1. shows carbide drills and the workpiece cut out of above welded structure parts. When this workpiece is drilled with carbide drills, chips are obtained as shown in Fig.2. As said above, two kinds of materials different in hardness have to be drilled in a single sequence, so the cutting condition was decided properly for the harder material(High-tension Steel). As the result, it is noted that the chips

Fig.1 Carbide drills and workpiece

obtained from another material became longer. However chips did not have to be cut shortly by step feed as before, so machining time of this parts became about a quarter.

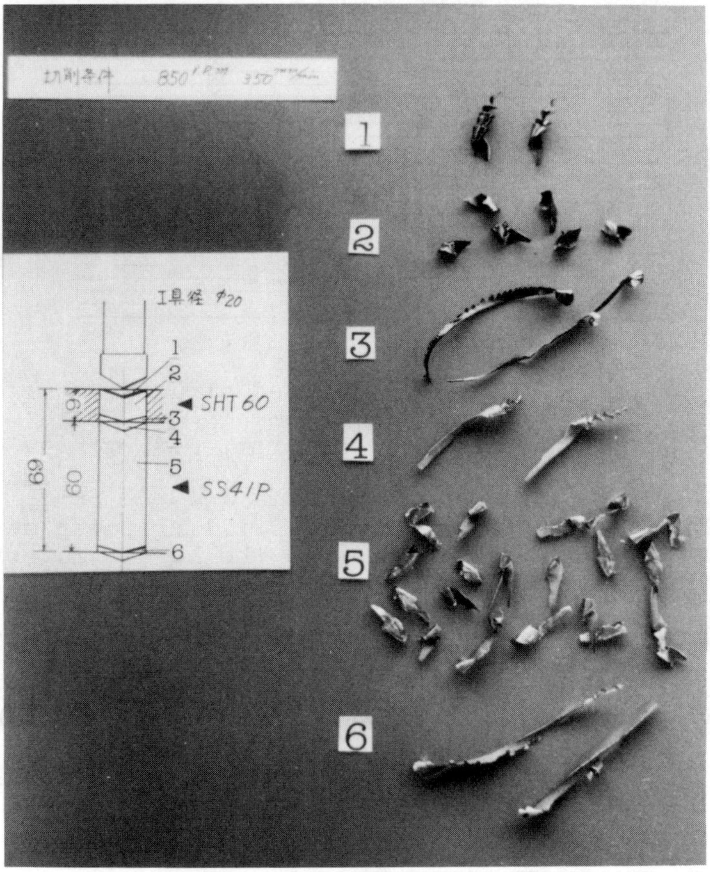

Fig.2. Chips obtained from welded structure parts

Workpiece	Welded structure parts
Drill diameter	20 mm
Spindle speed	850 rpm
Feed rate	0.41 mm/rev
Depth of hole	69 mm

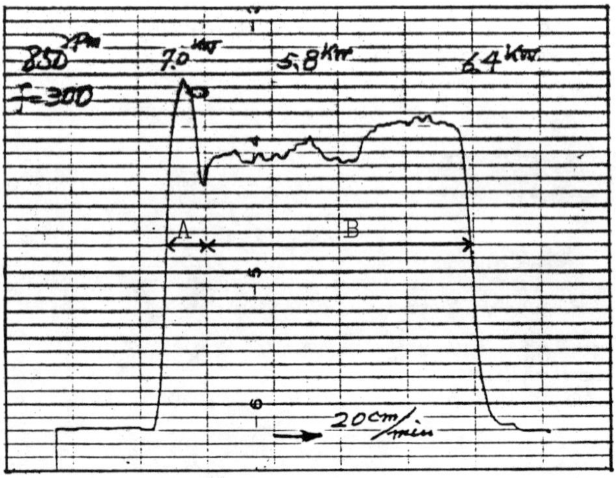

Fig.3 Drilling power

Work : Welded structure parts
Drill Diameter : 20 mm
Cutting condition : 850 rpm, 0.35 mm/rev

Fig.3 shows the result of drilling power measurement. Part A corresponds to the harder material, and part B to another material.

In application carbide drills for the welded structure parts, comparing with conventional HSS drills as shown in Table 1, productivity was four or five times higher and tool cost became about half.

Table 1. Comparing carbide drills with HSS drills

Item	Carbide drill	HSS drill
Diameter	20 mm	20 mm
Spindle speed	890 rpm	215 rpm
Feed rate	0.22 mm/rev	0.54 mm/rev
Machining time(ratio)	100	470
Tool life	about 400holes	about 28holes

2-2 Case Study 2,
Application for the Caterpiller Links

A caterpiller link is a forging in 331-352 Brinell hardness. In order to apply carbide drills, hole accuracy and cutting forces were measured and compared with those of the conventional HSS drills. Cutting tests were performed under the condition in Table 2.

Fig.4-a and b show the results of the hole accuracy measurement. As shown in the figures, hole oversize value by carbide drills decreased gradually with the number of drilled holes. However, the required tolerance of hole diameter was satisfied enough. The tolerance of the pitch error was satisfied too.

Table 2. Cutting condition used in the test.

Machine	NC drilling machine (Hosoi Machine Works Co. Ltd.)
Drill diamerer	12 mm
Workpiece	link Brinell-hardness 331-352
Spindle speed	1350 rpm (51m/min)
Feed	0.2 mm/rev
Depth of cut	36 mm (through hole)
Cutting fluid	sulpho-chlorinated water-miscible oil

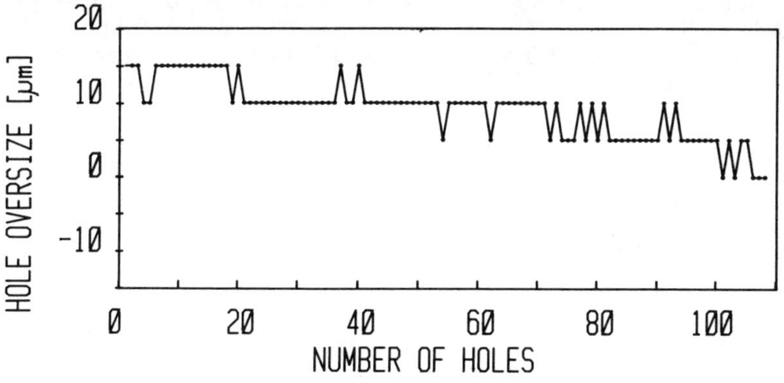

Fig.4-a. Hole oversize vs. Number of holes

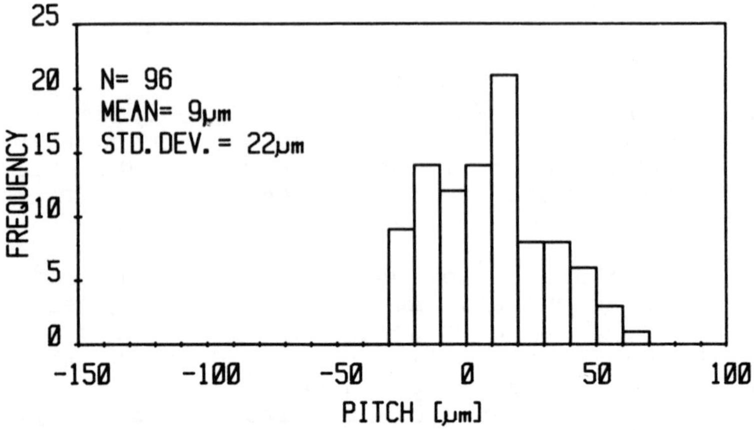

Fig.4-b. Histgram of pitch error.

Drill diamerer	12 mm
workpiece	link
Spindle speed	1350 rpm
Feed rate	0.2 mm/rev
Depth of hole	36 mm

Fig. 5. shows the results of the drilling force measurement at each number of hole. The range of thrust was from 256 kg to 337 kg and torque was from 132 kg-cm to 172 kg-cm.

Comparing these results with those of conventional HSS drills in Table 3. It was ascertained that carbide drills had higher productivity and higher hole accuracy than HSS drills.

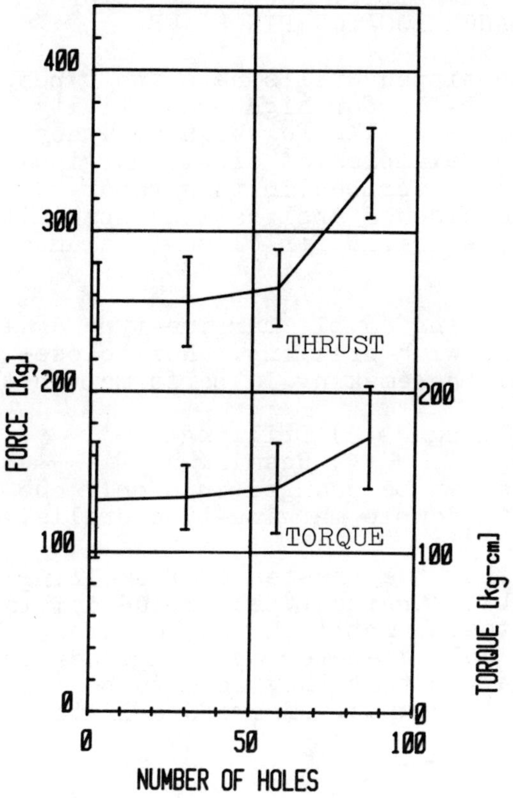

Fig.5. Drilling forces vs. Number of holes

Drill diameter 12 mm

Workpiece link

Spindle speed 1350 rpm

Feed rate 0.2 mm/rev

Depth of hole 36 mm

Table 3. Comparing carbide drills with HSS drills

Item	Carbide drill	HSS drill
Spindle speed	1350 rpm	200 rpm
Feed rate	0.2 mm/rev	0.15 mm/rev
Hole oversize	Mean=0.009 mm Std.dev.=0.004mm	Mean=0.07 mm Std.dev.=0.06mm
Pitch	Mean=0.009mm Std.dev.=0.022mm	Mean=0 mm Std.dev.=0.08mm

3. APPLICATION CAPABILITY FOR DRILLING OF CLOSE-TOLERANCE LOCATOR-PIN HOLES

Hosoi's carbide-tipped drills have two types, single margine-type drills for high productivity and double margine-type drills for high accuracy (Fig.6.) Application examples of single margine-type drills have been described in this paper. Next, we will try to discuss application capability of double margine-type drills for close-tolerance locator-pin holes.

The purpose of using double margine-type drills is to replace reaming with drilling. Now, close-tolerance locator-pin holemaking is performed in the following process.

(1) Center drilling (2) Drilling
(3) End milling (4) Reaming

These four processes can be changed into only one drilling process with double margine-type drills.

Fig.7-a and b show the pre-tests of drilling cast steel with double margine-type carbide drills. According to those tests, until the number of drilled holes is about 200, hole diameter and roughness are very variable, but after that they tend to become steady and keep the tolerance. Therefore if there

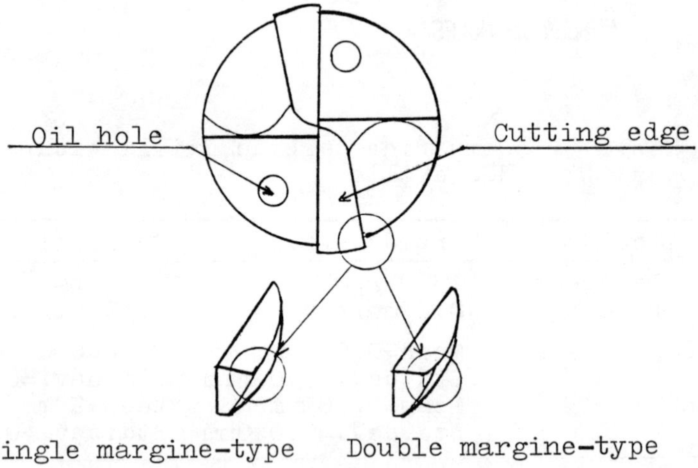

Fig.6. Single margine-type & double margine-type

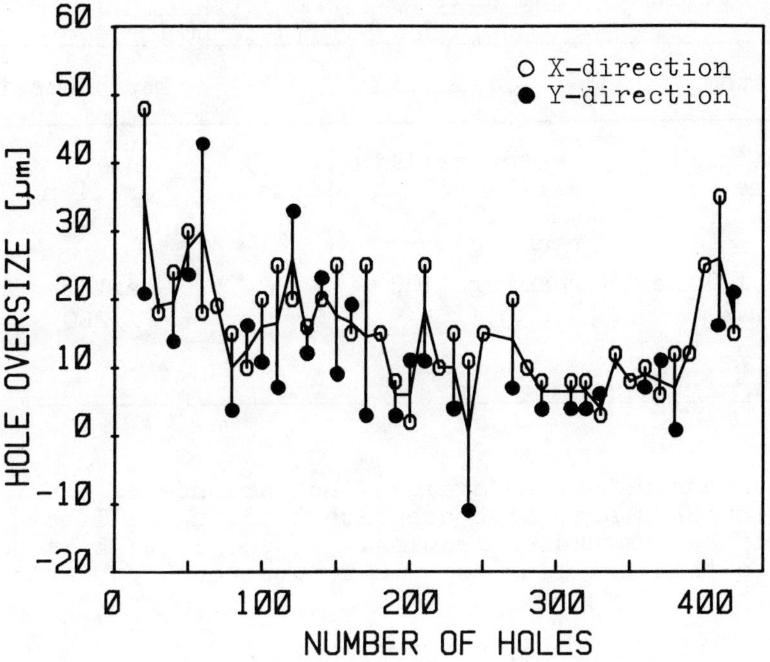

Fig.7-a. Hole oversize vs. Number of holes

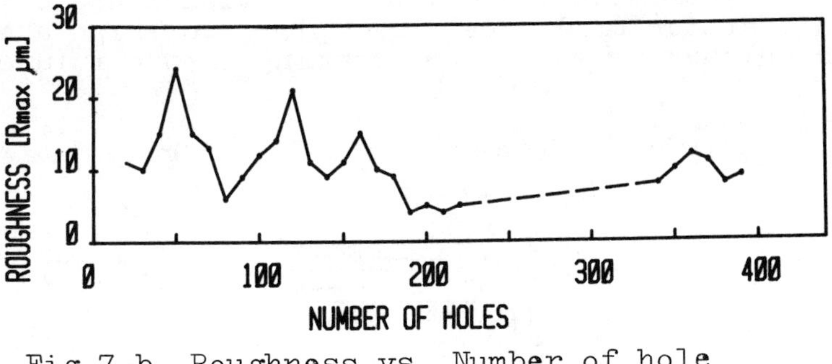

Fig.7-b. Roughness vs. Number of hole

Drill diameter	18 mm
Workpiece	cast steel
Spindle speed	1250 rpm (71 m/min)
Feed rate	0.34 mm/rev
Depth of hole	20 mm

Table 4. Cost analisys

Item	Conventional		Improvement
Cutting time	center drilling drilling end milling reaming	total 100	drilling 30
Tool life	reaming 100		carbide drill 260
Total tool cost (ratio)	100		70

is some counterplan against the variance of the beginning stage, high accurate drilling will be realized instead of reaming. If so, it will be possible that drilling cost is reduced such as Table 4.

4. CONCLUSION

The applications of Hosoi's carbide tipped twist drills were described in this paper. Steel drilling productivity with carbide drills can be four or five times higher than that with conventional HSS drills. Also, carbide drills can be applied for high accurate holemaking instead of reaming in the future.

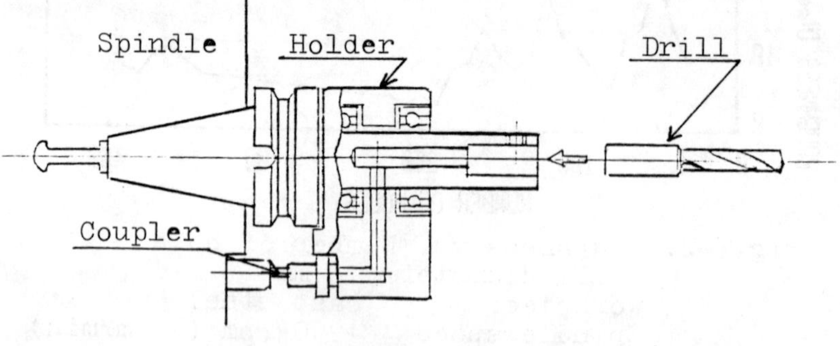

Fig.8. Special drill helder for the seandard NC Machining Centers.

However, there are two requirements to be satisfied. One is requirement for machines. It is necessary that the main spindle power and stiffness of the machines are high enough. The other is for the drill holders. The carbide drills have to be used with cutting fliud. As cutting fliud is supplied through drill body, special holders shown as Fig.8 are needed, if carbide drills are used on the standard NC machining centers.